Conservation Agriculture in Subsistence Farming

Case Studies from South Asia and Beyond

FSC
www.fsc.org
MIX
Paper from responsible sources
FSC® C013604

Conservation Agriculture in Subsistence Farming
Case Studies from South Asia and Beyond

Edited by

Catherine Chan

University of Hawai'i at Mānoa

and

Jean Fantle-Lepczyk

University of Hawai'i at Mānoa

www.cabi.org

CABI is a trading name of CAB International

CABI
Nosworthy Way
Wallingford
Oxfordshire OX10 8DE
UK

Tel: +44 (0)1491 832111
Fax: +44 (0)1491 833508
E-mail: info@cabi.org
Website: www.cabi.org

CABI
38 Chauncy Street
Suite 1002
Boston, MA 02111
USA

Tel: +1 800 552 3083 (toll free)
E-mail: cabi-nao@cabi.org

A catalogue record for this book is available from the British Library, London, UK.

Library of Congress Cataloging-in-Publication Data

Conservation agriculture in subsistence farming : case studies from South Asia and beyond / edited by: Catherine Chan, University of Hawaii at Manoa, Jean Fantle-Lepczyk, University of Hawaii at Manoa.
pages cm
Includes bibliographical references.
ISBN 978-1-78064-423-3 (hbk : alk. paper) 1. Agricultural conservation--South Asia--Case studies. 2. Soil conservation--South Asia--Case studies. 3. Sustainable agriculture--South Asia--Case studies. I. Chan-Halbrendt, Catherine, 1949- editor. II. Fantle-Lepczyk, Jean, editor.

S604.64.S64C66 2015
631.4′5--dc23

2014041668

ISBN-13: 978 1 78064 423 3

Commissioning editors: Rachel Cutts and Nicki Dennis
Editorial assistant: Emma McCann
Production editors: Tracy Head and Emma Ross

Typeset by SPi, Pondicherry, India.
Printed and bound in the UK by CPI Group (UK) Ltd, Croydon, CR0 4YY.

Contents

Contributors

Ares, Adrian, Feed the Future Innovation Lab for Collaborative Research on Sustainable Agriculture and Natural Resources Management (SANREM), Virginia Tech, 526 Prices Fork Rd, Blacksburg, VA 24061. E-mail: aresa@vt.edu

Chan, Catherine, University of Hawai'i-Mānoa, Department of Natural Resources and Environmental Management, College of Tropical Agriculture and Human Resources, 1910 East West Road, Sherman 104, Honolulu, HI 96822, USA. E-mail: chanhalb@hawaii.edu

Crow, Susan, University of Hawai'i-Mānoa, Department of Natural Resources and Environmental Management, 1910 East West Road, Honolulu, HI 96822, USA. E-mail: crows@hawaii.edu

Dash, Satya Narayan, Sustainable Management of Agroecological Resources for Tribal Societies, Orissa University of Agriculture and Technology, Bhubaneswar-751003, Odisha, India. E-mail: satya1962@yahoo.co.in

Eash, Neal S., University of Tennessee, Department of Biosystems Engineering and Soil Science, Room 206, Knoxville, TN 37996, USA. E-mail: eash@utk.edu

Fantle-Lepczyk, Jean, University of Hawai'i-Mānoa, Department of Natural Resources and Environmental Management, 1910 East West Road, Honolulu, HI 96822, USA. E-mail: fantle@hawaii.edu

Halbrendt, Jacqueline, University of Hawai'i-Mānoa, Department of Natural Resources and Environmental Management, 1910 East West Road, Honolulu, HI 96822, USA. E-mail: jhalbren@hawaii.edu

Himmelstein, Jennifer, Feed the Future Innovation Lab for Collaborative Research on Sustainable Agriculture and Natural Resources Management (SANREM), Virginia Tech, 526 Prices Fork Rd, Blacksburg, VA 24061, USA. E-mail: jhimmel1@vt.edu

Idol, Travis, University of Hawai'i-Mānoa, Department of Natural Resources and Environmental Management, 1910 East West Road, Honolulu, HI 96822, USA. E-mail: idol@hawaii.edu

Lai, Cynthia, University of Hawai'i-Mānoa, Department of Natural Resources and Environmental Management, 1910 East West Road, Honolulu, HI 96822, USA. E-mail: laicynth@hawaii.edu

Maher, Jane, Department of Geography, School of Natural Sciences, The Museum Building, Trinity College Dublin, College Green, Dublin 2, Ireland. E-mail: maherj8@tcd.ie

Mishra, Kshitendra Narayan, Sustainable Management of Agroecological Resources for Tribal Societies, Orissa University of Agriculture and Technology, Bhubaneswar-751003, Odisha, India. E-mail: khiturajprav@yahoo.co.uk

O'Brien, John, Concern Worldwide, 52–55 Lower Camden Street, Dublin 2, Ireland. E-mail: john.obrien@concern.net

Pande, Keshav Raj, Agriculture and Forestry University, Department of Soil Science and Agricultural Engineering, Rampur, Chitwan, Nepal. E-mail: keshabrajpande@yahoo.com

Paudel, Bikash, University of Hawai'i-Mānoa, Department of Natural Resources and Environmental Management, 1910 East West Road, Honolulu, HI 96822, USA. E-mail: paudel@hawaii.edu

Pradhan, Aliza, University of Hawai'i-Mānoa, Department of Natural Resources and Environmental Management, 1910 East West Road, Honolulu, HI 96822, USA. E-mail: apradhan@hawaii.edu

Pudasaini, Roshan, Local Initiatives for Biodiversity, Research and Development, PO Box 324, Pokhara, Kaski, Nepal. E-mail: rospud@gmail.com

Radovich, Theodore, University of Hawai'i-Mānoa, Department of Tropical Plant and Soil Science, 3190 Maile Way, Honolulu, HI 96822, USA. E-mail: theodore@hawaii.edu

Ray, Chittaranjan, University of Hawai'i-Mānoa, Department of Civil and Environmental Engineering, 2540 Dole Street, Holmes 383, Honolulu, HI 96822, USA. E-mail: cray@hawaii.edu

Ray, Plabita, Sustainable Management of Agroecological Resources for Tribal Societies, Orissa University of Agriculture and Technology, Bhubaneswar-751003, Odisha, India. E-mail: plabita.ray@gmail.com

Reed, Brinton Foy, University of Hawai'i-Mānoa, Department of Natural Resources and Environmental Management, 1910 East West Road, Honolulu, HI 96822, USA. E-mail: beefreed@yahoo.com

Reyes, Manuel, Department of Natural Resources and Environmental Design, North Carolina Agricultural and Technical State University, Greensboro, USA. E-mail: reyes@ncat.edu

Roul, Pravat Kumar, Sustainable Management of Agroecological Resources for Tribal Societies, Orissa University of Agriculture and Technology, Bhubaneswar-751003, Odisha, India. E-mail: roulpvt@yahoo.com.sg

Rust-Smith, Jessica, Hove, East Sussex, UK. E-mail: j.rust.smith@gmail.com

Schubert, Olivia, University of Hawai'i-Mānoa, Department of Natural Resources and Environmental Management, 1910 East West Road, Honolulu, HI 96822, USA. E-mail: oliviass@hawaii.edu

Tamang, Bir Bahadur, Local Initiatives for Biodiversity, Research and Development, PO Box 324, Pokhara, Kaski, Nepal. E-mail: btamang@libird.org

Thapa, Keshab, Local Initiatives for Biodiversity, Research and Development, PO Box 324, Pokhara, Kaski, Nepal. E-mail: kthapa@libird.org

Thierfelder, Christian, CIMMYT – International Maize and Wheat Improvement Center, PO Box MP163, Mount Pleasant, Harare, Zimbabwe. E-mail: c.thierfelder@cgiar.org

Wagstaff, Paul, Concern Worldwide, 52–55 Lower Camden Street, Dublin 2, Ireland. E-mail: paul.wagstaff@concern.net

Foreword

Under the constant pressures posed by climate change and a growing global population, striving for improved food self-sufficiency is not just a good idea, it is of critical importance. However, we need to increase the food supply in an environmentally and economically sound manner. Therefore, it is paramount for those who produce food around the world to have access to cost-saving, yield-increasing and environmentally sound crops and practices. And this is particularly true for people who farm on a small scale and thus have limited resources for learning about, and applying, new methods and technologies.

That is precisely why conservation agriculture is vital in subsistence agriculture. The key advantages of conservation agriculture over other technological advances are that its input requirements are not prohibitive, and resource-poor farmers can adopt them easily with adequate economic incentive.

This book is timely, and illustrates key examples of conservation agriculture that are inherently site specific due to consumption preferences, soil types, climatic zones, and general crop adaptability. Providing a comprehensive evaluation of conservation agriculture with a multidisciplinary approach, including economics, agronomics, soil ecology, gender implications, and technological transfer, this book is a must-read for those interested in the future of food production in the most vulnerable areas of the world.

Maria Gallo, PhD
Dean and Director of Research and Cooperative Extension
College of Tropical Agriculture and Human Resources
University of Hawai'i at Mānoa, Honolulu, Hawaii, USA

Preface

When the Sustainable Management of Agroecological Resources for Tribal Societies (SMARTS) project began in 2009, we had very optimistic plans and a strong desire to promote conservation agriculture (CA) practices to help tribal villagers in India and Nepal. Our goals were to prioritize local resource-use preferences, develop improved conservation practices, and identify market opportunities to enhance livelihood options for farmers living on marginal lands and struggling to get by. At the same time, we wished to work with local governments and non-profit organizations to build human capacity for conducting research and extension on CA and developing CA practices appropriate to the local context. We at the University of Hawai'i at Mānoa (UHM) were assisted in these endeavors through the generous support of a USAID grant funded through the Feed the Future Innovation Lab for Collaborative Research on Sustainable Agriculture and Natural Resources Management (SANREM) at Virginia Tech. We were eager to partner with local non-governmental organizations like Local Initiatives for Biodiversity, Research and Development (LI-BIRD) in Nepal, the faculty and staff at Orissa University of Agricultural Technology (OUAT) in India, and the Institute of Agriculture and Animal Science (IAAS) of Tribhuvan University in Nepal.

One of the highlights of this project has been the Frontiers in Agriculture in South Asia and Beyond Conference held in Nepal in March 2013. The conference was the culmination of 4 years of cooperation and interdisciplinary study by a diverse, international group of CA experts and practitioners, as well as regional researchers and extension agents. It was conceived as a way to highlight recent findings and current CA research in South Asia and other developing countries, with an emphasis on presenting interdisciplinary scientific knowledge that incorporated agricultural, economic, and social sciences.

This book is an extension of that conference. It focuses on the latest research in CA, with an emphasis on the applicability of results worldwide. Using Southeast Asia as a case study, it will examine the history and current state of CA regionally and globally, as well as explore the long-term impacts the adoption of CA practices has on the livelihoods, agricultural production, gender equity, adoption potential, and regional economic development of rural societies. This inclusive framework is achieved via interdisciplinary analysis at scales ranging from the household level to regional and national levels, and contributions by multidisciplinary and multinational CA researchers and experts. Using innovative assessment tools from multiple disciplines, this book provides a comprehensive analysis of the social, environmental, and economic factors that impact CA practice, and estimates the magnitude of such impacts, over the long term.

Of course, we did not get to this point, the production of this book, without the support and participation of many colleagues along the way. None of this would have been possible without the support of USAID and SANREM. The faculty and administration of OAUT and the IAAS, as well as the LI-BIRD staff, have invested heavily in this project, and their enthusiasm and willingness to share, learn, and grow together professionally has been an inspiration. We have had much support stateside as well. We are very thankful to all the professionals and students at UHM and other US institutions who have donated their time generously, both with the SMARTS project and with the book. Overall, this project and this book have both been an incredible, and enjoyable, learning experience for us. We hope you discover as much from reading the book as we have from editing it.

Catherine Chan and **Jean Fantle-Lepczyk**
10 September 2014

1 A Brief History of Conservation Agriculture

Travis Idol*

University of Hawai'i at Mānoa, Honolulu, Hawaii, USA

1.1 Introduction

"Conservation agriculture (CA) aims to achieve sustainable and profitable agriculture and subsequently aims at improved livelihoods of farmers through the application of the three CA principles: minimal soil disturbance, permanent soil cover and crop rotations. CA holds tremendous potential for all sizes of farms and agro-ecological systems, but its adoption is perhaps most urgently required by smallholder farmers, especially those facing acute labour shortages. It is a way to combine profitable agricultural production with environmental concerns and sustainability and it has been proven to work in a variety of agroecological zones and farming systems. It is been perceived by practitioners as a valid tool for Sustainable Land Management (SLM)" (FAO, 2014a).

This modern definition of conservation agriculture embodies almost a century of academic and public concern over the negative effects of agriculture on soils and other natural resources and a much longer recognition that the quality of these resources is essential for the sustainability of agricultural production and the well-being of the surrounding natural and human communities. The main culprit has been, and continues to be, the plowing of the soil. Tillage has been a part of the development of agriculture since its beginnings in North Africa, the Middle East, and the Indus River Valley of present-day Pakistan and northern India. These practices, while detrimental to soil quality, sustained the rise of the earliest recorded civilizations. The benefits of tillage are numerous, including weed control, seed bed preparation, acceleration of organic nutrient mineralization, and incorporation of soil amendments. At the same time, by breaking up the soil structure and leaving it exposed to wind and rain, tillage accelerates the natural processes of erosion.

*E-mail: idol@hawaii.edu

Conservation practices to protect the soil undoubtedly developed alongside tillage, including terraced hillsides, mulching, green manures, *terra preta*, seasonal and interannual crop rotations, intercropping, relay cropping, and others (Harrison and Turner, 1978; Ofori and Stern, 1987; Johansen, 1999; Mann, 2005; Faiella, 2006; Larkin *et al.*, 2012). Thus, CA as a practical matter has been around as long as intensified agriculture itself. However, degradation of the soil and the land base more generally has been a part of many ancient civilizations and likely is a significant contributing factor to their decline (Diamond, 2004).

1.1.1 Reduced tillage

While the purpose of any type of tillage is to disturb the soil in order to improve conditions for crop planting and early growth, it is the moldboard plow, i.e. a plow that turns the soil rather than simply cutting through it, that is associated most strongly with large increases in soil erosion and degradation. Variations on a soil-turning plow likely arose independently in several locations around the world. In Europe, an "iron share" plow that inverted the soil was in use during the Roman Empire of the 5th century AD. In the USA, Thomas Jefferson is credited with designing an improvement over the common European design in 1784. With the development of steam-powered tractors around 1910, heavier and larger plows could be utilized to farm larger areas. By 1940, there were 2 million tractors in the USA alone (Lal *et al.*, 2007).

This expansion of agricultural production left soils vulnerable to wind and water erosion. In the early 1930s, the western USA experienced several years of drought conditions, leading to massive wind erosion. Continent-wide dust storms (the "Dust Bowl") carried soil from the dry interior west all the way to the eastern coast, raising public (and political) awareness of the problems of erosion and the need to develop conservation practices and policies. Not long after the Dust Bowl, Fred Hoeme, a farmer in Oklahoma, developed a soil cultivator that acted like a ripper, bringing soil clods up to the surface that were more resistant to wind erosion than the smaller soil aggregates generated by turning over the soil. The Hoeme cultivator is considered the forerunner to the modern chisel plow. Another Midwestern farmer, C.S. Noble, was inspired by a machine that undercut rows of carrots to ease harvesting and developed a similar implement to control weeds without turning over the soil. This is considered the forerunner of the stubble-mulch tiller, allowing for greater residue retention on the soil surface.

Implements to allow farming without tillage ("no-till") were not explored until the late 1940s and 1950s with the development of chemical herbicides. Early results were disappointing, as weed control was less than desirable (Unger and Baumhardt, 2001). The development of 2,4-D, paraquat, and atrazine in the late 1950s and early 1960s provided farmers more effective weed control. Commercial no-till (NT) production first occurred in the USA in 1962. Home-made planters were used initially. Commercial planters were later developed by universities and agricultural implement companies. Combined with the reduced cost of the broad-spectrum herbicide, glyphosate, from approximately

US$40/l in the 1970s to less than US$10/l in the 1990s, NT became much more practical and economically feasible.

Outside of the USA, conservation tillage was also promoted as a response to accelerated erosion and general land degradation due to large-scale mechanized agriculture. While the USA is the world leader in area under NT at approximately 35 million hectares (Mha) as of 2008, globally there is more than 100 Mha under NT, including 19 countries with greater than 100,000 ha each. The International Maize and Wheat Improvement Center (CIMMYT) has been a global leader in the development and promotion of conservation tillage systems in major cereal production areas.

1.1.2 Crop rotation and diversification

While reducing tillage is the primary practice of CA, soil quality also benefits from crop diversification or rotation. Planting the same crop year after year tends to deplete certain soil nutrients, reduces soil microbial diversity and functionality, and promotes the build-up of crop-specific pests and diseases. As with reduced tillage, crop rotation has likely been a common practice in agriculture from its very beginnings. Instructions for crop production in a Sumerian farmer's almanac (1700 BC) included proper use of a seeder plow and the need for rotating legume and grain crops. The rise of intensive farming in China more than 2,500 years ago included crop rotation along with plowing, irrigation, and fertilization. At the same time in Egypt, agriculture had similarly been intensified, with soil fertility maintained not just by annual flooding of the Nile river but also by land management practices, including crop rotation and fallowing. In the Andean highlands, the pre-Columbian civilization of the Inca utilized crop rotation and fallow periods in potato production to restore soil fertility and reduce soilborne pests and diseases.

In the USA, Native Americans practiced some crop rotation as well, but the more common practice may have been fallowing of fields and moving entire settlements (Johansen, 1999). With no draft animals, and thus only hand tools, native farmers would burn fallow fields to kill weeds and till in the roots prior to sowing. European settlers continued this practice until the arrival of livestock and plows. Fields were then rotated between cropping and pasture. In the 18th and 19th centuries, agricultural systems such as tobacco (*Nicotiana* spp.), maize (*Zea mays*), and wheat (*Triticum* spp.) were regularly rotated with grasses and legumes to conserve and restore soil fertility. The rise of agriculture in the Midwest of the USA included regular rotations between grain and legume crops, primarily to restore nitrogen fertility. By the 1950s and 1960s, the mass production of chemical fertilizers led some researchers and agricultural leaders to question the need for crop rotation. However, by the 1990s, the multiple benefits of rotation for crop yield and sustainability were widely recognized.

Intercropping has a similar long history in agriculture and was practiced traditionally in Africa, Asia, and Latin America. Perhaps most famous is the *milpa* system of maize, beans, and squash practiced by the Mayans of Mexico, Belize, and Guatemala. By taking advantage of different timing of growth and habits of the various species, a mixed cropping system can generate yields of several crops from a single plot of land. Most interest has focused on cereal–legume intercropping (Ofori and Stern, 1987). Various yield and competition indices have been developed to quantify the relative performance of intercropped versus sole-cropped fields (e.g. Dhima *et al.*, 2007). The success of these systems often depends on proper spacing and timing to avoid competition but maximize use of available space and resources (e.g. Davis *et al.*, 1987).

Crop rotation can also include "relay cropping", in which a second crop is planted and grown after the first crop. In some cases, the second crop is not planted until the first crop is harvested, but in others, the second crop is planted while the first crop is still growing, taking advantage of the differential growth and maturation rates of the species. In cereal–legume intercropping, delaying planting of the legume for 10 days or so is common to minimize competitive effects on the cereal crop. Relay cropping is not necessarily a CA practice, though. In monsoonal climates such as India and Nepal, the second crop may be planted in order to take advantage of the residual soil water and postharvest rain events. This increases demands on soil nutrients and is often accompanied with additional tillage to prepare the field for the second crop along with cultivation to control weeds.

In the USA, intercropping and polyculture, like crop rotation, were common prior to the 1940s, before the increase in the availability of larger machinery and inexpensive fertilizers and pesticides. As machinery became more specialized for individual crops, intercropping became impractical. The global spread of these Green Revolution technologies and practices further reduced traditional intercropping systems. However, intercropping, especially with legumes, does offer similar advantages as crop rotations for modern agriculture, namely reduced nitrogen (N) fertilizer requirements, disruption of pest and disease cycles, weed suppression, increased microbial diversity, and greater crop residue cover. For smallholders, intercropping allows for continuous cultivation of important cereal or grain crops.

1.1.3 Organic soil cover

Exposure of the soil surface to water and wind makes it particularly vulnerable to erosion. It also increases exposure to gains or losses of heat, accelerates evaporation, and thus drying, increases risk of compaction from foot or machine traffic, and increases opportunities for germination and growth of weeds. Besides protecting the soil surface, organic soil cover also increases the cycling of organic matter and nutrients. Thus, keeping the soil covered is an essential component of CA. A minimum of 30% coverage is the standard for CA as defined by the United Nations Food and Agriculture Organization (FAO, 2014a).

Again, ancient descriptions of and recommendations for agricultural production include use of cover crops or organic residues to keep the soil surface protected. The ancient Roman poet, Virgil (70–19 BC) wrote of the use of cover crops in the epic poem, *Georgics* (Virgil, 1994). Prior to the widespread use of inorganic fertilizers, cover crops, especially nitrogen-fixing legumes, were used to restore fertility in agricultural soils. Cover crops have been applied to many types of production systems, from annual crops to orchard or vineyard crops. In perennial crop systems, cover crops are often maintained continuously under the main crops, while in annual cropping systems, they may be used as a relay crop; for example, as a "winter crop" in temperate systems (Hartwig and Ammon, 2002) or a post-rainy season crop in tropical and subtropical monsoonal systems (Venkateswarlu *et al.*, 2007). Winter wheat is a common relay crop following maize or rice (*Oryza sativa*) in subtropical and temperate areas (Derpsch *et al.*, 2010; Duiker and Thomason, 2014; Hongwen *et al.*, 2014; Lafond *et al.*, 2014).

While one of the purposes of cover cropping is to maintain vegetative cover of the soil, in temperate or seasonally dry crop systems, continuous live plant cover is not expected. However, maintaining soil cover with plant or crop residues is still important for achieving this third component of CA. Globally, more than half of all dry matter produced by crops is not part of the measured yield, and thus is considered crop residue. In NT or reduced-tillage systems, these residues naturally provide protection of the soil surface, and the crop root systems help maintain soil structure and porosity. In the majority of the land area managed in CA systems, soil cover is maintained primarily, if not exclusively, through crop residue management. Residues from cover crops also are important for soil cover. Less has been published about crop residue practices in modern agricultures, and nations do not normally keep statistics on residue production (Smil, 1999). Thus, harvest indices are typically used to estimate residue production. For cereals, 1–3 t of residue production per hectare is typical, which should provide reasonable levels of soil cover, i.e. 30% or more, to protect against wind and water erosion (Smil, 1999).

However, in many seasonally dry tropical and subtropical areas, traditional small-scale agriculture involves tending of livestock as well as growing crops. In the dry season after harvest of the final crop, fields are often open for communal grazing, as crop residues are an important source of dry season fodder. Indeed, the FAO held a workshop in 1987 on ways to improve the use of crop residues for animal fodder (Reed *et al.*, 1988). Use of residues for fuel, building materials, e.g. making bricks or roof thatching, and even for making paper were historically important and continue to be so in rural and low-income areas of developing countries. Integrated crop–livestock management is an active and important area of research and development for enabling smallholders to maintain organic soil cover in seasonally dry areas.

1.1.4 Integration of practices in CA systems

Conservation agriculture is ideally an integrated set of practices that complement each other and provide synergies for soil protection and quality enhancement (e.g. Villamil *et al.*, 2006). One of the benefits of reduced tillage, or ideally NT, is the maintenance of crop residues on the soil surface. Relay and intercropping provide additional crop residues and live plant cover, and crop rotations can include a fallow period using a cover crop. Finally, good soil cover and crop rotations over time suppress weed germination and growth and improve soil quality, reducing the need for tillage.

While the integration of CA practices can support low-input or organic agriculture, many farmers in low-input systems value tillage for weed control. Thus, historically, maintaining soil cover in reduced tillage systems has been facilitated by the development of effective and low-cost herbicides. However, long-term integrated CA systems can reduce annual weed pressure, allowing for reduced herbicide use. Combined with higher system biodiversity and increased organic matter and nutrient cycling, CA can promote reduced requirements for chemical inputs and labor (mechanical or manual) to generate similar productivity. The historical, and even current, association of CA systems with high chemical inputs is most likely due to transitional needs for weed control and fertilization until the system can generate the benefits of higher biodiversity and soil quality. Other reasons include a lack of full integration of practices or a lack of confidence by farmers that CA can replace the need for such inputs.

These potential synergies and complementarities of CA practices have been recognized in the modern history of CA development and promotion. CIMMYT's focus on NT rather than reduced-tillage systems was deliberately a way to maximize retention of crop residue cover. In the USA, tillage, crop rotation, and soil cover were promoted separately and customized to various cropping systems and farmer preferences and technical capabilities. However, their complementarity was noted, and combining practices has been promoted to reduce runoff and soil erosion better, maintain soil organic matter, and reduce the incidence of pests, diseases, and even specific weeds.

While large-scale mechanized production is responsible for most of the total area in CA, the origins, as well as much potential for future application, are in small-scale systems. In *The One Straw Revolution*, author and Japanese farmer, Masanobu Fukuoka, describes and promotes a NT rice–wheat system in which the residues from both crops provide significant soil cover to control weeds and maintain soil organic matter and nutrients (Fukuoka, 1978). This system is not widespread in Japan or elsewhere, but it relies on the integration of CA practices to maintain high productivity in a non-mechanized, low-input agricultural system.

1.2 Development of CA Around the World

Globally, there are more than 150 Mha currently classified as CA by the FAO (FAO, 2014b). This includes 27 countries that each has more than 100,000 ha (Table 1.1). Several countries, including Argentina, Australia, Brazil, Canada, Paraguay, and Uruguay, have at least one-third of their total arable land under CA.

Growth in CA has occurred primarily in the last 30 years. In the 1984–1987 period, there were only 6 Mha reported in CA (Fig. 1.1), three-quarters of which were in the USA. During that period, only eight countries reported any significant area in CA. Currently, there is documented CA in more than 50 of the 200 sovereign states reporting to the FAO. Presented below are summaries of the history of CA development and adoption from various regions of the world, with a focus on tropical and subtropical regions. For a more comprehensive global review of the development and current state of CA, see Jat *et al.* (2014).

Table 1.1. Countries with at least 100,000 ha in conservation agriculture (CA).

Country	Area under CA (2005–2014)[a] (1,000 ha)	Arable land (%)
USA	35,613	22
Brazil	31,811	44
Argentina	27,000	71
Canada	18,313	39
Australia	17,695	36
China	6,670	3
Russia	4,500	4
Paraguay	3,000	62
Kazakhstan	2,000	7
India	1,500	nd
Uruguay	1,072	36
Spain	792	5
Ukraine	700	2
Italy	380	1
Zimbabwe	332	3
Venezuela	300	11
Finland	200	7
France	200	1
Germany	200	nd
Zambia	200	06
Chile	180	14
New Zealand	162	34
Mozambique	152	3
UK	150	2
Colombia	127	6

[a]Period over which the latest data were reported. nd, no data.

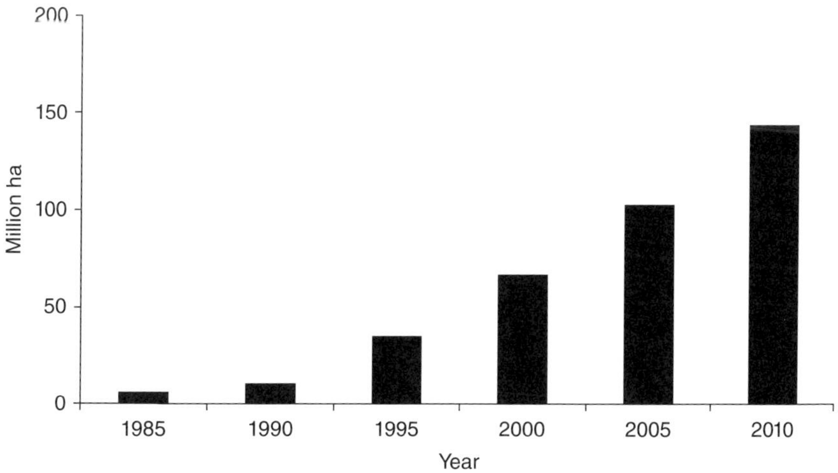

Fig. 1.1. Growth of global area (million ha) in conservation agriculture, 1985–2010.

1.2.1 USA

While most land area in the continental USA lies in the temperate zone, this country has been central to the development and thus modern history of CA. As a pioneer in the rise of mechanized agriculture and Green Revolution technologies, it experienced rapid agricultural expansion, the use of inorganic fertilizers to support multi-year monocropping, and the use of pesticides to control weed and pest infestations. The Dust Bowl in the 1930s made apparent the huge problems of accelerated soil erosion. Subsequent efforts to promote conservation resulted in the development of reduced tillage and NT implements, herbicides to support reduced tillage, and other practices that were compatible with large-scale mechanized agriculture. Currently, the USA is the world's leader in NT area cultivated at more than 35 Mha (FAO, 2014b).

While the problems of soil erosion were recognized early in the 20th century, it was the efforts of a few public champions, such as Hugh Bennett, a soil scientist with the US Department of Agriculture (USDA), that raised public and important political awareness of the need to take action. In 1928, he co-authored a government circular with W.R. Chapline, a grazing inspector with the US Forest Service, entitled *Soil Erosion: A National Menace* (Bennett and Chapline, 1928). The US Soil Erosion Service was created in 1933, with Bennett appointed as its head. In March of 1935, two separate major storms carried clouds of dust from the Great Plains over Washington, DC, darkening the skies just as Congress was holding hearings on a proposed soil conservation law. Bennett used these events to write editorials in the press and testify to Congress on the need for the creation of a permanent soil conservation agency. By April of that year, the Soil Conservation Service (SCS) was created (NRCS, 2014).

The Dust Bowl not only led to the development of the SCS but also spurred the development of conservation tillage practices and implements. As mentioned previously, tillage equipment was created to reduce turning of the soil and thus maintain better soil structure to resist erosion. Chemical herbicides developed in the late 1950s and early 1960s provided farmers more effective weed control. Commercial NT production began in the 1960s. By the early 1990s, the reduced cost of broad-spectrum herbicides made true NT planting economically feasible, which facilitated the spread of integrated CA systems.

Land grant universities have been central to research on and development of CA systems in the USA, as well as for extension and education. The SCS also supported NT research and demonstration efforts. Coordinated programs to promote NT began in the 1980s and 1990s, e.g. the mandate in the 1985 US Farm Bill for producers to reduce erosion on marginal lands to qualify for funding programs (Brock *et al.*, 2000). Finally, producer-led CA organizations have developed in various states and regions of the USA, often as a partnership among producers, government or university agriculture extension, and industry, e.g. equipment and input suppliers.

Continued innovation in equipment and experimentation with different cropping systems has made conservation tillage a viable option for most farmers in the USA. Currently, the largest areas under CA production reflect the major cropping systems, including maize; soybean (*Glycine max*); grain sorghum (*Sorghum bicolor*); small grain crops such as barley (*Hordeum vulgare*), oats (*Avena sativa*), and wheat; cotton (*Gossypium* spp.); and forage crops such as lucerne (*Medicago sativa*); and grasses grown for hay. Only soybeans, though, have more than 30% of total production in NT or conservation tillage systems (Duiker and Thomason, 2014).

1.2.2 South America

In Brazil, NT and CA arose in response to major agricultural expansion in the 1960s (Ekboir, 2002). As in the USA, the expansion of conventional agriculture led to massive increases in accelerated erosion, which in some cases resulted in crop failures and defaults on agricultural loans. In the 1970s in southern Brazil, farmers began to see the damaging effects of continuous tillage and, conversely, the benefits of NT and CA research. They partnered with farm equipment and chemical suppliers to test and promote these new practices on large and medium-sized farms. The spread of CA was inhibited by skeptical politicians who thought it was only relevant for large-scale farms, which limited research and extension efforts. However, the success of CA in the field eventually won over politicians and professionals, and state and national agriculture service agencies have now made CA an integral part of their research and support programs (Pieri *et al.*, 2002).

Promotion of NT for smallholders in Brazil began in earnest in the 1990s. This required the development of downsized planters and training of extension agents (Derpsch *et al.*, 2010). In a survey, 90–100% of farmers cited time and labor savings as reasons for adopting NT (Melo, 1997, in Pieri *et al.*, 2002). Short-term subsidies for the purchase of equipment were important for adoption by smaller farms (Pieri *et al.*, 2002). Also important has been the integration of livestock into CA systems, especially the development of rotational grazing and cropping patterns. Finally, farmer organizations have been important not only for promotion but also for innovation and adaptation of CA to new areas and cropping systems. As a result, Brazil has developed CA for a wide variety of crops and agricultural systems, as well as farm sizes and sophistication (Pieri *et al.*, 2002). Brazil currently has 31 Mha in CA, second only to the USA (FAO, 2014b).

The sustainability of NT in Brazil was found to depend on an integrated package of CA practices, as NT alone did not provide the expected benefits. As a result, most NT farmers also incorporate green manure/cover crops and crop rotation (i.e. the entire CA package). As well, a majority of Brazilian CA farmers practice "permanent no-till", meaning once they begin NT, they never plow again. In the USA, most farmers practicing NT will plow the field every few years.

In Argentina, the central plains are naturally grasslands growing on deep loess soils, known as the Pampas, much like the prairies of the central USA and Canada. In the 20th century, agriculture expanded rapidly through exploitation of these fertile soils, but erosion concerns led to efforts to develop soil and water conservation measures. Early experiences with NT farming in the 1970s were disappointing due to poor performance of herbicides, machinery, and technical knowledge (Derpsch *et al.*, 2010). This changed with the development of an NT farmers' association in 1989, La Asociación Argentina de Productores en Siembra Directa (AAPRESID). Coupled with this was an increase in overall production costs but a decrease in the cost of glyphosate, as in the USA. Mainly through the efforts of AAPRESID, NT and CA production expanded, especially in wheat, maize, and soybean systems. This promotion helped change the belief that plowing was necessary to grow crops. Equipment manufacturers responded to increased demand with large and small NT seeders. Government-sponsored research on CA followed this initial expansion; however, there are still no significant government policies favoring or promoting CA. Adoption of cover crops along with NT lagged until recently, when research demonstrated that cover crops could enhance water-use efficiency. Major benefits to farmers have included a 95% reduction in wind erosion, better soil water infiltration, reduced water use per unit crop yield, reduced production costs, and more stable yields (Albertengo *et al.*, 2014). From 1995 until 2010, the area under CA in Argentina grew from 500,000 ha to 27 Mha, trailing only Brazil and the USA. Argentina is the world's leader in percent of arable land in CA, at 71% (FAO, 2014b).

Other countries in Latin America also have seen large increases in the adoption of CA, including Uruguay, Bolivia, Venezuela, Chile, and Colombia, all with more than 100,000 ha currently in CA production. Paraguay currently has 3 Mha in CA and is second only to Argentina in percent of cultivated land in CA (62%). In Paraguay, CA was developed in response to the introduction of soybean and wheat to southern and eastern parts of the country in the early 1970s. Conventional tillage caused accelerated erosion and land degradation. The CA systems being developed in neighboring southern Brazil were imported and adapted to local conditions (Sorrenson, 1997). Besides the well-known benefits to reducing soil loss and savings of time and labor related to plowing, NT allowed for earlier planting of a second crop after harvesting of the first crop, extending the total cropping season by several weeks to months. An economic study in Paraguay noted that CA required skill development in managing a larger number of crops and managing weeds through the proper use of herbicides, cover crops, and crop rotations (Sorrenson, 1997). Typical CA crop rotations include wheat, oats, soybean, maize, and sunflower (*Helianthus* spp.). Common cover crops include *Crotolaria* and *Vicia* spp.

1.2.3 Africa

In terms of agricultural productivity and gains from Green Revolution technologies, Africa as a whole continues to lag far behind other regions of the world. While many areas of the continent experience distinct dry seasons and low overall precipitation (<1,000 mm annually), the old, highly weathered soils and lack of rural infrastructure and development are major factors constraining the adoption and success of these technologies. While CA may seem to hold much promise to sustain or even improve poor soils and allow farmers to realize the benefits of Green Revolution technologies, CA is most readily adopted where equipment, seeds, and other inputs are available and affordable for producers. This is simply not the case in many countries of Africa.

That said, CA has been promoted and developed throughout the African continent in the past 20 years. The success of CA in southern Brazil, Argentina, and Paraguay in the 1990s led to interest from researchers and professionals in Africa. Study tours and workshops led to the development of CA programs in Zambia, Tanzania, and Kenya (Ekboir, 2002). In southern and eastern Africa, more than 100,000 farmers are participating in NT and CA trials. However, the vast majority of producers are small. Thus, even though the adoption rate is high in some places, the total and percent area under NT or CA is generally small (Derpsch *et al.*, 2010; FAO, 2014a).

In North Africa, agricultural research organizations such as the International Center for Agricultural Research in Dry Areas (ICARDA) in Morocco and Tunisia and the French Agricultural Research Centre for International Development (CIRAD) in Algeria began working with farmers to develop CA systems, primarily for wheat production (Boulal *et al.*, 2014). Crucial to this was the importation and adaptation of a Brazilian-made NT seed planter. Research and on-farm trials have shown increases in soil organic matter and wheat yields after 2–3 years, reduced production costs, and better water infiltration and retention. Continuing challenges include the traditional use of crop residues for dry-season animal fodder, management of increased weeds in the first few years, and availability of affordable equipment, primarily NT seed planters. Conversion to CA requires a new set of knowledge and skills, so farmer training and local capacity building are important. Currently, CA adoption is quite low, <10,000 ha in Tunisia and <5,000 ha in Morocco, which represents <1% of the cultivated area (FAO, 2014a).

In sub-Saharan Africa, agriculture still largely consists of small farms, primarily farmed with manual labor. Much of the area is semi-arid to arid, with a prolonged dry season. Major crops grown include maize, rice, sorghum, and millet. Fields are generally open for grazing by livestock in the dry season. Farmers have practiced traditional soil and water conservation measures associated with low-input and small-scale agriculture, including zaï pits, half-moon depressions, stone bunds along sloping lands, direct seeding, and shifting cultivation, including a several-year fallow period to restore soil quality. Given the small size and large diversity of farmers and farming systems in this region, the development of CA has also been diverse, and the adoption of CA practices has often been partial and transitional (Nyamangara *et al.*, 2014).

In Ghana, there was early experimentation with NT systems in the 1960s and 1970s. However, it was not until the 1990s that CA was promoted to farmers through a partnership between the Sasakawa Africa Association based in Switzerland and the Monsanto Corporation. Direct seeding and use of herbicides for weed control were the primary practices encouraged through this effort. In the 1980s, CA was introduced on a commercial farm in Zimbabwe in order to reduce soil erosion and stabilize crop yields (Nyamangara *et al.*, 2014). Since then, CA has been promoted widely in the region through projects funded by European and US international development agencies (Nana *et al.*, 2014). The adaptation of indigenous practices, such as creating small depressional planting basins, has been developed to concentrate resources, capture water, and reduce labor while achieving the goals of reduced soil disturbance. Where animal-drawn power is available, modified conservation tillage implements have been developed and promoted (Nyamangara *et al.*, 2014).

The adoption of CA in sub-Saharan Africa has been greatest in countries with larger, mechanized farms, such as South Africa (368,000 ha), Zimbabwe (332,000 ha), Zambia (200,000 ha), and Mozambique (152,000 ha) (FAO, 2014b). This also tends to coincide with an integration of outside expertise, government policies, and farmer organizations to promote CA, as in Zambia (Baudron *et al.*, 2007; Nyamangara *et al.*, 2014). Overall, smallholder adoption remains a challenge in this and other regions. In a study in Burkina Faso where CA was promoted, adoption of the entire CA package was 64%, but on average, farmers were devoting only 15% of their land to CA (Nana *et al.*, 2014). In general, small farm size and high proportion of degraded land discourage the investments needed to transition to CA. By contrast, among smallholders, having a large number of livestock provides financial support to invest in CA. Outside financial support for soil conservation and attendance at farmer field schools also improved adoption.

As in other areas, CA has resulted in labor and fuel savings related to land preparation and weeding. However, intercropping or crop rotations have increased labor requirements. Over time, yields of staple crops like maize tend to increase. For smallholders, the labor for intercropping or crop rotation often generates a positive return to total yield, labor productivity, and household income. For low-input farms, though, low crop yields may not generate sufficient crop residues to meet the 30% requirement of organic soil cover. Cover crops have been promoted as one solution, but farmers in some cases have not considered this a high priority (Baudron *et al.*, 2007). Part of this is likely due to the tradition of communal grazing during the dry season. If farmers cannot control access to their lands during the dry season effectively, maintaining organic soil cover will be a challenge. More importantly, CA promotion and development for smallholders has largely been initiated and supported by external donors and international agriculture research organizations. Only recently have Africa-based organizations been created to support CA and sustainable agriculture, and these are mostly multinational associations (Nana *et al.*, 2014). The promotion of CA through regional policies is also limited and recent. Again, in areas where such support is available, such as Zambia, adoption and spread is much greater. However, it is likely that such support and integration are enabled by the generally better economic status of the farmers and the country.

1.2.4 South and Southeast Asia

Despite their importance to the world population and, increasingly, economic output, surprisingly little has been published about the history or status of CA in South and Southeast Asia. In China and India, CA systems mostly involve wheat or rice–wheat systems. In the warmer and more humid countries of Southeast Asia, CA has been developed for an array of cropping systems, including paddy rice, vegetable production, and adaptations for steeply sloping lands. Despite the relative lack of published information, valuable lessons can be learned from the history of CA development and continuing challenges for adoption.

In China, NT technologies were first developed and promoted in the 1970s, but the NT seed planters at the time were designed for large-scale farms. China was dominated by very small farm sizes (averaging 0.5 ha) and high food production demands. It was not until the 1990s that NT seeders were adapted for smaller systems, including manual seeders, implements suitable for two-wheel tractors, and more typical two- or four-row seeders and conservation tillage equipment designed for four-wheel tractors. In the 2000s, CA expanded into the larger-scale maize–wheat growing areas of the North China Plain, and CA equipment was developed for effective strip or chisel tillage systems. The government promoted conservation tillage and CA systems in this region due to its vulnerability to wind erosion and degradation. Thus, by 2005, when the FAO first reported CA cultivation in China, there were already 100,000 ha under CA. This grew to 3.1 Mha in 2010, and is currently estimated at 6.7 Mha (FAO, 2014b).

Conservation agriculture has resulted in improved soil water infiltration and storage, reduced wind erosion, higher soil organic matter and total nitrogen, and higher available phosphorus. As a result, crop yields, for example winter wheat, tend to be higher and more stable than under conventional tillage. Concerted efforts by the government since the 1990s have resulted in model systems adapted to a variety of local conditions. Problems are dealt with as challenges to be overcome, rather than as reasons not to adopt or promote CA, and the rapid growth in CA area over the past 10 years is a testament to its potential to provide multiple benefits to farmers and the natural resource base. Perhaps as importantly, China's innovation with CA equipment and manufacturing capabilities has allowed them to export these technologies to surrounding countries, enabling CA for other small and medium-sized farms in the region.

To the south of China, in the Indo-Gangetic plain of northern India and encompassing parts of Pakistan, Nepal, and Bangladesh, there is a mixture of cropping systems based on rainfall. The western region is dry and contains a diversity of cropping systems. The eastern region is wetter, and natural flooding is common. Here, rice followed by wheat is the dominant cropping system. Rice is planted during the monsoonal wet season, followed by wheat, which depends on less frequent rain and residual soil water. Chemical weed control had been developed previously, so NT packages have been available since the 1990s, mainly for wheat. Technologies have been developed for two- and four-wheel tractors, and even animal-drawn planters. Systems have been developed for establishing rice in puddled and unpuddled flat and raised beds for both direct seeding and transplantation (Hobbs *et al.*, 2003). Herbicide and water management are necessary to control weeds, but the labor savings of NT mean overall production costs are similar. Networking and support from several agricultural development organizations has helped with research, development, training and extension. Unofficial estimates of the area of NT rice–wheat in the region range from 2 to 5 Mha (Derpsch *et al.*, 2010). The FAO estimates 1.5 Mha of CA for all of India (FAO, 2014b).

As later chapters in this book describe, CA systems for other smallholder producers in India and Nepal are only now being developed and promoted through a project initially funded by the US Agency for International Development (USAID). These upland systems are based on maize or maize and millet, and they follow the monsoonal rains in the late spring and summer. Leguminous crops like cowpea (*Vigna unguiculata*) and horse gram (*Macrotyloma uniflorum*) are common in the area, so integration in CA systems as intercrops or relay crops is generally accepted by farmers. Because of the extended dry season and associated communal livestock grazing, farmers are limited in their ability to grow cover crops or maintain organic soil cover. In India, one solution is the management of mustard (*Brassica juncea*) residues, which are generally avoided by cattle. The crop is normally harvested and threshed for the seeds, so in experimental CA plots, the remaining stems are collected and returned to the fields rather than being burned, which is the traditional practice.

As in other regions, CA faces many challenges for the poorest farmers in India and Nepal. Tillage helps control weeds and incorporate fertilizers, primarily farmyard manure, so in Nepal, experiments with strip tillage have led to lower manure incorporation and thus lower crop yields. While intercropping of cowpea allowed for greater overall yields and labor productivity, labor associated with sowing and harvesting the crop offset gains from reduced tillage. Given the increased yields and income from the CA system in India, farmers in surrounding villages have readily adopted the CA package, and the state government of Orissa has committed to demonstrating the system on 500 ha to encourage further adoption. In Nepal, farmers are less encouraged by conservation tillage and thus are engaged in partial adoption only. Village isolation also limits farmer-to-farmer dissemination of experiences and the promotion of technologies by local non-governmental organizations (NGOs) and universities.

In the Philippines, sloping agricultural land technologies (SALT) have been pursued since the 1970s to reduce soil erosion and increase the sustainability of agriculture in these vulnerable areas. A focus was on establishing narrow terraces (3–5 m wide) bordered by hedgerows of nitrogen-fixing shrubs. Pruning of the shrubs was intended to reduce competition with the crops and provide green manure and soil cover for the cropped area. While successful in terms of soil quality and reduced erosion, the sustainability of these systems is questionable, as it becomes increasingly difficult to manage competition from the hedgerow shrubs. As well, economic pressures to increase yields and competing demands for labor have led to a reduction in the use of hedgerow species (Lienhard *et al.*, 2014). In the 1990s, a variety of projects in Southeast Asian countries focused on the development of improved fallows (2 years or less) using leguminous cover crops. As in other smallholder systems, communal grazing in the dry season and a dependable seed supply were important challenges for the sustainability and dissemination of these practices.

As the definition of CA coalesced in the early 2000s, a number of projects have been initiated in South and Southeast Asia, especially by CIRAD, USAID, and other international development and agricultural research organizations. The cropping systems being evaluated range from rubber (*Hevea brasiliensis*) and tea (*Camellia sinensis*) to maize, cassava (*Manihot esculenta*), soybean, rice, and forages (Lienhard *et al.*, 2014). The variety of systems reflects the various agroecological situations, ranging from sloping uplands where farmers are engaged in subsistence-based agriculture using shifting cultivation to intensive mixed cropping systems on rainfed plains. In Cambodia, Laos, Vietnam, and the Philippines, various combinations of relay and rotational systems of cassava, maize, soybean, rice, forages, and/or cover crops are being evaluated on small but generally mechanized farms. In Cambodia, mulching of the cover crop is followed by direct seeding of the main crop, providing excellent soil cover and weed suppression (Lienhard *et al.*, 2014). The equipment used in the project has been imported from Brazil and is appropriate for two- or four-wheel tractors common in the area. However, with equipment innovations and development in places like China and Thailand, costs and availability should be more suitable for broader adoption of these CA systems.

The outcomes of CA adoption on labor requirements and crop yields vary by agroecological situation. In rainfed plains, there is a general reduction in land preparation and weed control costs along with similar or increased yields for major crops, resulting in positive economic returns. For subsistence agriculture in uplands, moving away from shifting cultivation to settled agriculture increases labor requirements and production costs, but generally with the benefit of increased yields. The net effects on economic returns range from very high to near neutral. Increases in labor required for mulching and direct seeding were cited as being an important disincentive for CA adoption in Laos (Lienhard *et al.*, 2014). By contrast, in India and Nepal, farmers ranked increased labor demands as the lowest priority in selecting various CA practices or systems (Lai *et al.*, 2013).

1.3 Lessons from History

The modern history of CA development offers valuable lessons for consideration as CA is introduced, developed, promoted, and disseminated to new farmers and areas. Perhaps the most obvious lesson is that the development of appropriate equipment and inputs greatly facilitates NT planting and management of weeds and cover crops. Farmers, philosophers, politicians, and the general public have been aware of and concerned about the problems of soil erosion associated with tillage practically since farming began, but the rise of modern CA systems did not occur until innovations in farm equipment allowed for efficient and effective land preparation and sowing. This is evident not only from the history in early adopting countries like the USA and Brazil but also it is clear from the recent history of CA in China and sub-Saharan Africa. This does not mean CA is appropriate for or likely to be adopted only by larger mechanized farms, but it is the case that these farms have often been the early adopters, and due to their large size, have a disproportionate impact on the total area in CA. Downscaling and adaptation of equipment for two-wheel tractors, animal draft power, and even hand-held planters, have enabled the implementation of CA for smallholders in various regions of the world. That said, affordability and availability of such equipment remain major constraints to the adoption and spread of CA for smallholders, especially in South and Southeast Asia and sub-Saharan Africa. Short-term subsidies for equipment purchase have been important for adoption by smaller farms in places like Brazil (Pieri *et al.*, 2002).

The rise of effective, low-cost herbicides was another key development in the history of CA adoption and spread. Chemical rather than mechanical control of weeds reduces soil disturbance and can lower production costs, improving the economic returns of CA adoption. For medium- and large-scale agriculture, chemical weed control will continue to be an essential component of agricultural systems not dedicated to organic production. However, maintaining good soil cover in CA systems can and should reduce weed pressure. For smallholders not already using herbicides, this means CA systems do not generally require the adoption of chemical weed control to realize net economic benefits and a reduction in weed management effort.

In the seasonally dry tropics and subtropics, maintaining good soil cover in CA systems means integrating crops with livestock. In smallholder communities, crop residues are an important source of animal fodder, and farmyard manure is an important source of nutrient inputs for crops. The tradition of communal grazing during the dry season is common in Africa and South and Southeast Asia, and farmers wishing to adopt CA cannot simply exclude their fields without making other changes to cropping and grazing systems. Where livestock are an important source of wealth, however, farmers may be more capable of investing in the transition to CA and have options other than grazing in crop fields to maintain good animal nutrition and health. While a thorough discussion of this topic is beyond the scope of the current chapter, it is an active area of research and development, e.g. the Africa Rising program (Africa RISING, 2014). This will have to be an essential component of the further growth of CA among smallholders.

Where CA has been most successful, there is usually a history of a few influential farmers innovating, demonstrating, and promoting CA practices to other farmers. The spread of CA, though, is facilitated greatly by strong local to national farmer-led organizations. This can occur in the absence of strong government policies supporting or favoring CA, but partnerships among farmers, researchers, and government extension and agriculture service agencies are the most effective organizations to meet farmer needs for CA development, knowledge-sharing and skills development, and financial and policy support to invest in CA equipment and practices. In most countries, farmer-motivated development of CA has preceded shifts in policy and action by governments, but China offers a successful if somewhat unique model of how concerted efforts by the government can initiate and facilitate the rapid adoption and spread of CA for specific areas and cropping systems.

For farmers, successful CA implementation requires the acquisition of new knowledge and the development of new skills. A shift in perspective from agriculture as a set of prescriptive practices to a complex managed ecosystem is a necessary foundation for developing, adapting, and optimizing CA systems, especially considering crop–livestock integration. Farmers are generally aware of the negative effects of tillage and continuous monocropping on long-term yields and natural resource quality. Indeed, in counties such as Brazil and Argentina, farmer concern about land degradation was the motivation for the development and spread of NT and CA practices. However, the adoption of CA requires significant investments of time and resources by the farmer, and the transition process will likely require several years to begin realizing the long-term benefits. Thus, it is no surprise that CA adoption has been much more widespread where farmers have the resources to make these investments and stay committed through the transition period.

Just as important for farmers is the development of analytical and communication skills (Giller *et al.*, 2011). Especially for smallholders, using evidence-based approaches to address complex and context-specific problems is generally more effective than applying standard prescriptions. Effective farmer-to-farmer communication and sharing of experiences and findings are critical for CA dissemination. The benefits of such skills often extend beyond the existing project to other challenges of and opportunities for sustainable production, such as integrated pest and soil fertility management (Vanlauwe *et al.*, 2010). These skills, therefore, help build the capacity of farmers and farm communities to adopt and adapt new technologies and approaches to existing and changing agroecological conditions.

Facilitating this change in perspective and the acquisition of knowledge and skills are major tasks for CA development. There is a growing consensus on the need for collaboration among CA stakeholders and for participatory technology development (e.g. Pieri *et al.*, 2002; Giller *et al.*, 2009). Farmer field schools and farmer participation in experimental trials are common approaches to engaging them in the development process and for facilitating training (Van den Berg and Jiggins, 2007). This requires researchers and extension professionals to adapt strategies, practices, and technologies to farmer knowledge, capabilities, and preferences (e.g. Giller *et al.*, 2011). Promoting CA to smallholders means facilitating their participation in all aspects of the development process, from identifying and prioritizing needs, to selecting and implementing CA strategies and practices, to analyzing and interpreting outcomes, to sharing of knowledge and experiences, and training of other farmers.

This participation by farmers in the education process, and conversely by researchers and professionals in the learning process, helps to build both local and institutional capacity for continued innovation, adaptation, and dissemination of CA. Further spread of CA is then accomplished more easily through the creation or strengthening of farmer-led organizations and appropriate institutional support. Government and industry can address technical and financial constraints better when farmers and CA researchers and professionals have clearly identified the needs, strategies, and approaches to improving and sustaining agricultural production systems. Advocates of participatory appraisal and agricultural development have known this for several decades (Rhoades and Booth, 1982; Haverkort *et al.*, 1991), and it is essential for effective CA development projects involving smallholders.

Finally, one of the emerging lessons largely ignored in studies or reports on the history of CA development is that gender matters in agriculture. Again, sociologists and anthropologists have studied and written about this topic for several decades, but it is only recently that CA development projects have considered and analyzed gendered knowledge and roles in agriculture, and the impacts of changes in production practices and systems. There is real concern that for smallholders, CA may increase the labor burden for women, who are often responsible for weeding (Giller *et al.*, 2011), or fail to consider their role in the overall production system, such as where women primarily tend livestock (Erenstein, 2011). Conversely, decision making in CA systems development may involve primarily men. For example, since men generally own or operate tractors, draft animals, and the associated agricultural implements (Hassanein, 1997), women's issues and concerns can be marginalized. The emphasis of CA on staple crop production is certainly laudable, but in some cases it may disenfranchise women, as men tend to control land for staple crop production as opposed to land dedicated to vegetable production or "garden" crops (Hassanein, 1997). The converse – supporting garden plots and household vegetable production – can provide benefits for women specifically, as well as household nutrition, food security, and income more generally. While there are published methods for conducting gender analysis in agriculture (Feldstein and Jiggins, 1994), this must be accompanied by gender-specific approaches for building knowledge and skills in practices and cropping systems relevant for women, and in supporting gender-specific, farmer-to-farmer communication and training. Fortunately, understanding and promoting benefits and equality for women in rural development generally and CA projects specifically is becoming a central goal for development agencies like USAID, which is reflected in current CA projects (e.g. Lai *et al.*, 2012; Harman Parks *et al.*, 2014). The future success of CA adoption and dissemination will benefit from gendered analysis and the promotion of benefits to women specifically, as well as farm households more generally.

The following chapters on the experiences of CA development and promotion in the South Asia region reflect much of the history of CA globally, and the continuing and emerging issues with the development and adoption of CA among smallholders. It is hoped that these case studies and ongoing research and development projects will inform and engage readers in this critically important area of human and environmental as well as agricultural development, and perhaps inspire a new generation of scientists, professionals, service providers, and even a few farmers to continue the long history of the innovation and implementation of CA around the world.

References

Africa RISING (2014) Africa research in sustainable intensification for the next generation. Available at: http://africa-rising.net/ (accessed 1 August 2014).

Albertengo, J., Belloso, C., Giraudo, M.B., Peiretti, R., Permingeat, H., *et al.* (2014) Conservation agriculture in Argentina. In: Jat, R.A., Sahrawat, K.L. and Kassam, A. (eds) *Conservation Agriculture: Global Prospects and Challenges*. CAB International, Wallingford, UK, pp. 352–374.

Baudron, F., Mwanza, H.M., Triomphe, B. and Bwalya, M. (2007) *Conservation Agriculture in Zambia: A Case Study of Southern Province*. FAO, Rome.

Bennett, H.H. and Chapline, W.R. (1928) *Soil Erosion: A National Menace*. US Department of Agriculture, Circular No 33. US Government Printing Office, Washington, DC.

Boulal, H., El Mourid, M., Ketata, H. and Nefzaoui, A. (2014) Conservation agriculture in North Africa. In: Jat, R.A., Sahrawat, K.L. and Kassam, A. (eds) *Conservation Agriculture: Global Prospects and Challenges*. CAB International, Wallingford, UK, pp. 293–310.

Brock, B., Canterberry, J. and Naderman, G. (2000) Ten milestones in conservation tillage: History and role in the North Carolina conservation program. In: Sutherland, J. (ed.) *Proceedings of the 43rd Annual Meeting of the Soil Science Society of North Carolina*. Soil Science Society of North Carolina, Raleigh, North Carolina, pp. 13–18.

Davis, J.H.C., Roman, A. and Garcia, S. (1987) The effects of plant arrangement and density on intercropped beans (*Phaseolus vulgaris*) and maize II. Comparison of relay intercropping and simultaneous planting. *Field Crops Research* 16, 117–128.

Derpsch, R., Friedrich, T., Kassam, A. and Hongwen, L. (2010) Current status of adoption of no-till farming in the world and some of its main benefits. *International Journal of Agricultural and Biological Engineering* 3, 1–25.

Dhima, K.V., Lithourgidis, A.S., Vasilakoglou, I.B. and Dordas, C.A. (2007) Competition indices of common vetch and cereal intercrops in two seeding ratio. *Field Crops Research* 100, 249–256.

Diamond, J. (2004) *Collapse: How Societies Choose to Fail or Succeed.* Viking Penguin, New York.

Duiker, S.W. and Thomason, W. (2014) Conservation agriculture in the USA. In: Jat, R.A., Sahrawat, K.L. and Kassam, A. (eds) *Conservation Agriculture: Global Prospects and Challenges.* CAB International, Wallingford, UK, pp. 26–53.

Ekboir, J. (2002) Part 1: Developing no-till packages for small-scale farmers. In: Ekboir, J. (ed.) *CIMMTY 2000–2001 World Wheat Overview and Outlook: Developing No-Till Packages for Small-Scale Farmers.* CIMMYT, Mexico, DF, pp. 1–38.

Erenstein, O. (2011) Cropping systems and crop residue management in the Trans-Gangetic Plains: Issues and challenges for conservation agriculture from village surveys. *Agricultural Systems* 104, 54–62.

Faiella, G. (2006) *The Technology of Mesopotamia.* Rosen Publishing Group, New York.

FAO (Food and Agriculture Organization) (2014a) Conservation Agriculture. Available at: http://www.fao.org/ag/ca/ (accessed 28 March 2014).

FAO (2014b) AQUASTAT Database. Available at: http://www.fao.org/nr/water/aquastat/main/index.stm (accessed 10 July 2014).

Feldstein, H.S. and Jiggins, J. (1994) *Tools for the Field: Methodologies Handbook for Gender Analysis in Agriculture.* Kumarian Press, West Hartford, Connecticut.

Fukuoka, M. (1978) *The One Straw Revolution: An Introduction to Natural Farming.* Rodale Press, Emmaus, Pennsylvania

Giller, K.E., Witter, E., Corbeels, M. and Tittonell, P. (2009) Conservation agriculture and smallholder farming in Africa: The heretics view. *Field Crops Research* 114, 23–34.

Giller, K.E., Tittonell, P., Rufino, M.C., van Wijk, M.T., Zingore, S., *et al.* (2011) Communicating complexity: Integrated assessment of trade-offs concerning soil fertility management within African farming systems to support innovation and development. *Agricultural Systems* 104, 191–203.

Harman Parks, M., Christie, M. and Bagares, I. (2014) Gender and conservation agriculture: Constraints and opportunities in the Philippines. *GeoJournal* 2014, 1–17.

Harrison, P.D. and Turner, B.L. (1978) *Pre-Hispanic Maya Agriculture.* University of New Mexico Press, Albequerque, New Mexico.

Hartwig, N.L. and Ammon, H.U. (2002) Cover crops and living mulches. *Weed Science* 50, 688–699.

Hassanein, N. (1997) Networking knowledge in the sustainable agriculture movement: Some implications of the gender dimension. *Society and Natural Resources* 10, 251–257.

Haverkort, B., Kamp, J.v.d. and Waters Bayer, A. (1991) *Joining Farmers' Experiments: Experiences in Participatory Technology Development.* Intermediate Technology, London.

Hobbs, P.R., Gupta, R., Malik, R.K. and Dhillon, S.S. (2003) Conservation agriculture for the rice–wheat systems of the Indo-Gangetic plains of South Asia: A case study from India. In: Garcia-Torres, L., Benites, J., Martinez-Vilela, A. and Holgado-Cabrera, A. (eds) *Conservation Agriculture: Environment, Farmers Experiences, Innovations, Socio-economy, Policy.* Kluwer Academic, Dordrecht, the Netherlands, pp. 169–174.

Hongwen, L., Jin, H. and Huangwen, G. (2014) Conservation agriculture in China. In: Jat, R.A., Sahrawat, K.L. and Kassam, A. (eds) *Conservation Agriculture: Global Prospects and Challenges.* CAB International, Wallingford, UK, pp. 202–222.

Jat, R.A., Sahrawat, K.L. and Kassam, A. (2014) *Conservation Agriculture: Global Prospects and Challenges.* CAB International, Wallingford, UK.

Johansen, E. (1999) *The Encyclopedia of Native American Economic History*. Greenwood Press, Westport, Connecticut.

Lafond, G.P., Clayton, G.W. and Fowler, D.B. (2014) Conservation agriculture on the Canadian Prairies. In: Jat, R.A., Sahrawat, K.L. and Kassam, A. (eds) *Conservation Agriculture: Global Prospects and Challenges*. CAB International, Wallingford, UK, pp. 89–107.

Lai, C., Chan-Halbrendt, C., Halbrendt, J., Shariq, L., Roul, P., *et al.* (2012) Comparative economic and gender, labor analysis of conservation agriculture practices in tribal villages in India. *International Food and Agribusiness Management Review* 15, 73–86.

Lai, C.C., Chan-Halbrendt, C., Halbrendt, J., Naik, D. and Ray, C. (2013) Farmers' preference of conservation agricultural practices in Kendujhar, Odisha using the analytic hierarchy process. In: Mulvaney, M.J., Reyes, M.R., Chan-Halbrendt, C., Boulakia, S., Jumpa, K., Sukvibool, C. and Sobatpanit, S. (eds) *Conservation Agriculture in Southeast Asia and Beyond*. WASWAC Special Publication No 7. Funny Publishing, Bangkok, pp. 85–98.

Lal, R., Reicosky, D.C. and Hanson, J.D. (2007) Evolution of the plow over 10,000 years and the rationale for no-till farming. *Soil and Tillage Research* 93, 1–12.

Larkin, R.P., Honeycutt, C.W., Olanya, O.M., Halloran, J.M. and He, Z. (2012) Impacts of crop rotation and irrigation on soilborne diseases and soil microbial communities. In: He, Z., Larkin, R.P. and Honeycutt, C.W. (eds) *Sustainable Potato Production: Global Case Studies*. Springer New York, pp. 23–42.

Lienhard, P., Boulakia, S., Legoupil, J.-C., Gilard, O. and Séguy, L. (2014) Conservation agriculture in South-east Asia. In: Jat, R.A., Sahrawat, K.L. and Kassam, A. (eds) *Conservation Agriculture: Global Prospects and Challenges*. CAB International, Wallingford, UK, pp. 180–201.

Mann, C.C. (2005) *1491: New Revelations of the Americas Before Columbus*. Knopf, New York.

Melo, I.J.B. (1997) Validação de semeadoras tração animal em Sistema Plantio Direto. In: *Anais do II Seminário Internacional do Sistema Plantio Direto*. Embrapa – CNPT, Passo Fundo, Rio Grande do Sul, Brazil.

Nana, P.D., Dugué, P., Mkomwa, S., Da Sansan, J.B., Essecofy, G., *et al.* (2014) Conservation agriculture in West and Central Africa. In: Jat, R.A., Sahrawat, K.L. and Kassam, A. (eds) *Conservation Agriculture: Global Prospects and Challenges*. CAB International, Wallingford, UK, pp. 311–338.

NRCS (Natural Resources Conservation Service) (2014) 75 Years Helping People Help the Land: A Brief History of the NRCS. Available at: http://www.nrcs.usda.gov/wps/portal/nrcs/detail/national/about/history/?cid=nrcs143_021392 (accessed 27 March 2014).

Nyamangara, J., Chikowo, R., Rusinamhodzi, L. and Mazvimavi, K. (2014) Conservation agriculture in Southern Africa. In: Jat, R.A., Sahrawat, K.L. and Kassam, A. (eds) *Conservation Agriculture: Global Prospects and Challenges*. CAB International, Wallingford, UK, pp. 339–351.

Ofori, F. and Stern, W. (1987) Cereal–legume intercropping systems. *Advances in Agronomy* 41, 41–90.

Pieri, C., Evers, G., Landers, J., O'Connell, P. and Terry, E. (2002) *No-Till Farming for Sustainable Rural Development*. The International Bank for Reconstruction and Development, Washington, DC.

Rhoades, R.E. and Booth, R.H. (1982) Farmer-back-to-farmer: A model for generating acceptable agricultural technology. *Agricultural Administration* 11, 127–137.

Smil, V. (1999) Crop residues: Agriculture's largest harvest. *BioScience* 49, 299–308.

Sorrenson, W.J. (1997) *Paraguay – Financial and Economic Implications of No-Tillage and Crop Rotations as Compared to Conventional Cropping Systems*. FAO Investment Centre, Food and Agriculture Organization of the United Nations, Rome.

Unger, P.W. and Baumhardt, R.L. (2001) Historical development of conservation tillage in the southern Great Plains. In: Stiegler, J.H. (ed.) *24th Annual Southern Conservation Tillage Conference for Sustainable Agriculture*. Oklahoma Cooperative Extension Service, Oklahoma City, Oklahoma, pp. 1–18.

Van den Berg, H. and Jiggins, J. (2007) Investing in farmers – the impacts of farmer field schools in relation to integrated pest management. *World Development* 35, 663–686.

Vanlauwe, B., Bationo, A., Chianu, J., Giller, K.E., Merckx, R., *et al.* (2010) Integrated soil fertility management: Operational definition and consequences for implementation and dissemination. *Outlook on Agriculture* 39, 17–24.

Venkateswarlu, B., Srinivasarao, C., Ramesh, G., Venkateswarlu, S. and Katyal, J.C. (2007) Effects of long-term legume cover crop incorporation on soil organic carbon, microbial biomass, nutrient build-up and grain yields of sorghum/sunflower under rain-fed conditions. *Soil Use and Management* 23, 100–107.

Villamil, M.B., Bollero, G.A., Darmody, R.G., Simmons, F.W. and Bullock, D.G. (2006) No-till corn/soybean systems including winter cover crops: Effects on soil properties. *Soil Science Society of America Journal* 70, 1936–1944.

Virgil (1994) *Georgics*. Mynors, R.A.B. and Mynors, R.A. (eds). Oxford University Press, Oxford, UK.

2 Global Perspectives on Conservation Agriculture for Small Households

ADRIAN ARES,[1*] CHRISTIAN THIERFELDER,[2] MANUEL REYES,[3] NEAL S. EASH[4] AND JENNIFER HIMMELSTEIN[1]

[1]Virginia Tech, Blacksburg, Virginia, USA; [2]International Center for Maize and Wheat Improvement (CIMMYT), Harare, Zimbabwe; [3]North Carolina Agricultural and Technical State University, Greensboro, North Carolina, USA; [4]University of Tennessee, Knoxville, Tennessee, USA

2.1 Introduction

Conservation agriculture (CA) entails minimizing soil disturbance, maintaining year-round soil cover, and utilizing crop rotations or mixtures (Kassam *et al.*, 2009; Dubreil, 2011). Worldwide, the area under CA (or at least under one CA component: zero or minimum tillage) has increased vastly in the last 30 years. In 2011, no-till farming was practiced on almost 125 million hectares (Mha) (Friedrich *et al.*, 2012), mainly in USA, Brazil, Argentina, Australia, and Canada. Often, CA is implemented in large commercial estates (Bolliger *et al.*, 2006) such as the Bon Futuro farm in the state of Mato Grosso, Brazil, where genetically modified soybean is grown on 230,000 ha. Reductions in operating costs and decreased erosion were the main drivers for the spread of CA in the Americas (Tyler *et al.*, 1983; Langdale *et al.*, 1991). In the USA, no-till systems provided similar yields to those in conventional tillage-based systems, but at lower fixed and variable costs, as well as greatly reduced soil erosion rates (Pimentel *et al.*, 1995). The success in reducing soil degradation and production costs sparked interest among research and extension organizations to adapt CA to the needs and circumstances of smallholder farmers in developing countries (Benites *et al.*, 1998; Ekboir, 2002). The adoption of CA by small rural households in disadvantaged regions of the world, however, is still fairly limited, due to agronomic, economic, social, and technological constraints (Kassam *et al.*, 2012). In Africa, there are

*Corresponding author; e-mail: aresa@vt.edu

about 1 Mha under CA, of which 40% are in South Africa (Jat *et al.*, 2012), where CA is often implemented on large farms.

The development of CA by small households is being addressed by several programs focused on research, education, and technology transfer under different biophysical and socio-economic conditions. Among these are the Agroecology-based Aggradation-Conservation agriculture (ABACO) project funded by the European Union, the Conservation Agriculture Program of the International Maize and Wheat Improvement Center (CIMMYT), the Conservation Agriculture and Engineering Program of Centre de Coopération Internationale en Recherche Agronomique pour le Développement (CIRAD), and the Feed the Future Innovation Lab for Collaborative Research on Sustainable Agriculture and Natural Resources Management (SANREM) sponsored by the US Agency for International Development. Projects with the International Crops Research Institute for the Semi-Arid Tropics (ICRISAT), the International Institute of Tropical Agriculture (IITA), and the New Partnership for Africa's Development (NEPAD) have generated knowledge and transfer information on site-specific CA practices for small farm households (Haggblade and Tembo, 2003; Erenstein and Laxmi, 2008; Ares *et al.*, 2012; Lestrelin *et al.*, 2012; Tittonell *et al.*, 2012; Lienhard *et al.*, 2013). The Food and Agriculture Organization of the United Nations has also been a strong promoter of CA worldwide.

In this chapter, we examine the potential biophysical and socio-economic benefits and constraints for scaling up CA. We also discuss the relationships of CA with the popular topics of sustainable intensification (SI) and resilience. Case studies focused on CA for small landholders in southern Africa and peninsular Southeast Asia are presented.

2.1.1 Potential benefits of conservation agriculture

Most studies on the beneficial effects of CA for small landholders have addressed responses in yield, soil characteristics, and erosion (Table 2.1). There are also a substantial number of studies that deal with the effects of CA practices (e.g. intercropping) on integrated pest management (IPM). Socio-economic aspects related to CA have been less studied.

One of the most desirable benefits garnered from farmers, increased crop yield, often does not arise just after implementing CA practices. However, long-term studies have demonstrated steady increases in yield over time. In the highlands of Mexico, 10 years of wheat and maize rotations with residue conservation increased yields whereas interannual variability decreased (Govaerts *et al.*, 2005, 2009; Fuentes *et al.*, 2009; Verhulst *et al.*, 2011). In this system, the full potential of yield increases was realized after about 5 years, and when optimum inputs and cultivars were used (Govaerts *et al.*, 2005).

Evidence of statistically significant increases in soil organic carbon after switching from plow-based agriculture to CA may take even more time to be realized (e.g. 7–10 years), although changes can occur sooner in tropical soils than in temperate soils (Six *et al.*, 2002). Improved soil physical characteristics and reduced erosion with CA have been widely documented and tend to occur relatively

Table 2.1. Biophysical and socio-economic benefits of conservation agriculture.

Biophysical benefits	
Increased crop yields	Kouyaté *et al.*, 2000; Larbi *et al.*, 2002; Bado *et al.*, 2006; Ngigi *et al.*, 2006; Kosgei *et al.*, 2007; Rockström *et al.*, 2009; Paudel *et al.*, 2012; Thierfelder *et al.*, 2013b
Increased soil organic carbon and nutrients	Buerkert and Lamers, 1999; Muhr *et al.*, 2002; Chivenge *et al.*, 2007; Gwenzi *et al.*, 2009; Moussa-Machraoui *et al.*, 2010; Naresh *et al.*, 2012
Increased mycorrhizal fungi, microbial activity, soil mesofauna, and nitrogen fixation	Rebafka *et al.*, 1993; Rusinamhodzi *et al.*, 2006; Govaerts *et al.*, 2007b, 2008; Formowitz *et al.*, 2009; Thierfelder and Wall, 2010b
Improved soil structure, infiltration, and water content; reduced erosion and sedimentation	Gicheru *et al.*, 2004; Ngigi *et al.*, 2006; Kosgei *et al.*, 2007; Ouattara *et al.*, 2007; Munodawafa and Zhou, 2008; Govaerts *et al.*, 2009; Thierfelder and Wall, 2009, 2010b, 2012; Naudin *et al.*, 2010; Castellanos-Navarrete *et al.*, 2012; Ngwira *et al.*, 2013
Reduced greenhouse gas emissions	Patiño-Zúñiga *et al.*, 2009
Improved integrated management of weeds, insects, diseases, and others	Shenk and Saunders, 1984; Akobundu, 1987; Singh *et al.*, 2005; Govaerts *et al.*, 2007a; Sikirou and Wydra, 2008; Nyasani *et al.*, 2012; Muoni *et al.*, 2013
Socio-economic benefits	
Increased household income; increased gross margins	Ngwira *et al.*, 2012; Bisangwa, 2013; Nguema *et al.*, 2013
Decreased labor burden/ production costs	Bishop-Sambrook *et al.*, 2004; Khan and Hashmi, 2004; Erenstein and Laxmi, 2008; Erenstein *et al.*, 2012
Empowered women	Bishop-Sambrook *et al.*, 2004; Norwegian Agency for Development Cooperation, 2011; Owenya *et al.*, 2011

quickly. Keeping plant residues on site contributes greatly to improved physical and chemical soil characteristics, although residues may delay seed germination because of cooler temperatures (Aulakh *et al.*, 2012).

The effects of CA on greenhouse gases have been scarcely addressed in tropical and subtropical small farming systems. In western Kenya and eastern Uganda, emissions of nitrous oxide (N_2O) in maize–mucuna (*Mucuna pruriens*) cropping systems were 10 times greater at low-elevation sites than at high-elevation locations, but there were no differences with tillage treatments (deep and shallow hoeing and no-tillage). Assimilation of methane (CH_4) was higher in strip intercropping than in relay and farmer practices (J. Odhiambo, Laramie, Wyoming, 2013, personal communication). Laboratory studies with samples taken from CA field trials in the semi-arid, subtropical highlands of Mexico indicated that N_2O and carbon dioxide (CO_2) emissions were lower in permanent raised beds with crop residue retention than in conventionally tilled raised beds, while the opposite effect was found for nitrate (NO_3^-) (Patiño-Zúñiga *et al.*, 2009).

Minimizing soil disturbance, one of the three key factors of CA, can impact pest population levels directly or indirectly. Soil tillage transports seeds of competing vegetation to more superficial soil horizons, where increased light can cause them to germinate (Hobbs, 2007). Therefore, a no-till or minimum-tillage approach might reduce weed pressure. This can also preclude the incorporation of disease-infested plant foliage into the soil, which, depending on the pathosystem, could decrease soil pathogen populations (Pell *et al.*, 2010). No-till methods can also be advantageous for biological control practices. The biocontrol fungus, *Beauveria bassiana*, infected significantly more *Ostrinia nubilalis*, a pest of maize, in no-till fields compared to conventionally tilled fields (Bing and Lewish, 1993). Alternatively, CA could increase the presence of pathogens that have difficulty competing with soil saprophytes and so are more able to survive on the surface debris. An example of this is the higher incidence of the stem rot produced by *Sclerotinia sclerotiorum* of soybean in no-till treatments than in plowed plots (Mueller *et al.*, 2002).

Crop rotations and intercropping can also serve to support IPM-specific goals. Incorporating cover crops such as marigold (*Tagetes patula*) or Sudan grass (*Sorghum* × *drummondii*) into a crop rotation assists in root knot nematode management, as these cover crops produce chemicals that are toxic to the nematode (Widmer, 2000; Ploeg, 2002). The isothiocyantes found in *Brassica* cover crops such as *Brassica rapa*, which is grown in the tropics, are also known to be toxic to a number of plant pathogens (Mazzola *et al.*, 2001). A push–pull cropping system incorporates a trap plant, which pulls the pest away from the main crop, as well as a legume intercrop, which repels the target pest away from the principal crop and also acts as a weed suppressant (Khan *et al.*, 2011). Herbicides are often needed to attain control of competing vegetation in the early transition years of CA adoption. Ideally, herbicides are no longer needed once the management of residues and cover crops is mastered by farmers.

Economic studies indicate that reduced labor costs and increased gross margins are common benefits that arise from CA implementation (Fowler and Rockström, 2001; Erenstein *et al.*, 2012). Lower planting costs because of reduced plowing have been verified in the northwest Indo-Gangetic plains (Erenstein *et al.*, 2008), Tanzania (Owenya *et al.*, 2011), and the Andes region (Nguema *et al.*, 2013). The possibilities of reducing costs diminish when minimum or no-tillage increases the proliferation of weeds (Erenstein, 2002).

Often, farmers do not adopt all CA practices initially, and adoption becomes a step-wise process (Byerlee and Hesse De Polanco, 1986; Silici *et al.*, 2011). In South Asia, no-tillage and residue cover became widely used after a 10- to 15-year period of testing that led to a gradual increase in farmers' knowledge and availability of adequate implements (Jat *et al.*, 2012). Partial adoption of CA practices (e.g. no-till without crop residue retention), however, may lead to less synergistic outcomes, because of the reduced impacts on water availability and soil fertility (Thierfelder *et al.*, 2012, 2013b).

2.2 Constraints for Scaling Up Conservation Agriculture

Despite all its promises, scaling up CA practices for smallholders in developing countries has proved to be challenging because of a variety of constraints (Table 2.2). These include the limited exposure of farmers to CA practices because of undeveloped extension services, unsecure land tenure, lack of microfinance mechanisms, and shortage of suitable farming implements. Although many of these limitations are widespread, some constraints are more localized. In West and southern Africa, crop residues are used to provide animal shelter, feed for livestock, heat, and storage (e.g. silos made of maize and sorghum stalks in Mali). Therefore, the possibility of achieving the 30% residue retention proposed by the FAO is limited (Valbuena *et al.*, 2012), and may only apply to some regions (Erenstein, 2003).

Farmers are exposed to diverse information (e.g. from extension agents, agro-dealers, farmers' groups, religious leaders) that sometimes can be misleading. In Uganda, a company that manufactures ox plows advertises the "benefits of inverting and completely loosening the soil" (A. Ares, personal observation). A study on local agricultural networks in Lesotho indicated that one of the key sources of information for farmers was the tractor owner who rented his services to farmers to till the fields. The tractor owner was skeptical about the fact that tillage resulted in soil degradation, whereas the majority of the main community

Table 2.2. Agronomic, socio-economic, technological, and policy/institutional constraints for adopting conservation agriculture practices.

Agronomic
Limited experience with cover crops Control of competing vegetation (cost and labor requirements) Crop–livestock conflicts
Socio-economic
Strong culture of plowing and other mindset traditions Competitive use of residues Unsecure land tenure/farm size Unavailability of finance mechanisms Lack of market for legume crops Risk aversion Misleading information on suitable farming tools/agronomic practices
Technological
Lack of farming implements/animals Unavailability/high cost of herbicides Limited knowledge of CA
Policy/institutional
Lack of extension services Misleading policies Reluctant donor agencies

participants (subcounty chief, women and farmers' group leaders, district agricultural officer) acknowledged the impact of tillage on soils (Lamb *et al.*, 2013). This example illustrates that understanding the beliefs and expectations of the members of local communities is important and should be considered to develop efficient strategies for CA dissemination.

2.3 Sustainable Intensification

Recently, there has been an increased interest in SI, which consists of producing more output per unit of input (land, nutrient, and water), while reducing environmental impacts and increasing natural capital and environmental services (Cassman, 1999; Ruben *et al.*, 2006; Kassam *et al.*, 2011; Pretty *et al.*, 2011; Tilman *et al.*, 2011; Smith, 2013). The impetus for encouraging SI is based on the sheer need to increase food production greatly in some regions of the world (e.g. West Africa; Godfray *et al.*, 2010) in response to vast population growth, the limited possibilities for extensification (i.e. allocating additional land to agriculture; Young, 1999; Smith *et al.*, 2010), and the lack of long-term sustainability of agricultural models that were successful in several regions of the world for some time but were not able to sustain productivity and income gain in the long run. Agricultural intensification, in contrast to extensification, could also reduce tropical deforestation (Matthews and De Pinto, 2012) and mitigate the effects of global warming (Smith *et al.*, 2010). Sustainable intensification is the main theme for programs like the USAID Africa Research in Sustainable Intensification for the Next Generation (Africa RISING), the European Community IntensAfrica, and the ACIAR Sustainable Intensification of Maize–Legume Cropping Systems for Food Security in Eastern and Southern Africa (SIMLESA), aimed at increasing food security and providing other benefits for resource-poor smallholders.

Some of the characteristics of SI systems are a high production:input ratio, wise use of inputs, boosted ecological processes such as nutrient cycling and biological nitrogen fixation, limited adverse effects on human health (or even positive outcomes such as improved nutrition), good use of human/social capital, and minimal negative externalities. Conservation agriculture can indeed contribute to foster several of these components in agroecosystems.

One of the facets that can be explored to promote SI is reducing yield gaps (Mueller *et al.*, 2012). In West Africa, yields of maize, sorghum/millet, and soybean could be increased by 80–400%, 40–400%, and 30%, respectively, by using readily available technologies such as improved crop genotypes, integrated soil fertility management, water harvesting, crop rotations, and cover crops (Kpotor, 2012; Bonsu and Asibuo, 2013; V. Prasad, Manhattan, Kansas, 2013, personal communication). Nonetheless, a multi-wedged approach aimed to improve value chains, access to markets, women's rights, and other aspects are clearly needed to reach measurable gains in SI.

The benefits of SI practices for smallholder farmers can be diverse (Table 2.3) (Pretty *et al.*, 2011; The Montpellier Panel, 2013). Conservation agriculture and IPM practices are both considered integral components of SI (Pretty *et al.*, 2011).

Table 2.3. Benefits of sustainable intensification practices.

Beneficial goals	Practice
Increase productivity and income	Farming system improvement, integrated nutrient management, agricultural support services (input supply, credit, extension, output marketing)
Utilize improved crop varieties and livestock breeds	Crop/livestock breeding
Conserve soil and water; reduce external inputs	Conservation agriculture, farming system improvement, agroforestry, organic agriculture
Improve agroecological processes such as nutrient and water cycling, biological nutrient fixation, and biological pest management	Conservation agriculture, integrated pest management, agroforestry
Make productive use of human capital, and build social capital	Training, extension
Boost human nutrition	Crop breeding, farming system improvement, biofortification
Adapt to climate change	Conservation agriculture, water harvesting technologies
Decrease food waste	Postharvest technologies
Reduce impact on the environment and human health	Water quality improvement, dietary enhancement

IPM focuses on maintaining agricultural pest populations at levels that will not cause economic or environmental harm via an understanding of food-web and ecosystem processes, the reduction of pesticide use, and the utilization of multiple growing practices that are considered to have low levels of environmental impact (Puente *et al.*, 2011). On the other hand, CA emphasizes the enhancement and maintenance of soil quality. Both management systems share the same fundamental goals of increasing crop yield per unit area of cultivated land and decreasing negative environmental externalities arising from agricultural production (Fuglie and Kascak, 2001). The two approaches often complement one another, resulting in a synergistic effect that enhances each methodology's efficacy in achieving their more specialized goals. Unsurprisingly enough, the three primary components of CA are extremely relevant to IPM methodologies.

Sustainable intensification and agroecology are both paradigms that espouse biological diversity, which can be key to preventing pathogen epidemics and developing biological control (Jackson *et al.*, 2007; Pretty *et al.*, 2011). In Honduras, the activities of a diverse natural enemy complex prevent major destruction of maize fields by the fall armyworm (*Spodoptera frugiperda*) (Wyckhuys and O'Neil, 2010). Together, all three components of CA enhance the presence of beneficial organisms. The reduction of anthropogenic soil disturbance through minimized soil tillage supports the formation of a complex soil environment, where beneficial organisms–enemies of pests can flourish. Continuous plant cover promotes the presence of aboveground organisms that are natural enemies of pests, and also encourages the growth of favorable organisms that live within the soil environment via inputs such as nutritive root exudates and the build-up of soil organic matter (Kowalchuk *et al.*, 2002). The utilization of crop rotations and intercropping elevates the presence of beneficial organisms as plant species cultivate their own unique rhizosphere communities, and also differentially attract and foster aboveground beneficial insects (James, 2003). Promoting beneficial soil microorganisms is a form of IPM, since these organisms can induce systemic resistance in the plants, compete with the plant pathogens, or directly inhibit pathogen growth by producing antimicrobial compounds.

The adoption of IPM and CA technologies would be facilitated by public policy incentives or regulations, reduced costs and increased availability of inputs, financial credit resources, improved value chains and markets, adaptive management with both growers and scientists, and human and institutional capacity building (Jackson *et al.*, 2007; Wyckhuys and O'Neil, 2010; Khan *et al.*, 2011). The level of farmer education can be a major constraint for both practices (Fuglie and Kascak, 2001; Wyckhuys and O'Neil, 2010). CA adopters tended to have a higher education level than non-adopters in Mozambique (McNair, 2013) and West Bengal, India (Krishna *et al.*, 2012). This corresponds to IPM adoption results, except that a college education additionally impacts the time it takes for farmers to adopt IPM, with farmers having some college education adopting IPM 6.5 years faster than farmers with just a high school education (Fuglie and Kascak, 2001). Both practices can be considered somewhat knowledge-intensive, but possess different levels of complexity.

2.4 Resilience

Considerable conceptual elaboration on resilience thinking has taken place since the initial work by Holling (1978) and others. Enhanced resilience has been increasingly cited as a desirable attribute of farming systems worldwide, because of increased stresses on resources caused by land degradation, climate change, and population growth (Darnhofer *et al.*, 2010; Lin, 2011). Resilience has been described as "the ability of a system to withstand stresses of 'environmental loading'... a fundamental quality found in individuals, groups, organizations, and systems as a whole", (Horne and Orr, 1998), and "the capacity to cope with unanticipated dangers after they have become manifest, learning to bounce back" (Wildavsky, 1991). Several other definitions on resilience have been put forth, specifically in reference to agroecological systems (Table 2.4). The characteristics of resilient systems include diversity, modularity, openness, savings, social capital (that relates to leadership, truth, and social networks), and tightness of feedbacks (Walker and Salt, 2012). For example, Cambodia's SAMREM project fields where farmers had implemented CA did not require replanting when drought occurred, although conventionally treated plots did, since CA practices had built system resilience, enhancing soil water-holding capacity and intercropping plants with dissimilar zones of water utilization in the rhizosphere (Boulakia *et al.*, 2012, unpublished report). Conservation agriculture implementation can trigger a cascade of feedbacks (e.g. changes in

Table 2.4. Definitions of resilience that can be applied to agroecological systems.

Definition	Source
The capacity of a system to absorb disturbance and reorganize to retain the same function, structure, and feedbacks	Walker and Salt, 2012
The capacity of a socio-ecological system to absorb perturbations and to sustain function, structure, identity, and feedbacks through recovery or reorganization	Chapin *et al.*, 2009
The ability of people, households, communities, countries, and systems to mitigate, adapt to, and recover from shocks and stresses in a manner that reduces chronic vulnerability and facilitates inclusive growth	USAID, 2013

nutrient dynamics; Jat *et al.*, 2011). These interactions should be identified from the bottom up in order to be brought to scale.

Relevant applications of resilience, however, remain limited, owing to context-specific meanings (i.e. different definitions in engineering, psychology, education, environmental science, and other fields), and the restricted understanding of approaches to measure it in real systems. Resilience analyses applied to agriculture and related disciplines have commonly focused on a given component of a system or subsystems; for example, soil resilience (Lal, 1993). This has been called specified resilience. More recently, there has been interest in addressing systems as complex adaptive entities because of the limitation of one-sided solutions, and therefore the concept of general resilience has emerged. Systems change in predictable or unpredictable ways when certain limits or thresholds are reached, and the systems move to another state of stability domain through a regime shift (Walker and Salt, 2012). A resilience analysis requires the characterization of a range of factors, including the biophysical (e.g. crops, water, soil) and socio-economic (e.g. beliefs, values, resources) environments, the possible disturbances of interest, the system drivers and trends, and the interactions among these components at different scales. An assortment of indicators for resilience has been identified (Table 2.5). However, thresholds for these factors are often difficult to determine and are also variable. For instance, a given low available soil phosphorus level can represent a threshold for plant growth. Yet, if a legume variety with high phosphorus use efficiency is introduced, the soil phosphorus threshold level can change. The above is a simple example; systems are more complex and changes trigger a variety of responses in different directions and at various scales. The dynamics of these systems and their capacity for resilience can differ drastically from one another due to differences in the variables that are inherent in a resilience analysis. The resilience of different locales when faced with similar challenges will vary depending on the types of crops grown or the soil type (biophysical factors), cultural preferences in food, or access to microloans (socio-economic factors), and more. This is why strategies for enhancing resilience need to be adapted to local landscapes and societies.

Table 2.5. Some indicators of resilience. (Adopted from van Oudenhoven *et al.*, 2011.)

Acquisition, use, and transmission across generations and among communities of traditional ecological knowledge
Existence of traditional land tenure systems, indigenous governance, and customary laws
Number of generations interacting with the landscape and experienced farmers
Use of traditional exchange and reciprocity systems (e.g. seed exchange)
Availability and use of traditional foods, seeds, and medicines in local production system
Food sovereignty and self-sufficiency
Low levels of threat from illegal encroachment, land grabbing, privatization, government expropriation, and forced resettlement
Crop yield stability
Conservation of resources and biodiversity
Wise use of fertilizers, insecticides, and/or herbicides on agricultural land
Diversity of landscape components, agricultural systems, cultivated crops, and varieties and breeds

2.5 Regional Perspectives

While CA systems across the world share common principles, there are also important differences among geographic locations because of climatic and soil conditions, farming system types, crop–livestock interactions, access to resources by farmers, and other factors. The case studies below illustrate the effects of such differential conditions on the development of CA systems.

2.5.1 Conservation agriculture in southern Africa

Conservation agriculture became attractive for farmers in southern Africa as a response to increased soil degradation and fertility decline (Sanginga and Woomer, 2009), as well as the more unreliable climatic conditions characterized by frequent droughts and other potential negative impacts of climate change (Lobell *et al.*, 2008). Currently, CA systems practiced in southern Africa are distinguished by their planting techniques (Johansen *et al.*, 2012), which range from very simple manual systems such as planting with a pointed stick or a hoe, sometimes in planting basins (Mazvimavi and Twomlow, 2009), to more sophisticated systems based on mechanical jab and hoe planters, animal traction rip-line seeding, and direct seeding (Johansen *et al.*, 2012). Very few seeding options are currently available for the commercial farming sector in southern Africa, with the exception of South Africa.

Benefits of CA in southern Africa

Since 2004, there has been strong interest in documenting research results from CA systems in southern Africa. Despite views that CA would be a suitable technology for only a limited number of farmers (Giller *et al.*, 2009), CA has proven to perform better than the traditional tillage-based farming systems, and significant yield increases in paired plots have been reported in many cases (Ngwira *et al.*, 2012; Thierfelder *et al.*, 2012, 2013a,c). While it is indeed advantageous that CA can yield better than conventional tillage systems, the goal of CA is to provide yield stability while intensifying smallholder production. Many smallholders till their land in excess of the area they can manage effectively, leaving their fields subject to high erosion and low yields because of the lack of effective weed control. Conservation agriculture provides an opportunity to intensify subsistence production into smaller manageable areas (Thierfelder and Wall, 2011). In Malawi, the results from nine target communities, with six replicated paired plots in each location, showed marked yield gains of CA systems with an increasing yield trend (Figs 2.1, 2.2a and b). Figure 2.1 shows CA treatment plots that consistently yielded more maize grain than conventionally treated plots, from 2006 to 2012. Figure 2.2a and b demonstrate a similar pattern. At the regional level, the conclusion was that the greatest benefit could only be achieved if all principles were applied; i.e. leaving out mulching and/or rotations would lead to a very slow increase in productivity (Thierfelder *et al.*, 2013b).

The results from research in Malawi highlight that CA has immediate biophysical and socio-economic benefits. The retention of surface crop residue as mulch increased soil water infiltration, biological activity, and rainfall effectiveness (Thierfelder and Wall, 2009). Furthermore, residue mulching lowered water runoff and topsoil loss (Thierfelder and Wall, 2010a,b). Residues reduced

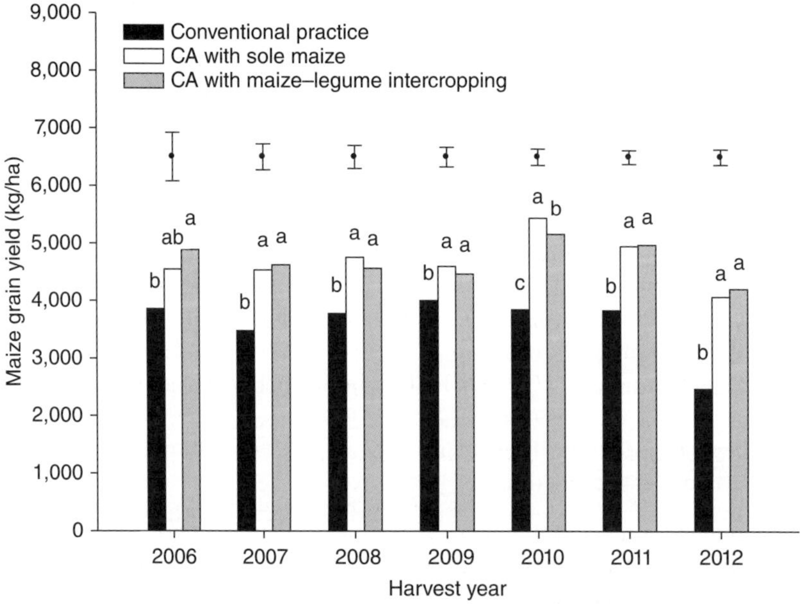

Fig. 2.1. Maize grain yields (kg/ha) of one conventional and two conservation cropping system in Malawi 2005/06–2011/12 (adapted from Thierfelder *et al.*, 2013a). The same letters above each mean bar graph indicate no significant difference ($P < 0.05$); error bars in each yearly comparison show the standard error of the difference.

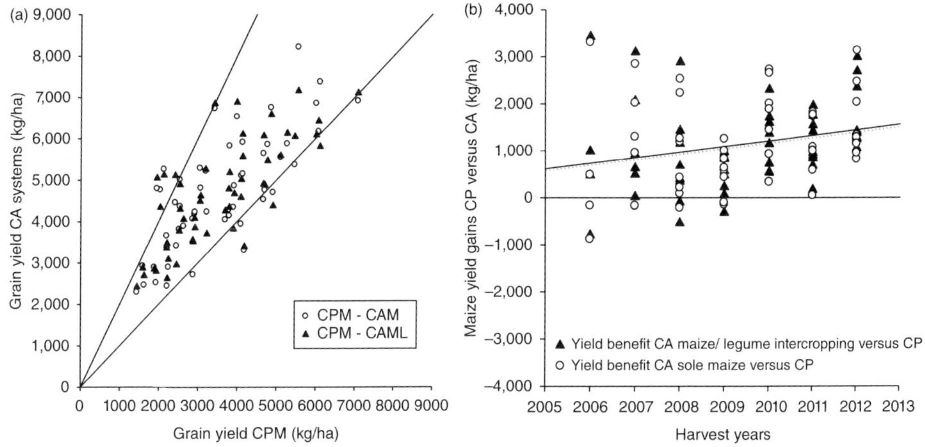

Fig. 2.2. Relative difference between conservation agriculture (CA) and conventional (CP) treatments with maize (M) and maize/legume intercropping (ML) (a) and yield gains (b) in Malawi from 2006 to 2012 (adapted from Thierfelder *et al.*, 2013a). Dots above the 1:1 line in graph (a) show a benefit towards CA, dots above the 1:2 line show double the yield on CA plots in comparison to CP.

evaporation losses, likely because of decreased solar radiation absorption at the soil surface (Lal, 1974), which improved crop water balance (Thierfelder and Wall, 2009; Mupangwa *et al.*, 2013) and resulted in crop moisture stress being less frequent and intense (Mupangwa *et al.*, 2008; Thierfelder and Wall, 2010a). Reported socio-economic benefits included reduced traction and labor requirements for land preparation and weeding if herbicides were used (hence saving the costs of manual labor), animal draft, and fuel, depending on the farming system (Mazvimavi *et al.*, 2008; Johansen *et al.*, 2012; Ngwira *et al.*, 2012, 2013). Over the long term, CA can increase soil organic matter, develop more resilient soil structure, improve nutrient availability, and increase water-holding capacity (Rusinamhodzi *et al.*, 2012; Thierfelder *et al.*, 2013b).

The positive effects of CA have translated into increased adoption of CA in southern Africa. In Malawi, more than 80,000 farmers implemented CA in 2013, compared with just 12 practicing farmers in 2005. There was also adoption of specific CA components by 371,000 farmers in Zimbabwe, and by 250,000 farmers in Zambia (Aagard, 2009; Derpsch and Friedrich, 2009).

Constraints to the adoption of CA in southern Africa

Despite the successes, there are still significant constraints to the adoption of CA that warrant additional research. Because of the pastoral activities common to the African continent, old crop residues (i.e. vegetal materials left to protect the soil surface as one of the three tenets of CA) are valuable in mixed crop–livestock systems and can be grazed during the winter season, leaving the soil surface with little protection (Mupangwa *et al.*, 2012; Valbuena *et al.*, 2012). If CA is to be sustainable and effective, weed suppression and control with and without herbicides need to be studied further (Mashingaidze *et al.*, 2012; Muoni *et al.*, 2013) and evaluated in parallel with crop rotations where landholding size is limited (Snapp *et al.*, 2002; Thierfelder and Wall, 2010b). Further, the often highly degraded African soils need inputs of fertilizers or sufficient quantities of manure and good seed, which are often difficult to obtain through the dysfunctional markets common throughout Africa (Morris *et al.*, 2007). Without adequate credit, markets for agronomic inputs, outputs, and machinery, CA success could be short-lived (Sanchez, 2010; Sims *et al.*, 2012). Often, the mindset of farmers to this new way of farming is critical, and in many instances, it is difficult for farmers to accept that farming can be very productive and environmentally benign if tillage is abandoned (Wall, 2007).

A case of successful promotion of CA in southern Africa

The development of CA in Malawi was facilitated via an innovative systems approach initiated in 2005 in Nkhotakota by key stakeholders. A range of partners were identified, providing complementary skills that included research and extension organizations, the private sector, and farmers. Discussions within target farm communities highlighted key constraints to crop production during the project implementation phase. Experiments enabled agronomists and soil scientists to assess the effects and performance of different CA interventions on long-term productivity and soil fertility. Trials also served as learning centers for the socio-economic and agronomic aspects of CA, and contributed to promoting farmer-to-farmer exchange activities.

In 2009, focus group discussions identified participants who were involved in the adoption of CA by using a pluralistic innovation support approach (Klerkx *et al.*, 2009; Spielman *et al.*, 2011). Most of the initial research and training was led by the government extension service and the regional non-governmental organization (NGO), total landcare (TLC), and focused on target communities. Organized discussions between local stakeholders and participating farmers promoted widespread community discussions and feedback that eventually provided a foundation for technology adaptation and adoption. Farmers and farmer groups near the validation trials that were implemented in farmers' fields were encouraged to participate in field days, discussion groups, and farmer-to-farmer exchange visits. In these activities, farmers were able to see first-hand the effects of practices such as relay crops, minimum tillage and residue conservation on soil characteristics, crop yields and farm income. Survey results suggested that the main catalyst for the innovation network was the growing interest of farmers in herbicides. Since input suppliers were initially not available in the study areas to supply herbicides, farmer access to herbicides was facilitated by TLC. Farmers interested in CA registered with the TLC field coordinators and paid a deposit of US$14.5 each, with a commitment to pay the cost of the supplied inputs within 9 months. This amount included improved maize seed varieties and herbicides that cost about US$50/0.4 ha of land. The amount of deposits was used to estimate the demand for maize seed and herbicides that were purchased by TLC and distributed to farmers. The repayment of the "soft" loans was estimated at about 90% and was used as a revolving fund to support CA farmers the following season.

Interaction and information exchange between all relevant stakeholders was crucial for adapting CA systems locally. The initial process was supported by international donor funds to create the necessary critical mass of successful CA examples. While innovation networks from previous studies were built around marketed commodities and the resulting markets (Brooks and Loevinsohn, 2011; Kilelu *et al.*, 2011; Spielman *et al.*, 2011), later experiences focused on the evaluation of a whole crop management system. The main drivers in this innovation network were knowledge transfer and capacity building, adaptive CA research, access and availability of critical inputs, and an enabling environment for CA in Malawi. In 2005, there were almost no farmers practicing CA in Central Malawi. By 2012, the number of farmers rose to about 18,000 on around 6,000 ha (Fig. 2.3).

Evolution of conservation agriculture in Monze, Zambia

During the mid-1990s, CA was introduced to Zambia by the Conservation Farming Unit (CFU), as well as a number of other players (Haggblade and Tembo, 2003), to address critical constraints to small-scale farmer food security. The Zambian efforts originated from the experiences of a Zimbabwean commercial farmer, Brian Oldrieve, who was invited to Zambia as a consultant in 1995. Based on his experiences with smallholder CA production systems in Zimbabwe, a system locally referred to as conservation farming (CF) was developed. It was based largely on the preparation of planting basins during the dry season and the use of an animal-drawn ripper, with direct planting into the ripped lines. Other practices such as residue retention and crop rotations were promoted alongside the basins and rip-line seeding system, but were applied to a more limited extent. In 2004, CIMMYT started to encourage animal traction to develop CA

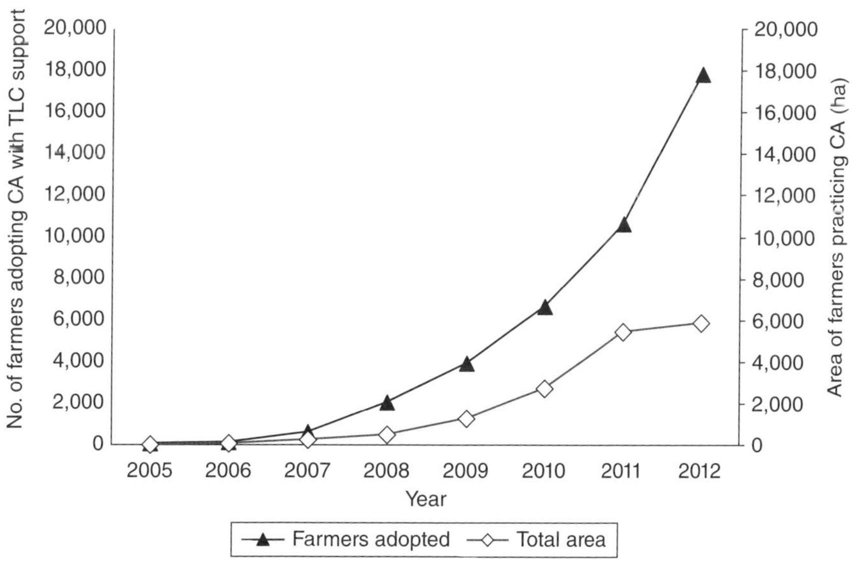

Fig. 2.3. Extent of CA adoption facilitated by Total LandCare (TLC) in Malawi, 2005–2012 (T. Bunderson, Total Land Care, unpublished data, 2012).

seeding systems in Monze. An innovation platform was established with the advent of three new projects in September 2009, namely Research into Use (RIU), CA Scaling up Production and Productivity (CASPP), and Farmer Input Support Response Initiative (FISRI). The innovation network in Monze has approximately 45 active members representing research and extension organizations, input suppliers, church groups, and the radio and local news channels. The network has meetings four times per year and joint field visits, and discusses and broadcasts CA promotion successes and challenges.

The CASPP and FISRI projects, along with other initiatives led by CFU, CIMMYT, the Archdiocese of Monze, Mantantala, Research into Use, Prevention against Malnutrition, and MRI seeds, targeted about 83,000 farmers in 2009 (about 70% of all farming households in the Monze district). The ambitious targets set in the initial project documents of CASPP and FISRI and other initiatives have not been met so far, and only about 25% of the target farmers have started to adopt CA. The main reason for the slow development has been lack of access to critical inputs, cash constraints, limited market for products (especially legumes), and competition for crop residues. In 2012, data showed that around 16,000 farmers in the Monze district practiced CA in their fields using their own inputs (Fig. 2.4).

2.5.2 Conservation agriculture in Southeast Asia

In many mountainous regions of Southeast Asia (SEA), traditional subsistence farming and agriculture with long fallows have been replaced by extensive agriculture development based on plowing, because of increased demand for cash crops. This trend has accelerated the mining of soil resources and caused land

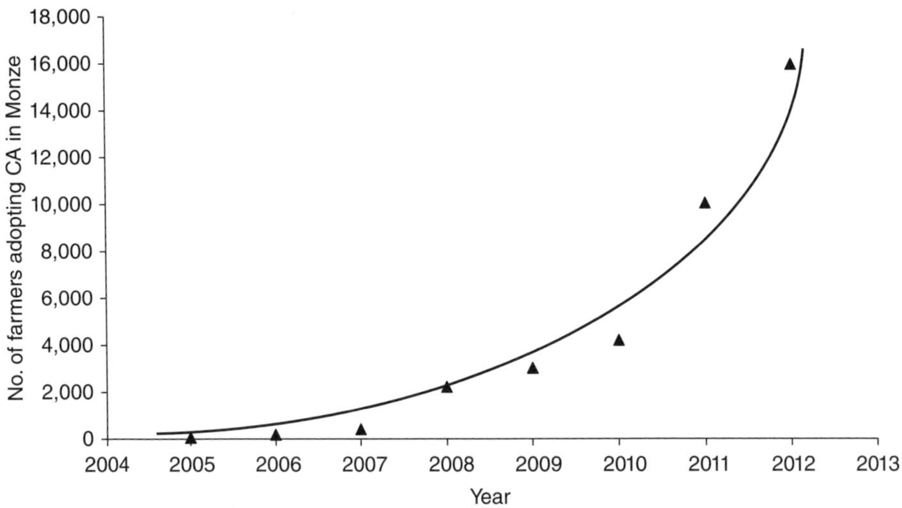

Fig. 2.4. Extent of adoption of conservation agriculture (CA) technologies in the Monze district from 2005 to 2012 (Zambia Ministry of Agriculture and Livestock, unpublished Monze District data, 2012).

degradation. If these processes continue, many biologically diverse mountainous regions in SEA will decline significantly in productivity, resulting in food insecurity. To address this problem, CA research in SEA commenced in Vietnam in 1998 through the Mountainous Agrarian Systems (SAM) project (Le and Ha, 2010) managed by CIRAD. Additional projects on CA and related topics were later initiated (Table 2.6). In addition, the Conservation Agriculture Network in SEA (CANSEA) began in 2009, with Cambodia, Indonesia, Vietnam, Thailand, and China as founding country members. The first CA research study applied to vegetable production started in 2013 in Cambodia. Most of the research dealt with crop yield, soil conservation, and the economic impacts of CA systems compared with plow-based systems. In Battambang and Siem Reap, Cambodia, and in the Philippines, the impact of CA on gender was explicitly included in the research.

Benefits of CA in Southeast Asia

The majority of the benefits derived from CA in SEA involved soil carbon sequestration, increased crop yield, and enhanced soil quality. As far as we know, the oldest experiment in SEA included no-tillage (NT), the first principle of CA, and began in 1987 in Lampung, Indonesia. In this study, after 23 years of NT, soil carbon storage at 0–20 cm depth was 43% higher in NT than in tilled areas (Utomo, 2013). NT sequestered some 4.4 Mg C/ha, while tilled systems depleted soil carbon by as much as 6.6 Mg C/ha. Microbial biomass carbon was 14.4% higher in NT. Soybean and maize yields were higher in NT than in tilled systems.

There are numerous examples of CA impacting farmer yield positively. Results from CIRAD's SAM project, one of the first CA projects in Cambodia, indicated that yields of upland crops (rice, maize, and cassava) increased 200%, and soil erosion decreased 96% in CA systems compared with plow-based systems in the Cho Don district of Bac Kan province of northern Vietnam (Le and Ha, 2010). In a CIRAD study in Kampong Cham, Cambodia, cassava yield for CA-practicing farmers averaged 8 t/ha compared with 6.3 t/ha in plowed fields, with gross profit margins of US$1,083/ha for CA and US$970/ha for plowed systems (Chabierski *et al.*, 2012). In a CIRAD study in Sayaboury province, the main maize production area in Laos, maize yield recorded under CA in a 2-year sequence of maize–rice bean and maize–*Brachiaria ruziziensis* was 50% higher compared to monocropped maize. Production costs in maize monocropping with tillage reached up to US$250/ha, whereas production costs were reduced by at least 50% in CA systems. The net income generated by maize production under CA increased 100%.

In 2010, the USAID-funded SANREM project, in partnership with CIRAD, initiated research and demonstration studies in Battambang, in northwestern Cambodia. The focus of this project has been CA research and development for smallholder upland cultivation. CA technologies successfully applied in Kampong Cham were applied to the socio-economic and agroecological conditions in Battambang for maize, cassava, soybean, and upland rice. The results of these many case studies which support CA's ability to enhance field production arise from a range of important factors, one notably being CA's impact on soil quality. In the CIRAD/SANREM Battambang project, maize yield in CA systems was higher

Table 2.6. Conservation agriculture experiments in Southeast Asia. (From Lienhard *et al.*, 2013, with additions by M. Reyes.)

Country	Province	Period	Project	Crops/ activities	Donor	Technical support
Cambodia	Kampong Chan	2004–2008	PHF	Rubber	AFD	MAFF, CIRAD
	Kampong Chan, Battambang	2008–2013	PADAC	Maize, cassava, soybean	AFD	MAFF, CIRAD
	Battambang	2010–2014	SANREM	Maize, cassava	USAID	NCA&T, CIRAD
Laos	Sayabouri	2001–2002	PRODESSA	Maize	AFD	NAFRI, CIRAD
	Sayabouri, Xien Khouang	2003–2008	PRONAE PASS	Maize, upland, forages	AFD	NAFRI, CIRAD
	National	2007–2011	PROSA	Scaling up	AFD	MAFF, CIRAD
Philippines	Claveria	2010–2014	SANREM	Maize	USAID	NCA&T, WAC
Thailand	Sakhon Nakhon	2005–	Soil biology laboratory	Cover crops, upland rice	Thai gov.	KU, CIRAD
Vietnam	Bac Kan	1998–2004	SAM	Rice, maize, cassava, forages	French gov.	VAAS, CIRAD, IRD, IRRI
	Pleiku	1999–2004	ADP	Rubber	WB	NIR, CIRAD
	Phu To, Son la, Yen Bai	2008–2012	ADAM	Maize, tea	AFD	NOMAFSI, CIRAD
	Son la, Yen Bai	2009–2013	IME	Maize	ACIAR	NOMAFSI, UQ
Regional (six countries)		2009–	CANSEA	Research/ training	AFD	CIRAD
Regional (four countries)		2009–2013	PAMPA	Impact studies	AFD	CIRAD, IRD

Notes: ACIAR = Australian Center for International Agricultural Research; ADAM = Support to Conservation Agriculture Extension in Mountainous Areas of Vietnam; ADP = Agricultural Diversification Project; AFD = Agence Française de Développement; CANSEA = Conservation Agriculture Network in Southeast Asia; CIRAD = French Agriculture Research Center for Development; IME = Improved Market Engagement Project; IRD = French Research Institute for Development; IRRI = International Rice Research Institute; KU = Kasetsart University, Thailand; MAFF = Ministry of Agriculture, Forestry and Fisheries, Cambodia; NAFRI = National Agriculture and Forest Research Institute, Laos; NIR = National Institute of Rubber, Vietnam; NOMAFSI = Northern Mountainous Agricultural and Forestry Science Institute, Vietnam; NCA&T = North Carolina Agricultural and Technical State University, USA; PADAC = Projet d' Appui au Dévelopment de l' Agriculture du Cambodge; PAMPA = Programme d' Appui Multi-Pays pour l' Agroécologie; PASS: Development Project for the South of Sayabouri Province; PHF = Rubber for Smallholders Project; PRODESSA: Projet de Développement Rural de Sayabour; PRONAE: Programme National En Agroécologie; PROSA: Programme Sectoriel en. Agroécologie; SAM = Mountainous Agrarian Systems Project; SANREM = Sustainable Agriculture and Natural Resource Management Innovation Lab; UQ = University of Queensland, Australia; USAID = United States of America International Development Agency; VAAS = Vietnamese Academy of Agricultural Sciences; WAC = World Agroforestry Center; WB = World Bank.

than in plow-based systems, irrespective of the number of years in CA (Boulakia *et al.*, 2012, unpublished report) (Table 2.7). Furthermore, maize yield in plow-based systems decreased, while maize yield in CA systems increased from 2009 to 2011. Plots with maize grown under CA and plow-based systems are shown in Fig. 2.5. The gross profit margin for CA in year 3 was US$581/ha compared with US$495/ha for the plow-based systems. Cover crops such as *Stylosanthes* could be established successfully in the acidic soils of Kampong Cham. This leguminous cover crop serves as a nitrogen source for the soil and generates large amounts of biomass, thereby providing quality soil cover. In the limestone-derived soils of Battambang, however, other leguminous cover crops like pigeon pea (*Cajanus cajan*) were a better fit (Boulakia *et al.*, 2012, unpublished report).

In the Plain of Jars in northeastern Laos, a 3-year rotation of rice/maize/soybean was tested in one conventional tillage-based (CT) system and three CA systems (Lienhard *et al.*, 2014). The results of studies conducted from 2007 to 2010 showed that compared with CT, CA systems led to higher grain production, increased profits, livestock system intensification, and higher labor productivity. Non-tillage also improved soil physical and chemical characteristics such as aggregate stability, organic carbon, and cation exchange capacity, as well as microbial abundance (total biomass, bacterial and fungal densities).

In 2013, the SANREM and HORTICULTURE Innovation Labs jointly started a study on vegetable production by women using CA and drip irrigation in Siem Reap, Cambodia. The goals of the projects were to enhance food security in a sustainable way by encouraging the application of CA methodologies, directly increasing women's income and the household availability of nutritious produce. Many women only have access to small tracts of land, close to home, which they can use for their own purposes. Drip irrigation is a way to maximize the usage of this space while also decreasing the labor needed to water the plants. Initial results indicated that yield of cucumber in CA and drip irrigation treatments were not significantly different than those with tillage and drip irrigation (Table 2.8) (Edralin and Reyes, 2013).

Knowledge on gendered participation in CA in SEA is still limited. A review of the literature in Cambodia and the Philippines, which were the two countries in SEA where the SANREM project was implemented, revealed that there was a paucity of published studies for these countries; hence, the review was expanded to cover CA literature in other parts of the world (Javier, 2013, unpublished report). The analysis showed variations in CA practices, although the

Table 2.7. Maize yield in conservation agriculture (CA) and plowed systems.

Cropping system	Yield (t/ha)	Plots (number)
Millet/maize + Stylosanthes//maize + Stylosanthes Year 1	3.9 ± 0.6	12
Maize + Stylosanthes//maize + Stylosanthes Year 2	3.5 ± 1.0	15
Maize + Stylosanthes//maize + Stylosanthes Year 3	4.1 ± 0.6	3
Maize with traditional management (plowing)	3.3 ± 0.7	20

Notes: Years 1, 2, and 3 indicate the number of years under the CA treatments. Yield values are means ± one standard deviation.

(a)

(b)

Fig. 2.5. Maize in conservation agriculture (a) and plow-based system (b) in Battambang, Cambodia.

three CA principles were usually observed. The reported CA benefits were often not sex-disaggregated, except in the case of two studies in Zambia that detailed the gendered benefits of reduced labor from CA. The review showed that adapting CA to local sociocultural and biophysical conditions brought about benefits and challenges to both female and male smallholder farmers in developing countries. Among the main benefits for women were reduced labor and

Table 2.8. Fresh yield of cucumber in conservation agriculture (CA) and tilled systems with and without drip irrigation in Siem Reap, Cambodia.

Treatment	Marketable fruits (number/ha)	Mean yield of marketable fruits (t/ha)
CA	148,889 b	13.1 b
CA with drip irrigation	185,556 ab	15.5 ab
Tilled	170,000 ab	14.9 ab
Tilled with drip irrigation	222,222 a	19.7 a

Note: Values followed by the same letter are not significantly different at $P \leq 0.05$.

time, owing to decreased tasks of raking and gathering vegetative debris/crop residues, weeding, and fetching irrigation water. This generates extra time for personal leisure and non-farm chores. Benefits for men were labor and time savings due to no or minimum tillage and non-burning, relief from physical stress, and more time to engage in other income-generating activities.

In the Philippines, SANREM began testing CA practices in replicated research and demonstration plots on the mountainous steep slopes of northern Mindanao in 2010. The main study included the following treatments:

T1: Maize + *Arachis pintoi* followed by maize planted alongside established *A. pintoi*.
T2: Maize + *Stylosanthes guianensis* followed by *S. guianensis* fallow.
T3: Maize + cowpea/upland rice + cowpea followed by maize + cowpea/upland rice/cowpea.
T4: Rice beans/maize followed by rice beans/maize.
T5: Cassava + *S. guianensis*.
T6: Control, which is farmers' traditional plow-based practice for growing maize.

Two years after applying the treatments, the cassava with *S. guianensis* treatment yielded the highest biomass and total sales of dried cassava chips, followed by the farmers' traditional practice (T6), which produced more maize grain and total dry matter compared with other maize-based CA practices (Mercado *et al.*, 2012). Maize with cowpea had the lowest yield, possibly because of the very close spacing between rows at 30 cm. This closer proximity likely resulted in increased competition for available nutrients and solar radiation. The intercropping of maize with either cowpea or rice beans did not produce better total grain and biomass yield, but provided higher sales due to relatively better prices of cowpea and rice beans. The maize + *A. pintoi* treatment appears the most promising of conservation agriculture production systems (CAPS) treatments (Mercado *et al.*, 2013, unpublished report). *A. pintoi* is perennial, outcompetes weeds, and contributes to control soil erosion. Furthermore, *A. pintoi* leaves and stems can also be fed to chickens, cattle, carabao, and pigs, and honeybees also frequent *A. pintoi* flowers. Yield of maize sowed with *Arachis* is greater than in plow-based treatment. *A. pintoi* and maize growing in fields in Claveria are featured in Fig. 2.6. There is also the prospect of having five maize crops biannually (instead of four) because of the time savings in the NT system. By year 3, the net return of maize grown in *A. pintoi* was US$1,576/ha compared with only US$352/ha in the plow-based maize

(a)

(b)

Fig. 2.6 Maize–*Arachis pintoi* practice (a) and maize in tilled field (b) in Claveria, Mindanao, Philippines.

system. Higher input costs with lower maize yields decreased net income from the plow-based practice significantly. There was a trend of increased soil organic carbon in some CA treatments, like the maize + *S. guianensis* followed by *S. guianensis* fallow (T2 treatment), although changes with time were not significant at $P \leq 0.05$ (Table 2.9) (Ella, 2013, unpublished report). The trend seems reversed

Table 2.9. Relationship between percent soil organic matter (SOM) at 0–5 cm depth and time (days) in conservation agriculture treatments in Claveria, Mindanao, Philippines.

Treatment		r^2	*P*
T1	SOM = 6.23 + 0.0001 time	0.01	0.90
T2	SOM = 5.63 + 0.0028 time	0.57	0.14
T3	SOM = 5.58 + 0.0008 time	0.27	0.36
T4	SOM = 5.41 + 0.0004 time	0.36	0.28
T5	SOM = 6.23 + 0.0004 time	0.05	0.71
T6	SOM = 5.61 – 0.0012 time	0.45	0.21

in the plow-based treatment (T6 treatment), but again, there were no significant differences with time. All treatments received 30 kg/ha of diphosphorus pentoxide (P_2O_5).

Conservation agriculture adoption constraints in Southeast Asia

Some of the main constraints to the adoption of CA in SEA are due to larger-scale infrastructural problems that limit the availability of inputs and agricultural training: (i) local unavailability of suitable CA implements; (ii) communal grazing after a main crop; (iii) absence of a suitable cover crop that can produce income after the main crop; (iv) lack of credit systems suited for CA; (v) limited skills in weed management in CA systems; (vi) a weak agricultural extension system; and (vii) highly specialized and entrenched plow-based agriculture (Lienhard *et al*., 2013). Boulakia *et al*. (2012, unpublished report) also observed that farmers with smaller landholdings and less capital resources in Kampong Cham and Battambang, Cambodia, were slower in adopting CA compared to farmers who were better educated and had bigger landholdings and more capital resources. This trend was also present in the Philippines (Mercado, Claveria, the Philippines, 2013, personal communication). As a response to some of these constraints (i, iv, v, and vi), the CIRAD/SANREM project introduced, tested and promoted NT planters and medium-sized sprayers imported from Brazil, and convinced a Thai machinery manufacturer to fabricate CA implements. No-till machinery became available at half the cost of the imported Brazilian model, and with readily available spare parts. In addition, a pilot microfinance approach to assist farmers in the transition from plow-based agriculture to CA was well received with very high repayment rates. In the Philippines, SANREM researchers tested several NT machines and tools and concluded that for smallholder farmers, a simple modified machete was the most appropriate tool for NT sowing of maize (Mercado *et al*., 2013, unpublished report). Additionally, Filipino farmers adopted the permanent living mulch, *A. pintoi*. It appears that decreased labor, savings in inputs, and increases in yields under CA systems will eventually lead to elevated adoption rates by SEA farmers.

Conservation agriculture case studies in Southeast Asia

In Sayaboury province, Laos, agriculture has expanded rapidly to cover more than 42,000 ha (Tran Quoc *et al*., 2010). While livelihoods have certainly been improved, agricultural intensification and expansion can have very negative social and ecological effects in the long term, including increased soil erosion, mining of soil resources, and chemical pollution of soil and water resources. The advent of these undesirable outcomes in SEA due to changes in land use has already been widely established (Foley *et al.*, 2005). In response to these negative externalities, CA has been tested in four districts of southern Sayaboury province. In 2008, more than 1,200 smallholders adopted CA, on a total area of 1,500 ha, with greater adoption rates occurring in most degraded areas. In the Plain of Jars, one of the more degraded areas in the western part of Xieng Khouang province, various CA-based systems that incorporate intercropping, reduced tillage, and continuous ground cover (cover crops) have been introduced and tested with small farmers as an alternative to tillage-based agriculture since 2006 (Lienhard *et al.*, 2014).

In Battambang, Cambodia, the area under CA increased from 32 ha with 26 households in 2010/11 to 165 ha with 64 households in 2013/14 (Fig. 2.7). In addition to the area managed by farmers participating in CA projects, spontaneous adoption extension occurred in 35 ha in 2013 (Kong *et al.*, 2013, unpublished report). Following this initial accomplishment, scaling up some CA practices in Cambodia looks promising. To back up this effort, large-scale demonstrations, replicated field trials, seed production areas, and germplasm collections with 183 rice, 61 soybean and 27 cassava varieties and 42 cover crop species and varieties are maintained at the Cambodian Ministry of Agriculture, Forestry and Fisheries (MAFF) station in Bos Knohr, Kampong Cham. Furthermore, the Conservation Agriculture Service Center under MAFF was started in partnership with the Royal University of Agriculture and the University of Battambang.

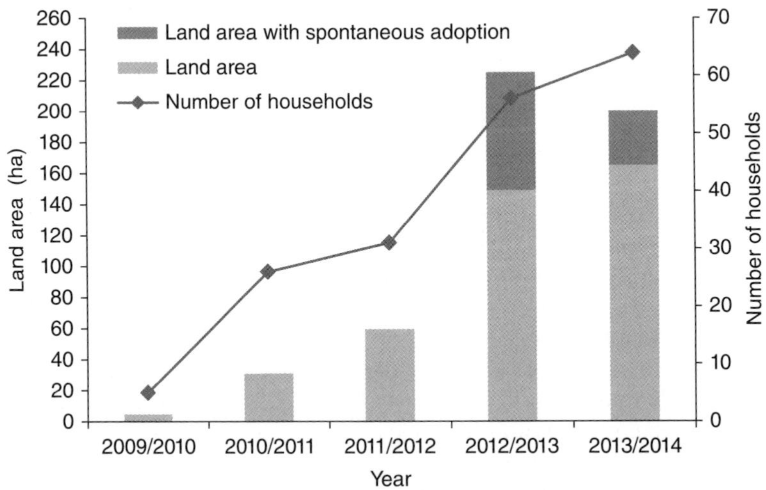

Fig. 2.7. Extent of adoption of conservation agriculture technologies from 2009 to 2013 in Battambang, Cambodia. (From Kong *et al.*, 2013, unpublished report.)

In the mountainous region of northern Mindanao in the Philippines, SANREM has been working on CA since 2010. In addition to crop production, tree planting can prevent land slippage and largely reduce erosion. More than 2,000 farmers have been trained at the experimental site in Claveria that includes demonstrations of agroforestry, cover crops, conservation agriculture, and rainwater ponds built with animal-based technology for rainwater harvesting and fish production. Among the demonstration plot cover crops, the perennial native grass adlai (*Coix lacryma-jobi*) produces high biomass amounts and grains that have multiple uses. Since demonstrations are inclusive of a diverse number of CA methodologies, farmers are implementing multiple CA techniques at once, making this site a prime example of a working model of ecological agriculture.

2.6 Conclusions

Substantial knowledge has been amassed recently on the potential benefits of CA for resource-poor, small householders and the limitations for widespread adoption. Studies from different regions reveal some commonality in findings, but also point to the situational nature of both partial and full application of CA principles. Like any other cropping system, CA is not a one-size-fits-all solution, and requires significant adaptation and fine-tuning to the needs of farmers in target areas. Site-specific applied research using participatory methods and innovation networks as drivers for adoption have proven to work well in many areas, and encouraging adoption trends have been reported from various regions. The CA adaptation process is likely most efficient when a local "innovation system" emerges and begins to acquire a self-sustaining dynamic (Harrington and Erenstein, 2005).

In southern Africa, the increasing need to address soil fertility decline and the negative impacts of climate variability and change has forced farmers to consider new ways of farming. One of the "greener" solutions at hand is CA, which has the potential to address many of the aforementioned constraints and challenges. Reported CA benefits mentioned in this chapter include soil quality improvements and increased soil water retention, greater productivity, and more stable cropping yields.

In SEA, CA adoption has not mirrored the most successful examples from Africa, such as Malawi. However, there seems to be a window of opportunity for further developments of CA in countries like Cambodia, where SANREM/CIRAD research has tested innovative concepts such as the use of cover crops and minimum tillage. Similar to the situation in southern Africa, the need to promote economic diversification and reduce land degradation can provide momentum for the adoption of CA in southern Asia. Towards this purpose, it will be crucial to develop pathways and strategies encompassing biophysical, socio-economic, and political domains and their interactions at different scales. Further analysis of the role of CA in promoting sustainable intensification and resilience in agroecosystems can contribute to more integral approaches to favor small landholders in disadvantageous conditions.

References

Aagard, P. (2009) Maize production and conservation farming in Zambia. An update. Conservation Farming System, Lusaka, Zambia. Available at: http://conservationagriculture.org/uploads/pdf/MAIZE-PRODUCTION-AND-CF-IN-ZAMBIA.pdf (accessed 22 October 2014).

Akobundu, I.O. (1987) *Weed Science in the Tropics: Principles and Practices*. Wiley, Chichester, UK.

Ares, A., Moore, K.M. and Mulvaney, M. (2012) Conservation agriculture systems to improve rural livelihoods: The Sustainable Agriculture and Natural Resources Management Collaborative Support Program. In: Hauswirth, D., Pham, T.S., Nicetic, O., Tivent, F., Le Quoc, D., *et al.* (eds) *Conservation Agriculture and Sustainable Upland Livelihoods. Innovations For, With and By Farmers to Adapt to Local and Global Changes*. Proceedings of the 3rd International Conference on Conservation Agriculture in Southeast Asia. Hanoi, Vietnam, 10–15 December 2012. CIRAD, Montpellier, France; NOMAFSI, Phu Tho, Vietnam; University of Queensland, Brisbane, Australia, pp. 346–347.

Aulakh, M.S., Manchanda, J.S., Garg, A.K., Kumar, S., Dercon, G., *et al.* (2012) Crop production and nutrient use efficiency of conservation agriculture for soybean–wheat rotation in the Indo-Gangetic Plains of Northwestern India. *Soil and Tillage Research* 120, 50–60.

Bado, B.V., Bationo, A. and Cescas, M.P. (2006) Assessment of cowpea and groundnut contributions to soil fertility and succeeding sorghum yields in the Guinean savannah zone of Burkina Faso (West Africa). *Biology and Fertility of Soils* 43, 171–176.

Benites, J., Chuma, E., Fowler, R., Kienzle, J., Molapong, K., *et al.* (eds) (1998) *Conservation Tillage for Sustainable Agriculture*. Proceedings International Workshop Part I (Workshop Report). Deutschegesellschaft, GTZ, Eschborn, Germany.

Bing, L.A. and Lewish, L.C. (1993) Occurrence of the entomopathogen *Beauveria bassiana* (Balsamo) Vuillemin in different tillage regimes and in *Zea mays* L. and virulence towards *Ostrinia nubilalis* (Hubner). *Agriculture, Ecosystems and Environment* 45, 147–156.

Bisangwa, E. (2013) *The Influence of Conservation Agriculture Adoption on Input Demand and Maize Production in Butha Buthe, Lesotho*. University of Tennessee, Knoxville, Tennessee.

Bishop-Sambrook, C., Kienzle, J., Mariki, W., Owenya, M. and Ribeiro, F. (2004) Conservation agriculture as a labour saving practice for vulnerable households. A study of the suitability of reduced tillage and cover crops for households under labor stress in Babati and Karatu Districts, Northern Tanzania. IFAD and FAO.

Bolliger, A., Magid, J., Amado, T.J.C., Scora Neto, F., Dos Santos Ribeiro, M.D.F., *et al.* (2006) Taking stock of the Brazilian "zero-till revolution": A review of landmark research an farmers' practice. *Advances in Agronomy* 91, 47–110.

Bonsu, P.O. and Asibuo, J.Y. (2013) Rotational effects of legumes on maize yield. *International Journal of Scientific and Technology Research* 2, 222–227.

Brooks, S. and Loevinsohn, M. (2011) Shaping agricultural innovation systems responsive to food insecurity and climate change. *Natural Resources Forum* 35, 185–200.

Buerkert, A. and Lamers, J.P.A. (1999) Soil erosion and deposition effects on surface characteristics and pearl millet growth in the West African Sahel. *Plant and Soil* 215, 239–253.

Byerlee, D. and Hesse De Polanco, E. (1986) Farmers' stepwise adoption of technological packages: Evidence from the Mexican Altiplano. *American Journal of Agricultural Economics* 68, 519–527.

Cassman, K.G. (1999) Ecological intensification of cereal production systems: Yield potential, soil quality, and precision agriculture. *Proceedings of the National Academy of Sciences* 96, 5952–5959.

Castellanos-Navarrete, A., de Goede, R.G.M., Kooistra, M.J., Sayre, K.D., Brussaard, L., *et al.* (2012) Earthworm activity and soil structural changes under conservation agriculture in central Mexico. *Soil and Tillage Research* 123, 61–70.

Chabierski, S., Rada, K., Sona, S., Tyneth, L., Penot, E., *et al.* (2012) A preliminary experience of conservation agriculture practices extension with smallholders in Cambodia: The study case of Kampong Cham Province. In: Mulvaney, M.J., Reyes, M.R., Chan-Halbrendt, C., Boulakia, S., Jumpa, K., *et al.* (eds) *Conservation Agriculture in Southeast Asia and Beyond.* Special Publication No 7, World Association of Soil and Water Conservation (WASWAC), Beijing, pp. 3–28.

Chapin, F.S. III, Folke, C. and Kofinas, G.P. (2009) A framework for understanding change. In: Chapin, F.S., Kofinas, G.P. and Folke, C. (eds) *Principles of Ecosystem Stewardship.* Springer, New York, pp. 3–18.

Chivenge, P.P., Murwira, H.K., Giller, K.E., Mapfumo, P. and Six, J. (2007) Long-term impact of reduced tillage and residue management on soil carbon stabilization: Implications for conservation agriculture on contrasting soils. *Soil and Tillage Research* 94, 328–337.

Darnhofer, I., Fairweather, J. and Moller, H. (2010) Assessing a farm's sustainability: Insights from resilience thinking. *International Journal of Agriculture Sustainability* 8, 186–198.

Derpsch, R. and Friedrich, T. (2009) *Global Overview of Conservation Agriculture Adoption.* Proceedings 4th World Congress on Conservation Agriculture: Innovations for improving efficiency, equity and environment. 4–7 February 2009, New Delhi. Print Process, New Delhi, pp. 429–438.

Dubreil, N. (2011) Targeting conservation agriculture innovations to combat soil degradation and food insecurity in semi-arid Africa: A literature review. MSc thesis, Wageningen University, the Netherlands.

Edralin, D.I.A and Reyes, M.R. (2013) Conservation agriculture with drip irrigation in Siem Reap, Cambodia. Poster presented at the 2013 Water Education Summit, Chattanooga, Tennessee, 24 – 26 September 2013.

Ekboir, J. (2002) *CIMMYT 2000–2001 World Wheat Overview and Outlook: Developing No-till Packages for Small-scale Farmers.* CIMMYT, Mexico, DF.

Erenstein, O. (2002) Crop residue mulching in tropical and semi-tropical countries: An evaluation of residue availability and other technological implications. *Soil and Tillage Research* 67, 115–133.

Erenstein, O. (2003) Smallholder conservation farming in the tropics and sub-tropics: A guide to the development and dissemination of mulching with crop residues and cover crops. *Agriculture, Ecosystems and Environment* 100, 17–37.

Erenstein, O. and Laxmi, V. (2008) Zero tillage impacts in India's rice–wheat systems: A review. *Soil and Tillage Research* 100, 1–14.

Erenstein, O., Farooq, U., Malik, R.K. and Sharif, M. (2008) On-farm impacts of zero tillage wheat in South Asia's rice–wheat systems. *Field Crops Research* 105, 240–252.

Erenstein, O., Sayre, K., Wall, P., Hellin, J. and Dixon, J. (2012) Conservation agriculture in maize- and wheat-based systems in the (sub) tropics: Lessons from adaptation initiatives in South Asia, Mexico, and Southern Africa. *Journal of Sustainable Agriculture* 36, 180–206.

Foley, J.A., DeFries, R., Asner, G.P., Barford, C., Bonan, G., *et al.* (2005) Global consequences of land use. *Science* 309, 570–574, doi:10.1126/science.1111772.

Formowitz, B., Joergensen, R.G. and Buerkert, A. (2009) Impact of legume versus cereal root residues on biological properties of West African soils. *Plant and Soil* 325, 145–156.

Fowler, R. and Rockström, J. (2001) Conservation tillage for sustainable agriculture: An agrarian revolution gathers momentum in Africa. *Soil and Tillage Research* 6, 93–107.

Friedrich, T., Derpsch, R. and Kassam, A. (2012) Overview of the global spread of conservation agriculture. Field Actions Science Reports, Special Issue 6.

Fuentes, M., Govaerts, B., De León, F., Hidalgo, C., Dendooven, L., *et al.* (2009) Fourteen years of applying zero and conventional tillage, crop rotation and residue management systems and its effect on physical and chemical soil quality. *European Journal of Agronomy* 30, 228–237.

Fuglie, K.O. and Kascak, C.A. (2001) Adoption and diffusion of natural-resource-conserving agricultural technology. *Review of Agricultural Economics* 23, 386–403.

Gicheru, P., Gachene, C., Mbuvi, J. and Mare, E. (2004) Effects of soil management practices and tillage systems on surface soil water conservation and crust formation on a sandy loam in semi-arid Kenya. *Soil and Tillage Research* 75, 173–184.

Giller, K.E., Witter, E., Corbeels, M. and Tittonell, P. (2009) Conservation agriculture and smallholder farming in Africa: The heretic's view. *Field Crops Research* 114, 23–34.

Godfray, H.C.J., Beddington, J.R., Crute, I.R., Haddad, L., Lawrence, L., *et al.* (2010) Food security: the challenge of feeding 9 billion people. *Science* 327, 812–818.

Govaerts, B., Sayreb, K.D. and Deckers, J. (2005) Stable high yields with zero tillage and permanent bed planting? *Field Crops Research* 94, 33–42.

Govaerts, B., Fuentes, M., Mezzalama, M., Nicol, J.M., Deckers, J., *et al.* (2007a) Infiltration, soil moisture, root rot and nematode populations after 12 years of different tillage, residue and crop rotation managements. *Soil and Tillage Research* 94, 209–219.

Govaerts, B., Mezzalama, M., Unno, Y., Sayre, K.D., Luna-Guido, M., *et al.* (2007b) Influence of tillage, residue management, and crop rotation on soil microbial biomass and catabolic diversity. *Applied Soil Ecology* 37, 18–30.

Govaerts, B., Mezzalama, M., Sayre, K.D., Crossa, J., Lichter, K., *et al.* (2008) Long-term consequences of tillage, residue management, and crop rotation on selected soil micro-flora groups in the subtropical highlands. *Applied Soil Ecology* 38, 191–210.

Govaerts, B., Sayre, K.D., Goudeseune, B., De Corte, P., Lichter, K., *et al.* (2009) Conservation agriculture as a sustainable option for the central Mexican highlands. *Soil and Tillage Research* 103, 220–230.

Gwenzi, W., Gotosa, J., Chakanetsa, S. and Mutema, Z. (2009) Effects of tillage systems on soil organic carbon dynamics, structural stability and crop yields in irrigated wheat (*Triticum aestivum* L.) – cotton (*Gossypium hirsutum* L.) rotation in semi-arid Zimbabwe. *Nutrient Cycling in Agroecosystems* 83, 211–221.

Haggblade, S. and Tembo, G. (2003) Early evidence of conservation farming in Zambia. Paper for the International Workshop on Reconciling Rural Poverty and Resource Conservation: Identifying Relationships and Remedies. Ithaca, New York.

Harrington, L. and Erenstein, O. (2005) Conservation agriculture and resource conserving technologies. A global perspective. In: Abrol, I.P., Gupta, R.K. and Malik, R.K. (eds) *Conservation Agriculture. Status and Prospects*. Centre for Advancement of Sustainable Agriculture, New Delhi, pp. 1–12.

Hobbs, P.R. (2007) Conservation agriculture: What is it and why is it important for future sustainable food production? *Journal of Agricultural Science* 145, 127–137.

Holling, C.S. (ed.) (1978) *Adaptive Environmental Assessment and Management*. Wiley, London.

Horne, J.F. III and Orr, J.E. (1998) Assessing behaviors that create resilient organizations. *Employment Relations Today* 24, 29–39.

Jackson, L.E., Pascual, U. and Hodgkin, T. (2007) Utilizing and conserving agrobiodiversity in agricultural landscapes. *Agriculture, Ecosystems and Environment* 121, 196–210.

James, D.G. (2003) Synthetic herbivore-induced plant volatiles as field attractants for beneficial insects. *Environmental Entomology* 32, 977–982.

Jat, M.L., Saharawat, Y.S. and Gupta, R. (2011) Conservation agriculture in cereal systems of south Asia: Nutrient management perspectives. *Karnataka Journal of Agricultural Science* 24, 100–105.

Jat, R.A., Wani, S.P. and Sahrawat, K.L. (2012) Conservation agriculture in the semi-arid tropics: Prospects and problems. *Advances in Agronomy* 117, 191–273.

Johansen, C., Haque, M.E., Bell, R.W., Thierfelder, C. and Esdaile, R.J. (2012) Conservation agriculture for small holder rainfed farming: Opportunities and constraints of new mechanized seeding systems. *Field Crops Research* 132, 18–32.

Kassam, A., Friedrich, T., Shaxson, F. and Pretty, J. (2009) The spread of conservation agriculture: Justification, sustainability and uptake. *International Journal of Agriculture Sustainability* 7, 292–320.

Kassam, A., Friedrich, T., Shaxson, F., Reeves, T., Pretty, J., *et al.* (2011) Production systems for sustainable intensification. Integrating productivity with ecosystem services. *Technikfolgenabschätzung. Theorie und Praxis* 20, 38–45.

Kassam, A., Friedrich, T., Shaxson, F. and Pretty, J. (2012) Opportunities for scaling up conservation agriculture: barriers, conditions, and strategies. In: Hauswirth, D., Pham, T.S., Nicetic, O., Tivent, F., Le Quoc, D., *et al.* (eds) *Conservation Agriculture and Sustainable Upland Livelihoods.*

Innovations For, With and By Farmers to Adapt to Local and Global Changes. Proceedings of the 3rd International Conference on Conservation Agriculture in Southeast Asia, Hanoi, Vietnam, 10–15 December 2012. CIRAD, Montpellier, France; NOMAFSI, Phu Tho, Vietnam; University of Queensland, Brisbane, Australia, pp. 308–321.

Khan, M.A. and Hashmi, N.I. (2004) Impact of no-tillage farming on wheat production and resource conservation in the rice–wheat zone of Punjab, Pakistan. In: Lal, R., Hobbs, P.R., Uphoff, N. and Hansen, D.O. (eds) *Sustainable Agriculture and the Rice–Wheat System*. Marcel Dekker, New York, pp. 219–228.

Khan, Z., Midega, C., Pittchar, J., Pickett, J. and Bruce, T. (2011) Push–pull technology: A conservation agriculture approach for integrated management of insect pests, weeds and soil health in Africa. *International Journal of Agricultural Sustainability* 9, 162–170.

Kilelu, C., Klerkx, L., Leeuwis, C. and Hall, A. (2011) Beyond knowledge brokerage: An exploratory study of innovation intermediaries in an evolving smallholder agricultural system in Kenya. Discussion Paper 13, Research into Use. Open University, UK.

Klerkx, L., Hall, A. and Leeuwis, C. (2009) Strengthening agricultural innovation capacity: Are innovation brokers the answer? *International Journal of Agricultural Resources, Governance and Ecology* 8, 409–438.

Kosgei, J.R., Jewitt, G.P.W., Kongo, V.M. and Lorentz, S.A. (2007) The influence of tillage on field scale water fluxes and maize yields in semi-arid environments: A case study of potshini catchment, South Africa. *Physics and Chemistry of the Earth* 32, 1117–1126.

Kouyaté, Z., Franzluebbers, K., Juo, A.S.R. and Hossner, L.R. (2000) Tillage, crop residue, legume rotation, and green manure effects on sorghum and millet yields in the semiarid tropics of Mali. *Plant and Soil* 225, 141–151.

Kowalchuk, G.A., Buma, D.S., de Boer, W., Klinkhamer, P.G.L. and van Veen, J.A. (2002) Effects of above-ground plant species composition and diversity on the diversity of soil-borne microorganisms. *Antonie van Leeuwenhoek* 81, 509–520.

Kpotor, P. (2012) Evaluation of newly released maize varieties in Ghana for yield and stability under three nitrogen application rates in two agroecological zones. MSc Thesis, Kwame Nkrumah University of Science, Ghana.

Krishna, V., Bhatia, M., Teufel, N. and Bashnoi, D. (2012) Characterizing the cereal systems and identifying the potential of conservation agriculture in NW India, Nepal Terai and NW Bangladesh. Baseline Village Survey Report: Cereal Systems Initiative for South Asia (CSISA). Socio-economics Working Paper 5.

Lal, R. (1974) Soil temperature, soil moisture and maize yield from mulched and unmulched tropical soils. *Plant and Soil* 40, 129–143.

Lal, R. (1993) Tillage effects on soil degradation, soil resilience, soil quality, and sustainability. *Soil and Tillage Research* 27, 1–8.

Lamb, J.N., Moore, K.M., Marake, M., Lambert, D., Wilcox, M., *et al.* (2013) Agricultural actors, networks, and farmer identity: Examining perspectives and adoption of conservation agriculture in Botha Bothe, Lesoto. SANREM Working Paper 01–13, Virginia Tech Office of International Research, Education and Development, Blacksburg Virginia.

Langdale, G., Blevins, R., Karlen, D., McCool, D., Nearing, M., *et al.* (1991) Cover crop effects on soil erosion by wind and water. Soil and Water Conservation Society, Ankeny, Iowa, pp. 15–22.

Larbi, A., Smith, J.W., Adekunle, I.O., Agyare, W.A., Gbaraneh, L.D., *et al.* (2002) Crop residues for mulch and feed in crop–livestock systems: Impact on maize grain yield and soil properties in the West African humid forest and savanna zones. *Experimental Agriculture* 38, 253–264.

Le Q.D. and Ha D.T. (2010) Conservation agriculture on sloping lands in northern mountainous regions of Vietnam. In: Chanphengxay, M.B., Khamhung, A., Panysiri, K., Chabanne, A., Julien, F., *et al.* (eds) *Investing in Sustainable Agriculture: The Case of Conservation Agriculture and Direct Seeding Mulch-based Cropping Systems.* Proceedings of the Regional Workshop on Conservation Agriculture, Phonsavan, Xieng Khouang Province, Lao PDR, 28 October–1 November 2008, Vientiane. Laos Ministry of Agriculture and Forestry, Phonsavan, Laos, pp. 27–36.

Lestrelin G., Nanthavong, K., Jobard, E., Keophoxay, A., Lienhard, P., *et al.* (2012) "To till or not to till?" Opportunities and constraints to the diffusion of Conservation Agriculture in Xieng Khouang province, Lao PDR. *Outlook on Agriculture* 41, 41–49.

Lienhard P., Boulakia, S., Legoupil, J.C., Gilard, O. and Séguy, L. (2013) Chapter 7. Development and prospects of conservation agriculture in South East Asia: A comprehensive review. In: Jat, R.A., Sahrawat, K.L. and Kassam, A. (eds) *Conservation Agriculture: Global Prospects and Challenges*, Chapter 7. CAB International, Wallingford, UK, pp. 180–201.

Lienhard, P., Panyasiri, K., Sayphoummie, S., Leudphanane, B., Lestrelin, G., *et al.* (2014) Profitability and opportunity of conservation agriculture in the acid savannah grasslands of Laos. *International Journal of Agricultural Sustainability* 12, 391–406.

Lin, B.B. (2011) Resilience in agriculture through crop diversification: Adaptive management for environmental management. *BioScience* 61, 183–193.

Lobell, D.B., Burke, M.B., Tebaldi, C., Mastrandrea, M.D., Falcon, W.P., *et al.* (2008) Prioritizing climate change adaptation needs for food security in 2030. *Science* 319, 607–610.

McNair, W.E. (2013) Assessing the influence of conservation agriculture on household wellbeing and maize marketing in Tete and Manica, Mozambique. MSc thesis, University of Tennessee, Knoxville, Tennessee.

Mashingaidze, N., Madakadze, C., Twomlow, S., Nyamangara, J. and Hove, L. (2012) Crop yield and weed growth under conservation agriculture in semi-arid Zimbabwe. *Soil and Tillage Research* 124, 102–110.

Matthews, R. and De Pinto, A. (2012) Should REDD+ fund "sustainable intensification" as a means of reducing tropical deforestation? *Carbon Management* 3, 117–120.

Mazvimavi, K. and Twomlow, S. (2009) Socioeconomic and institutional factors influencing adoption of conservation agriculture by vulnerable households in Zimbabwe. *Agricultural Systems* 101, 20–29.

Mazvimavi, K., Twomlow, S., Belder, P. and Hove, L. (2008) An assessment of the sustainable uptake of conservation farming in Zimbabwe International Crops Research Institute for the Semi-Arid Tropics: Global Theme on Agroecosystems. Report No 39, Bulawayo, Zimbabwe.

Mazzola, M., Granatstein, D.M., Elfving, D.C. and Mullinix, K. (2001) Suppression of specific apple root pathogens by *Brassica napus* seed meal amendments regardless of glucosinolate content. *Phytopathology* 9, 673–679.

Mercado, A.R., Reyes, M.R., Ella, V. and Boulakia, S. (2012) Conservation agriculture in Southeast Asia and beyond. In: Mulvaney, M.J., Reyes, M.R., Chan-Halbrendt, C., Boulakia, S., Jumpa, K., *et al.* (eds) *Conservation Agriculture Research in the Philippines*. Special Publication No 7. World Association of Soil and Water Conservation (WASWAC), Beijing, pp. 51–72.

Montpellier Panel, The (2013) Sustainable Intensification: A New Paradigm for African Agriculture, London.

Morris, M.L., Kelly, V.A., Kopicki, R.J. and Byerlee, D. (2007) *Fertilizer Use in African Agriculture: Lessons Learned and Good Practice Guidelines*. World Bank Publications, Washington, DC.

Moussa-Machraoui, S.B., Errouissi, F., Ben-Hammouda, M. and Nouira, S. (2010) Comparative effects of conventional and no-tillage management on some soil properties under Mediterranean semi-arid conditions in northwestern Tunisia. *Soil Tillage Research* 106, 247–253.

Mueller, D.S., Hartman, G.L. and Pedersen, W.L. (2002) Effect of crop rotation and tillage system on *Sclerotinia* stem rot on soybean. *Canadian Journal of Plant Pathology* 24, 450–456.

Mueller, N.D., Gerber, J.S., Johnston, M., Ray, D.K., Ramankutty, N., *et al.* (2012) Closing yield gaps through nutrient and water management. *Nature* 490, 254–257.

Muhr, L., Tarawali, S.A., Peters, M. and Schultze-Kraft, R. (2002) Soil mineral N dynamics and maize grain yields following *Centrosema macrocarpum* and *Stylosanthes guianensis*: Effects of different rotations and varying levels of N fertiliser. *Field Crops Research* 78, 197–209.

Munodawafa, A. and Zhou, N. (2008) Improving water utilization in maize production through conservation tillage systems in semi-arid Zimbabwe. *Physics and Chemistry of the Earth* 33, 757–761.

Muoni, T., Rusinamhodzi, L. and Thierfelder, C. (2013) Weed control in conservation agriculture systems of Zimbabwe: Identifying economical best strategies. *Crop Protection* 53, 23–28.

Mupangwa, W., Twomlow, S. and Walker, S. (2008) The influence of conservation tillage methods on soil water regimes in semi-arid southern Zimbabwe. *Physics and Chemistry of the Earth*, Parts A/B/C 33, 762–767.

Mupangwa, W., Twomlow, S. and Walker, S. (2012) Reduced tillage, mulching and rotational effects on maize (*Zea mays* L.), cowpea (*Vigna unguiculata* (Walp) L.) and sorghum (*Sorghum bicolor* L. (Moench)) yields under semi-arid conditions. *Field Crops Research* 132, 139–148.

Mupangwa, W., Twomlow, S. and Walker, S. (2013) Cumulative effects of reduced tillage and mulching on soil properties under semi-arid conditions. *Journal of Arid Environments* 91, 45–52.

Naresh, R.K., Singh, S.P. and Chauhan, P. (2012) Influence of conservation agriculture, permanent rasied bed planting and residue management on soil quality and productivity in maize–wheat system in western Uttar Pradesh. *International Journal of Life Science, Biotechnology and Pharma Research* 1, 27–34.

Naudin, K., Goze, E., Balarabe, O., Giller, K.E. and Scopel, E. (2010) Impact of no tillage and mulching practices on cotton production in north Cameroon: A multi-locational on farm assessment. *Soil and Tillage Research* 108, 68–76.

Ngigi, S.N., Rockstrom, J. and Savenije, H.H.G. (2006) Assessment of rainwater retention in agricultural land and crop yield increase due to conservation tillage in ewaso ng'iro river basin, Kenya. *Physics and Chemistry of the Earth* 31, 910–918.

Nguema, A., Norton, G.W., Alwang, J., Taylor, D.B., Barrera, V., *et al.* (2013) Farm-level economic impacts of conservation agriculture in Ecuador. *Experimental Agriculture* 49, 134–147.

Ngwira, A.R., Aune, J.B. and Mkwinda, S. (2012) On-farm evaluation of yield and economic benefit of short term maize legume intercropping systems under conservation agriculture in Malawi. *Field Crops Research* 132, 149–157.

Ngwira, A.R., Thierfelder, C. and Lambert, D.M. (2013) Conservation agriculture systems for Malawian smallholder farmers: long-term effects on crop productivity, profitability and soil quality. *Renewable Agriculture and Food Systems* 28, 350–363.

Norwegian Agency for Development Cooperation (2011) Report from a fact finding mission: Women, Gender and Conservation Agriculture in Zambia. Norad Report 5, Oslo.

Nyasani, J.O., Meyhöfer, R., Subramanian, S. and Poehling, H.M. (2012) Effect of intercrops on thrips species composition and population abundance on French beans in Kenya. *Entomologia Experimentalis et Applicata* 142, 236–246.

Ouattara, K., Ouattara, B., Nyberg, G., Sédogo, M.P. and Malmer, A. (2007) Ploughing frequency and compost application effects on soil infiltrability in a cotton–maize (*Gossypium hirsutum–Zea mays* L.) rotation system on a ferric luvisol and a ferric lixisol in Burkina Faso. *Soil and Tillage Research* 95, 288–297.

Owenya, M.Z., Mariki, W.L., Kienzie, J., Friedrich, T. and Kassam, A. (2011) Conservation agriculture (CA) in Tanzania: The case of the Mwangaza B CA farmer field school (FFS), Rhotia Village, Karatu District, Arusha. *International Journal of Agricultural Sustainability* 9, 145–152.

Patiño-Zúñiga, L., Ceja-Navarro, J.A., Govaerts, B., Luna-Guido, M., Sayre, K.D., *et al.* (2009) The effect of different tillage and residue management practices on soil characteristics, inorganic N dynamics and emissions of N_2O, CO_2 and CH_4 in the central highlands of Mexico: A laboratory study. *Plant Soil* 314, 231–241.

Paudel, B., Radovich, T., Crow, S., Halbrendt, J., Chan-Halbrendt, C., *et al.* (2012) On-farm performance evaluation of conservation agriculture production systems in the central middle hills of Nepal. In: Hauswirth, D., Pham, T.S., Nicetic, O., Tivent, F., Le Quoc, D., *et al.* (eds) *Conservation Agriculture and Sustainable Upland Livelihoods. Innovations For, With and By Farmers to Adapt to Local and Global Changes*. Proceedings of the 3rd International Conference on Conservation Agriculture in Southeast Asia, Hanoi, Vietnam, 10–15 December 2012. CIRAD, Montpellier; NOMAFSI, Phu Tho, Vietnam; University of Queensland, Brisbane, pp. 289–291.

Pell, J.K., Hannam, J.J. and Steinkraus, D.C. (2010) Conservation biological control using fungal entomopathogens. *Biocontrol* 55, 187–198.

Pimentel, D., Harvey, C., Resosudarmo, P., Sinclair, K., Kurz, D., *et al.* (1995) Environmental and economic costs of soil erosion and conservation. *Science* 267, 1117–1123.

Ploeg, A. (2002) Effects of selected marigold varieties on root-knot nematodes and tomato and melon yields. *Plant Disease* 86, 505–508.

Pretty, J., Toulmin, C. and Williams, S. (2011) Sustainable intensification in African agriculture. *International Journal of Agricultural Sustainability* 9, 5–24.

Puente, M., Darnall, N. and Forkner, R.E. (2011) Assessing integrated pest management adoption: Measurement problems and policy implications. *Environmental Management* 48, 1013–1023.

Rebafka, F., Ndunguru, B.J. and Marschner, H. (1993) Crop residue application increases nitrogen fixation and dry matter production in groundnut (*Arachis hypogaea* L.) grown on an acid sandy soil in Niger, West Africa. *Plant and Soil* 150, 213–222.

Rockström, J., Kaumbutho, P., Mwalley, J., Nzabi, A.W., Temesgen, M., *et al.* (2009) Conservation farming strategies in east and Southern Africa: Yields and rain water productivity from on-farm action research. *Soil and Tillage Research* 103, 23–32.

Ruben, R., Kruseman, G. and Kuyvenhoen, A. (2006) Strategies for sustainable intensification in East African highlands: labor use and input efficiency. *Agricultural Economics* 34, 167–181.

Rusinamhodzi, L., Murwira, H.L. and Nyamangara, J. (2006) Cotton–cowpea intercropping and its N_2 fixation capacity improves yield of a subsequent maize crop under Zimbabwean rain-fed conditions. *Plant and Soil* 287, 327–336.

Rusinamhodzi, L., Corbeelsc, M., Nyamangarad, J. and Giller, K. (2012) Maize–grain legume intercropping is an attractive option for ecological intensification that reduces climatic risk for smallholder farmers in central Mozambique. *Field Crops Research* 136, 12–22.

Sanchez, P.A. (2010) Tripling crop yields in tropical Africa. *Nature Geoscience* 3, 299–300.

Sanginga, N., and Woomer, P.L. (2009) *Integrated Soil Fertility Management in Africa: Principles, Practices and Developmental Process*. CIAT, Cali, Colombia.

Shenk, M. and Saunders, J.L. (1984) Vegetation management systems and insect responses in the humid tropics of Costa Rica. *Tropical Pest Management* 30, 186–193.

Sikirou, R. and Wydra, K. (2008) Effect of intercropping cowpea with maize or cassava on cowpea bacterial blight and yield. *Journal of Plant Diseases and Protection* 115, 145–151.

Silici, L., Ndabe, P., Friedrich, T. and Kassam, A. (2011) Harnessing sustainability, resilience and productivity through conservation agriculture: The case of likoti in Lesotho. *International Journal of Agricultural Sustainability* 9, 137–144.

Sims, B.G., Thierfelder, C., Kienzle, J., Friedrich, T. and Kassam, A. (2012). Development of the conservation agriculture equipment industry in sub-Saharan Africa. *Applied Engineering in Agriculture* 28, 813–823.

Singh, R., Dabur, K.R. and Malik, R.K. (2005) Long-term response of plant pathogens and nematodes to zero-tillage technology in rice–wheat cropping system. Technical Bulletin No 7, Regional Research Station, Kaul, Department of Nematology and Directorate of Extension Education, Hisar, CCS Haryana Agricultural University, India.

Six, J., Feller, C., Denef, K., Ogle, S.M., Moraes Sa, J.C., *et al.* (2002) Soil organic matter, biota and aggregation in temperate and tropical soils. Effects of no-tillage. *Agronomie* 22, 755–775.

Smith, P. (2013) Delivering food security without increasing pressure on land. *Global Food Security* 2, 18–23.

Smith, P., Gregory, P., van Vuuren, D., Obersteiner, M., Rounsevell, M., *et al.* (2010) Competition for land. *Philosophical Transactions of the Royal Society B* 365, 2941–2957.

Snapp, S.S., Rohrbach, D.D., Simtowe, F. and Freema, H.A. (2002) Sustainable soil management options for Malawi: Can smallholder farmers grow more legumes? *Agriculture, Ecosystems and Environment* 91, 159–174.

Spielman, D.J., Davis, K., Negash, M. and Ayele, G. (2011) Rural innovation systems and networks: Findings from a study of Ethiopian farmers. *Agriculture and Human Values* 28, 195–212.

Thierfelder, C. and Wall, P.C. (2009) Effects of conservation agriculture techniques on infiltration and soil water content in Zambia and Zimbabwe. *Soil and Tillage Research* 105, 217–227.

Thierfelder, C., and Wall, P.C. (2010a) Investigating conservation agriculture (CA) systems in Zambia and Zimbabwe to mitigate future effects of climate change. *Journal of Crop Improvement* 24, 113–121.

Thierfelder, C. and Wall, P.C. (2010b) Rotations in conservation agriculture systems of Zambia: Effects on soil quality and water relations. *Experimental Agriculture* 46, 309–325.

Thierfelder, C. and Wall, P.C. (2011) Reducing the risk of crop failure for smallholder farmers in Africa through the adoption of conservation agriculture. In: Bationo, A., Waska, B., Okeyo, J.M., Maina, F. and Kihara, J.M. (eds) *Innovations as Key to the Green Revolution in Africa*. Springer, the Netherlands, pp. 1269–1277.

Thierfelder, C. and Wall, P.C. (2012) Effects of conservation agriculture on soil quality and productivity in contrasting agro-ecological environments of Zimbabwe. *Soil Use and Management* 28, 209–220.

Thierfelder, C., Cheesman, S. and Rusinamhodzi, L. (2012) A comparative analysis of conservation agriculture systems: Benefits and challenges of rotations and intercropping in Zimbabwe. *Field Crops Research* 137, 237–250.

Thierfelder, C., Chisui, J.L., Gama, M., Cheesman, S., Jere, Z.D., *et al.* (2013a) Maize-based conservation agriculture systems in Malawi: Long-term trends in productivity. *Field Crops Research* 142, 47–57.

Thierfelder, C., Mwila, M. and Rusinamhodzi, L. (2013b) Conservation agriculture in eastern and southern provinces of Zambia: Long-term effects on soil quality and maize productivity. *Soil and Tillage Research* 126, 246–258.

Thierfelder, C., Mombeyarara, T., Mango, N. and Rusinamhodzi, L. (2013c) Integration of conservation agriculture in smallholder farming systems of southern Africa: Identification of key entry points. *International Journal of Agricultural Sustainability* 11, 317–330.

Tilman, D., Balzer, C., Hill, J. and Befort, B.L. (2011) Global food demand and the sustainable intensification of agriculture. *Proceedings of the National Academy of Sciences* 108, 20260–20264.

Tittonell, P., Scopela, E., Andrieu N., Posthumush, H., Mapfumoi, P., *et al.* (2012) Agroecology-based aggradation-conservation agriculture (ABACO): Targeting innovations to combat soil degradation and food insecurity in semi-arid Africa. *Field Crops Research* 132, 168–174.

Tran Quoc H., Lestrelin, G., Jullien, F., Kongthong, K., Khamxaykhay, C., *et al.* (2010) Adoption of conservation agriculture in Laos: A case study in the Mekong Corridor. Abstract. In: Chanphengxay, M.B. *et al.* (eds) *Investing in Sustainable Agriculture: The Case of Conservation Agriculture and Direct Seeding Mulch-based Cropping Systems.* Proceedings of the Regional Workshop on Conservation Agriculture, Phonsavan, Xieng Khouang Province, Lao PDR, 28 October–1 November 2008. Laos Ministry of Agriculture and Forestry, pp. 175–176.

Tyler, D., Overton, J. and Chambers, A. (1983) Tillage effects on soil properties, diseases, cyst nematodes, and soybean yields. *Journal of Soil and Water Conservation* 38, 374–376.

USAID (2013) The resilience agenda: Helping vulnerable communities emerge from cycles of crisis onto a pathway toward development. USAID, Washington, DC.

Utomo, M. (2013) Conservation tillage for improving soil health, sustaining crop productivity and mitigating greenhouse gas emission. Abstract. The Fourth International Conservation Agriculture Conference in Southeast Asia, Conservation Agriculture and Sustainable Crop Production Intensification: Producing More While Enhancing Ecosystem Services, 9–13 December 2013. The University of Battambang, Battambang, Cambodia.

Valbuena, D., Erenstein, O., Homann-Kee Tui, S., Abdoulaye, T., Claessens, L., *et al.* (2012) Conservation agriculture in mixed crop–livestock systems: Scoping crop residue trade-offs in Sub-Saharan Africa and South Asia. *Field Crops Research* 132, 175–184.

van Oudenhoven, F.J.W., Mijatovic, D. and Eyzaguirre, P.B. (2011) Social-ecological indicators of resilience in agrarian and natural landscapes. *Management of Environmental Quality* 22, 154–173.

Verhulst, N., Nelissen, V., Jespers, N., Haven, H., Sayre, K.D., *et al.* (2011) Soil water content, maize yield and its stability as affected by tillage and crop residue management in rainfed semi-arid highlands. *Plant and Soil* 344, 73–85.

Walker, B. and Salt, D. (2012) *Resilience Practice. Building Capacity to Absorb Disturbance and Maintain Function.* Island Press, Washington, DC.

Wall, P.C. (2007) Tailoring conservation agriculture to the needs of small farmers in developing countries: An analysis of issues. *Journal of Crop Improvement* 19, 137–155.

Widmer, T.L. (2000) Mechanism of suppression of *Meloidogyne hapla* and its damage by a green manure of Sudan grass. *Plant Disease* 84, 562–568.

Wildavsky, A. (1991) *Searching for Safety*. Transaction Publishers, New Brunswick, New Jersey.

Wyckhuys, K.A.G. and O'Neil, R.J. (2010) Social and ecological facets of pest management in Honduran subsistence agriculture: Implications for IPM extension and natural resource management. *Environment, Development and Sustainability* 12, 297–311.

Young, A. (1999) Is there really spare land? A critique of estimates of available cultivable land in developing countries. *Environment, Development and Sustainability* 1, 3–18.

3 Potential of Conservation Agriculture Production Systems (CAPS) for Improving Sustainable Food and Nutrition Security in the Hill Region of Nepal

Bikash Paudel,[1*] Theodore Radovich,[1] Catherine Chan,[1] Susan Crow,[1] Jacqueline Halbrendt,[1] Keshab Thapa[2] and Bir Bahadur Tamang[2]

[1]*University of Hawai'i at Mānoa, Honolulu, Hawaii, USA;* [2]*Local Initiatives for Biodiversity, Research and Development, Pokhara, Nepal*

3.1 Introduction

3.1.1 Food and nutritional security in Nepal

One in eight people in the world suffers from hunger (FAO/IFAD/WFP, 2013). Although the population considered malnourished has decreased from 1,015 to 842 million from 1990–1992 to 2011–2013, the rate of progress varies across different regions. During this period, the undernourished population has decreased from 319 to 295 million people in southern Asia, yet the region's contribution to the global malnourished population has increased from 31% to 35% in the same duration. The increase in southern Asia's proportion of the world's undernourished population is due mainly to comparatively slower progress in improving food security in the region. Although Nepal is included in southern Asia, its progress in fighting hunger is slightly better than other countries in the region. Nepal has reduced the prevalence of undernourishment successfully from 25.4% (in 1990–1992) to 16.0% (in 2011–2013). Thus, the country is well on track to meet the Millennium Development Goal (MDG) of reducing hunger to below 15% by 2015.

*Corresponding author; e-mail: paudel@hawaii.edu

Despite encouraging progress on reducing undernourishment, the prevalence of severe malnutrition remains very high in Nepal. Undernourishment refers to the condition of not having enough to eat, whereas malnutrition refers to a lack of the right balance of nutrition in food (FAO/IFAD/WFP, 2013). Generally, a reduction in the proportion of the undernourished in a population is associated with an improvement in the overall nutritional condition of people. However, Nepal has been an exception to this rule, because despite having only 16% of its population undernourished, the prevalence of underweight children (29%) and stunting (40%) in Nepal is still one of the highest in the world (FAO/IFAD/WFP, 2013). Since estimates indicate that 35–40% of Nepal's adult population consumes less than the required level of calories (CBS/WFP, 2006), malnutrition is still extremely high, particularly considering that the country is about to meet its food security target set by the MDG. Severe malnutrition problems requiring clinical treatments, such as *kwashiorkor* (dangerous swelling of the face, feet, and limbs caused by protein-energy deficiency) and *marasmus* (severe emaciation caused by extreme energy deficiency), are frequently observed among Nepalese children (Regmi *et al.*, 2004). Because of the very high prevalence of severe malnutrition incidences, milder forms of undernutrition are not even considered a problem in Nepal (Adhikari, 2010). In addition to the high prevalence of malnutrition, failure to reduce great regional disparity, due to extreme variation in topography and climate, is another failed objective in Nepal's race against hunger. In fact, the prevalence of undernourishment is as low as 4% in cities to as high as 40% in the mountain region (CBS/WFP, 2006).

3.1.2 Nepal's agroecological condition

Although Nepal is a small country with an area of 147,181 km^2, it extends from the Ganges river plains in the south, through the central hill region, to the Himalayas in the north. Within a short north–south width of 145–241 km, the country's cascading landscape rises from about 70 m to the world's highest peak, Mount Everest (8,848 m). The climate also varies from subtropical in the southern end to a cold (arctic-like) climate in the north. This extreme variation in topography and climate within a short distance has resulted in extreme differences in crops and agriculture systems in the country. For example, barley, buckwheat, and potato are the major staple food crops produced in the high hills, while maize, finger millet, and grain legumes are the major crops in the middle and lower hills; and rice, wheat, legumes, and oilseed crops are the major crops in the plain areas (Pariyar, 2005).

Despite great variation in topography, climate, and agriculture systems, Nepal is divided broadly into three east-/west-aligned ecological regions in order to simplify the planning process. The northern part of the country falling above 2,500 m is called the *mountain region*. The wide middle belt of the country, ranging from 500 to 2,500 m, is called the *hill region*. Finally, the *terai region* is in the southern part of the country, with elevation lower than 500 m (Pariyar, 2005). Mountain, hill, and *terai* regions cover approximately 35, 42, and 23%, respectively, of Nepal's area (MoAC, 2011). Although about half of the country's population live in the mountain and hill regions, agricultural lands are of more limited availability because of the sloping terrain. Additionally, available agriculture lands in the mountain and hill regions are less fertile than agriculture land in the *terai* region. As a result, production in the mountain and hill regions is not sufficient to secure food self-sufficiency in the area (MoAC/WFP/FAO, 2011).

3.1.3 Agriculture and food security

Although the low availability of agricultural land is a constraint to achieving food security throughout Nepal, it is a more serious problem in the hill and mountain regions. Since only about 16.7% of Nepal's total land area is arable, per capita arable land is less than 0.09 ha/person (World Bank, 2013), which is less than half the world's average (0.20 ha/person). Nepal's per capita arable land is even lower than that in India (0.13 ha/person), which is one of the most highly populated countries in the world (WORLDSTAT, 2012). Therefore, Nepali farmers hold fairly small land parcels, with an average 0.8 ha/household (CBS, 2002). Unfortunately, these parcels are not sufficient to produce enough food for each family.

Regardless of the limited land availability, agriculture is still the mainstay of the country's economy, contributing about one-third of national gross domestic product (35.1% in 2011/12), and employing about 70% of the country's employable people (CBS, 2008; MoF, 2013). Nepalese are highly dependent on the agriculture sector for their livelihood, because there is limited opportunity for off-farm jobs. As a result, increasing agriculture production is the foremost option to improve the country's food and nutritional security. Since increasing the arable land area is not possible, increasing food production depends mainly on increasing crop yields. The hill and mountain regions contain about two-thirds of Nepal's agriculture land (Partap, 1999), so increasing crop yields in these regions is crucial for the country's food and nutritional security. As existing crop yields in the hill and mountain regions are very low (compared to world averages, or even compared to the *terai* region), this presents an opportunity to harness potential yield to improve food self-sufficiency in these regions.

3.1.4 Historical trends of crop yields and food balance in Nepal

With the exception of Africa, crop yields in South Asia are lower than elsewhere in the world. In particular, Nepal's major food crop yields, such as rice, wheat, and maize, are even lower than South Asia's average. In 2012, Nepal's average rice yield was 3.31 t/ha, which was about 5% lower than the South Asian average (3.46 t/ha). Nepal's 2012 wheat yield (2.4 t/ha) was also about 15% lower than South Asia's average (2.85 t/ha). On a more optimistic note, however, rice and wheat yields are increasing much quicker, on average, in Nepal than in the rest of South Asia. During 1991–2012, rice and wheat yields increased by 38% and 75%, respectively, compared to 33% and 52% in South Asia. Because of satisfactory growth rates of rice and wheat yield in the *terai* region, Nepal has been able to produce sufficient food quantities to feed the population on average (MoAC/WFP/FAO, 2011). However, due to high regional imbalance, the hill and mountain regions of the country are still facing food deficit conditions. Most *terai* districts have a food surplus, while most districts in the hill and mountain regions have a food deficit. As food transport from *terai* to hill and mountain regions is difficult due to rough topography, it is essential to increase crop yields in the hills and mountains to maintain food security in these regions. Since maize and millets are the main food crops grown in the hill and mountain regions, slower growth rate of their yields has been the major obstacle for attaining food security in these regions. While maize yield increased by 82% in South Asia during 1991–2012, it increased by only 44% in Nepal (Fig. 3.1). In fact, Nepal's maize yield in 1991 (1.62 t/ha) was about 6% higher than the average yield in South Asia (1.53 t/ha). However, since South Asia's average maize yield surpassed Nepal's in 1996, Nepal's yield had remained lower than the regional average, with the gap between them increasing. By 2012, Nepal's maize yield (2.5 t/ha) was about 11% lower than that of South Asia (2.79 t/ha) (Fig. 3.1).

The productivity of millet through time has been even more depressing than that of maize. Although Nepal's millet yield (1.13 t/ha) was slightly higher than South Asia's average (1.11 t/ha) in 2012, similar to maize, the trend showed that Nepal's millet yield had actually decreased during 1991–2012 versus a slight increase in South Asia. Nepal's millet yield in 1991 (1.17 t/ha) was about 69% higher than South Asia's average (0.69 t/ha). By 2011, South Asia's millet yield had slightly surpassed Nepal's (Fig. 3.1). The trend of maize and millet yield in the hill region heavily influences the national trend, since about 69% of maize and 77% of millet are grown in the hills (Table 3.1).

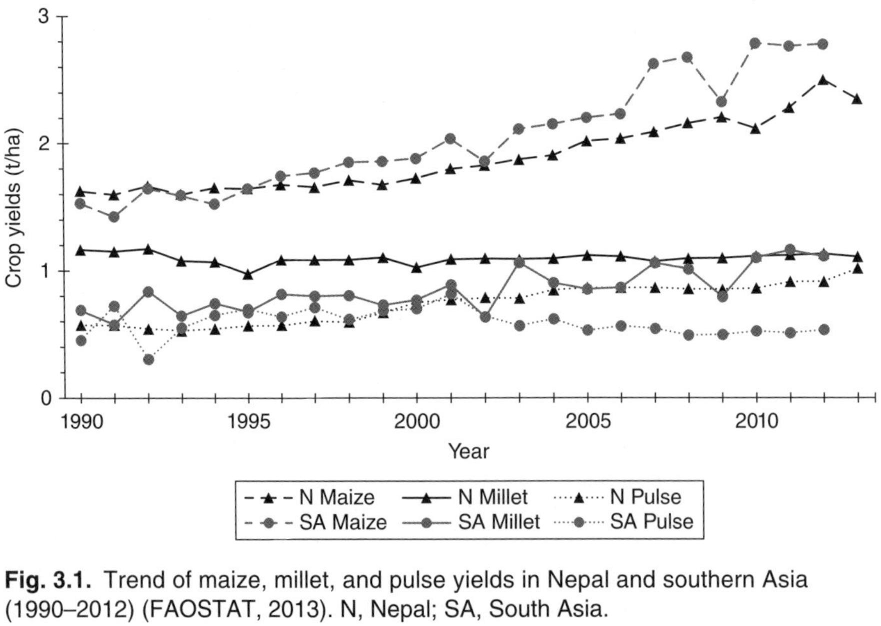

Fig. 3.1. Trend of maize, millet, and pulse yields in Nepal and southern Asia (1990–2012) (FAOSTAT, 2013). N, Nepal; SA, South Asia.

Table 3.1 Area (thousand ha) and yield (t/ha) of major crops (2011) by geographical region in Nepal. (From MoAC, 2011.)

	Mountain			Hill			Terai		
Crops	Area (thousand ha)		Yield	Area (thousand ha)		Yield	Area (thousand ha)		Yield
Rice	66.7	(4.50)[a]	2.06	407.0	(27.2)	2.77	1022.7	(68.3)	3.12
Maize	98.5	(10.9)	2.12	623.1	(68.8)	2.27	184.5	(20.4)	2.42
Wheat	52.1	(6.80)	1.51	265.4	(34.6)	1.99	449.9	(58.6)	2.53
Millet	53.8	(20.0)	1.04	206.3	(76.5)	1.15	9.6	(3.6)	1.07
Legume	12.5	(3.70)	0.85	69.0	(20.6)	0.90	252.8	(75.6)	0.97
Oilseed	3.5	(1.60)	0.73	45.2	(21.2)	0.81	164.9	(77.2)	0.83

[a]Indicates percent of the total area of the crop in the country.

In terms of pulses, Nepal's overall yields are better than those of southern Asia. But national trends cannot be used to generalize pulse yield trends in the hill and mountain regions because only about 24% of pulses are grown in these two regions (Table 3.1). Lentil, chickpea, and pigeon pea, which are grown in Nepal's *terai* region, comprise most of the pulses grown. Even though pulse yields look higher in Nepal, the yields of pulses such as black gram, soybean, and cowpea, which are grown in hill and mountain regions, are still lower than the southern Asian average. Therefore, slow or no yield growth for maize, millet, and pulses during the last two decades is a major contributing factor to the food deficit conditions in Nepal's hill region.

Crop yields from 2011 highlight the high yield gaps between the *terai* and the hill and mountain regions. In general, yields are highest in the *terai* region, followed by the hill region, with the lowest yield from the mountain region. On average, crop yields in the hill and mountain regions were 15% and 30% lower, respectively, than the *terai* region. Maize yields in the hill (2.28 t/ha) and mountain (2.12 t/ha) regions were approximately 6% and 12% lower than in the *terai* region (2.42 t/ha), respectively. Similarly, rice yield in the hill (2.77 t/ha) and mountain (2.06 t/ha) regions were approximately 11 and 34% lower than in the *terai* region (3.12 t/ha). Although legume and oilseed yields in the hill region (0.90 and 0.81 t/ha, respectively) were almost comparable to the *terai* region (0.97 and 0.83 t/ha, respectively), these crops were minor productions of the hill region (only about 21% for both crops); hence, they did not contribute much to food self-sufficiency. Millet was the only crop that had a slightly higher yield in the hill region (1.15 t/ha) than in the *terai* region (1.07 t/ha). Lower production in the mountain region was less of a concern to national food security because, by 2011, the region comprised only 7% of Nepal's total food crop area. Since the hill region contains about 27% of rice, 69% of maize, 35% of wheat, and 76% of millet cultivation area in Nepal, increasing the crop yields in this region is crucial for the country to reach food self-sufficiency.

Low agriculture productivity in the hill and mountain regions is not only a cause for high food insecurity (MoAC/WFP/FAO, 2011) but also is the reason for the areas' high incidence of poverty (CBS, 2005) and malnutrition (FAO, 2010).

In fact, in 2011, 47.1% of Nepal's poor lived in the hill region (MoF, 2012). More directly, the consequences of low agriculture productivity on poverty were shown by the fact that the poverty rate of Nepalese farmers was almost double (54%) of the country's overall poverty incidence rate (30.8%) in 2009 (Upadhyaya, 2010). Therefore, the eradication of undernourishment, poverty, and malnutrition in Nepal is very difficult, if not impossible, without increasing the agriculture productivity of the hill region.

That being said, increasing crop yields in the hill region is not easy. Several geophysical, technological, and socio-economic factors contribute directly to the lower crop yields in the region. Small landholdings, sloping terrain, less fertile soil, unavailability of irrigation resulting in high dependence on rainfall, use of local and traditional varieties, low or no use of external inputs (such as fertilizers), low or no use of plant protection technologies, and high rates of soil degradation are all important reasons for lower crop yields. The average landholding size of 0.65 ha/household in the hill region was lower than those seen in the *terai* (0.93 ha/household) and mountain (0.73 ha/household) regions. Additionally, irrigation facilities are better in the *terai* region compared to the hill and mountain regions. Since only 25% of agricultural land is irrigated in the hill region (Upadhyaya, 2010), crop yield depends greatly on rainfall. For example, there was a sharp decline in maize yield in 2010 due to inadequate rainfall. After producing normal yields in 2011 and 2012, maize yield again decreased sharply in 2013, due to inadequate rainfall (Fig. 3.1).

In addition to the lack of irrigation, the adoption rates of new agricultural technologies are also slower in the hill region compared to the *terai* region. Only about 7.1% of farmers in the hill region had adopted improved maize varieties compared to 14% in the *terai* region (CBS, 2013). In fact, only about 10% of households had adopted more than four new agricultural technologies, while about 30% of households did not adopt any kind of new agricultural technologies in the hill region of Nepal until 2003. Furthermore, 60% of households in the region had adopted two or less new agricultural technologies (Floyd *et al.*, 2003) up to the same time. Less than two technology adoptions for 60% of farmers was fairly low, because in that study, the adoption of an improved variety of a crop was counted as a full technology.

3.1.5 Intensification of hill agriculture and concerns for sustainability

Despite lower yield and slower growth rates of crop yields, food demand is increasing in the hill region, due to a population growth of about 47% during 1991–2012 (CBS, 2012). As a consequence, the hill agricultural system has gone through rapid intensification, including increased cropping frequency on a given land plot in an attempt to increase food production (Hall *et al.*, 2001). Currently, the most common cropping pattern in the hill region is a maize–legume/millet–fallow system. In this system, maize is cultivated during the spring–summer season, which is followed by millet or legume crops during the rainy–post-rainy season; fields are left fallow during the winter season. Although the cultivation of millet and legume crops in the post-rainy season is now a very common practice, up until 20 years ago, the majority of farmers used to leave their land fallow during that period. Although intensification has increased food production, it has also had negative repercussions. An increase of cropping intensity with a conventional system increases the frequency of plowing and soil disturbance. In the conventional system of farming, land is plowed thoroughly two to three times in winter and left fallow until the sowing of crops at the onset of spring. Additionally, land is again plowed at least once before crop sowing. This increased plowing, open fallowing, and harrowing in sloping lands increases water-induced erosion (Gardner and Gerrard, 2003). Although farmers in Nepal have traditionally employed terracing and agroforestry systems in some areas, a large portion of agricultural lands do not have these protections. Therefore, while intensification has produced a short-term increase in food production, it has compromised the long-term sustainability of regional systems. Indeed, higher rates of land degradation due to soil erosion and soil nutrition depletion show that greater attention is required to maintain the sustainability of the agricultural system.

3.1.6 Soil degradation status in the hill region of Nepal

While it seems clear that erosion is a real problem in Nepal, a wide range of estimates of the actual magnitude of losses have been reported. Shrestha (1997) estimated that 32 t/ha/year of soil were lost in rainfed maize–millet systems in the hill region of Nepal, while Maskey and Joshi (1991) suggested 14 t/ha/year. Moreover, Das and Bauer (2012) predicted soil erosion from the conventional maize cropping system to be around 21 metric t/ha/year, while Partap and Watson (1994) reported the average erosion rate to be 4–20 metric t/ha/year for the same region. Similarly, Gardner and Gerrard (2003) estimated an average of 2.7–8.2 metric t/ha/year and up to 12.9 metric t/ha/year of erosion in rainfed agricultural terraces in the hill region of Nepal. Acharya *et al.* (2007) reported about 8 metric t/ha/year for the same region. Officially, MoEST (2006) reported about 5–15 metric t/ha/year soil loss in well-managed *bari* land (dry terrace) in the hilly region. Though there is a wide range of predictions, all of these rates are very high considering 5 metric t/ha/year is the maximum benchmark in the USA, and an upper limit of 1 metric t/ha/year is suggested by the Sustainable Europe Research Institute (SERI). No such benchmark has been established for Nepal thus far, soil erosion in Nepal is already much higher than these international standards. Regardless of the actual level, soil erosion is a major challenge to the sustainable development of Nepal's hill region. Hence, among many factors contributing to reduce crop yields in the hill region, loss of fertile soil due to water-induced erosion is one of the most important (Tiwari *et al.*, 2004).

Higher cropping intensity not only increases soil erosion, but also increases nutrient uptake from soil. If these nutrients are not properly replenished, soil productivity is reduced. In the prevalent maize-millet cropping system in the hill region, both crops are extremely high nutrient exhaustive crops. Therefore, in addition to high soil erosion, continuous cropping of cereal crops (ex. the maize-millet system) has also negatively affected soil fertility and sustainability of the system (Neupane *et al.*, 2002; Manandhar *et al.*, 2009). Further exacerbating the degradation, the amount and frequency of nutrient replenishment has decreased over time despite higher cropping intensity. Crop residues are not recycled in villages because farmers harvest them to feed their domestic animals. Instead, farmers' sole method of nutrient recycling is putting farmyard manure (FYM) on agricultural land. However, the amount of FYM and frequency of its use has decreased in recent years because of reduction in Nepal's total livestock population (Hall *et al.*, 2001).

In short, the two simultaneous challenges of lower productivity and high soil degradation need to be addressed in order to achieve sustainable food and nutritional security in the hill region of Nepal. It is obvious that a technology that can control soil degradation and enhance soil fertility is required in the region. However, for the peoples' immediate need to increase production to address severe food insecurity, the new technology must be able to augment productivity from the beginning. Globally, conservation agriculture production systems (CAPS) have been found suitable for these requirements (FAO, 2013).

3.1.7 Conservation agriculture production systems (CAPS) as a method to increase productivity and enhance sustainability

Conservation agriculture production systems (CAPS) are an innovative system of farming in which three conservation practices are implemented simultaneously to enhance conservation and increase return. The three pillars of CAPS are: (i) minimizing soil disturbance; (ii) optimizing crop rotation; and (iii) maintaining year-round soil (live or dead) cover (Kassam *et al.*, 2009). CAPS help reduce soil erosion by minimizing soil disturbance and maintaining year-round soil cover. However, conservation is not the only objective of CAPS; it also maximizes production by adopting optimal crop rotation and intensification. The intensification in CAPS is done through relay cropping, intercropping, and strip cropping. CAPS also target crop rotations, which will ensure judicious utilization of plant nutrients in soil (Kassam *et al.*, 2009). Therefore, CAPS has the potential to achieve sustainable and profitable agriculture systems simultaneously (FAO, 2013). Furthermore, the adoption of CAPS also enables farmers to grow cash crops, which would help elevate their economic livelihoods, while improving the soil and environmental conditions (FAO, 2013).

Indeed, CAPS has been found to increase crop yield and profit worldwide (Hobbs *et al.*, 2003; Rockström *et al.*, 2009; Lienhard *et al.*, 2013; Thierfelder *et al.*, 2013). Kassam *et al.* (2009) summarized various studies conducted to evaluate the crop yields from CAPS, reporting a 20–120% yield increase compared to conventional systems. However, there are cases of lower crop yields from CAPS compared to traditional systems (Gangwar *et al.*, 2004; Nyamangara *et al.*, 2013). Furthermore, the yield advantage of CAPS depends on other technological factors such as the use of fertilizers (Nyamangara *et al.*, 2013) and herbicides (Grabowski and Kerr, 2013). It has also been shown that up to 7 years might be needed to realize the benefits of CAPS fully (Hobbs, 2007). Giller *et al.* (2009) have concluded that although CAPS improve soil quality and have the potential to increase crop yield in the long term, short-term yield loss is a possibility in subsistence agriculture. However, if the adoption of CAPS requires the compromise of a significant reduction of crop yields, it would be difficult for smallholder farmers in the hill region of Nepal to adopt CAPS due to their pressing need for immediate food security. Therefore, evaluation of the effects of CAPS on crop yields is necessary before wide-scale promotion can be considered.

The principles of CAPS (i.e. reduced tillage, intercropping, and soil cover) have been rarely investigated in the hill region of Nepal. However, the few studies that have been conducted concluded that crop yields from reduced tillage systems were not lower than those from the conventional system (Bharati *et al.*, 1986; Atreya *et al.*, 2006; Begum *et al.*, 2010). Additionally, the CAPS practices of crop rotation and the integration of nitrogen-fixing legume crops have been reported to be the most viable option for long-term sustainability in the hill region of Nepal (Schreier *et al.*, 1994). However, investigating the effect of only one principle of CAPS (as done by Bharati *et al.*, 1986) does not provide a complete picture of the benefits of CAPS because the objective of CAPS is to produce synergistic effects by implementing multiple conservation practices simultaneously (Bot and Benites, 2005). Moreover, a 1-year study (as undertaken by Atreya *et al.*, 2006, 2008; Begum *et al.*, 2010) is likely too short to evaluate the full effect of conservation practices. Additionally, Rusinamhodzi *et al.* (2011) summarized the published literature to indicate that about 63% of results showed that increased yields via CAPS were due to proper crop rotation. Thus, identification of the best crop combinations for CAPS is a key factor affecting the scope of its efficacy in Nepal. To this end, the "Sustainable Management of Agroecological Resources in Tribal Societies (SMARTS)" project was started in 2010, with the goal of assessing the potential of CAPS for ensuring food security, reducing poverty, and achieving sustainable agriculture development in three villages of Nepal's central mid-hill region. The effects of CAPS on crop yields, soil health, profit, and social aspects were investigated by implementing conservation agricultural practices with different crop rotations for multiple years in Nepal's hill farming system.

3.2 On-farm Evaluation of CAPS

3.2.1 Study villages

On-farm trials were conducted in three villages inhabited by the Chepang, one of Nepal's poorest and most marginalized tribal communities. The selected villages were Thumka, Hyakrang, and Kholagaun, which are in Gorkha, Dhading, and Tanahun districts, respectively, in central Nepal. The geophysical and climatic conditions of the study villages are representative of the country's hill farming system, although there are some differences in cropping patterns. These villages are typical of Nepal's maize-based hill farming systems in terms of the spring–summer season crop. In the study area, as with the rest of Nepal, maize is grown on about 95% of the rainfed upland area in the spring–summer season (Chan-Halbrendt *et al.*, 2011, unpublished results). However, in the rest of the hill region, the spring–summer maize is followed by post-rainy season millet, black gram, or a fallow period (38%, 23%, and 32%, respectively) (MoAC, 2011). In the study villages, 56%, 30%, and 7% of the spring–summer maize area is followed by post-rainy season cowpea, black gram, and millet. This indicates farmers in the study area have started to replace millet with cowpea. Additionally, the post-rainy season fallow area was lower (<5%) in the study villages. This difference in the post-rainy season crop was governed by recent changes in the farming system. All study villages are located near the highway joining the *terai* region to Kathmandu; hence, they have better access to market compared to the average hill region village in Nepal. Due to better market access, their farming systems have a higher percent of market-oriented crops such as cowpea than do other villages in the hill region.

3.2.2 Identification of suitable CAPS

Following the selection of study villages, a team of researchers and development workers met with local farmers to identify potential CAPS. It was expected that the exchange of ideas among these stakeholders would identify CAPS that had a sound scientific basis and high feasibility for adoption by farmers. Farmers were informed of different conservation agriculture principles and assisted in making an informed choice to implement those principles according to their requirements. For example, the key questions in the discussion included: What crops do the farmers currently grow? What are the crop rotations the farmers see possible in the villages? Can those crops be grown with reduced tillage? What can be done to maintain year-round soil cover? In most studies, experts identify research treatments based on their own knowledge. However, this study gave high value to farmers' opinions in identifying appropriate CAPS. Careful selection of CAPS was very necessary at this stage, as only a few treatments could be implemented on the small terraces available for establishing on-farm trials in the study villages. In the end, three CAPS treatments and one traditional production system were identified for the on-farm trial (Table 3.2). These CAPS treatments combined different tillage methods and the integration of legume crops through intercropping.

Obviously, maize was chosen as the crop for the spring–summer season (March–July). There was no feasible alternative crop to replace or to intercrop

Table 3.2. Summary of CAPS treatments of on-farm evaluation trials in the central part of the hill region of Nepal (2011–2012).

			Crop grown Year 1 (2011)		Crop grown Year 2 (2012)	
On-farm trial treatments	Tillage system	Cropping system	Spring–summer	Post-rainy	Spring–summer	Post-rainy
Traditional	Conventional	Sole crop	Maize	Millet (Mi)	Maize	Millet
CAPS1	Conventional	Sole crop	Maize	Cowpea (CP)	Maize	Blackgram (BG)
CAPS2	Conventional	Intercrop	Maize	Mi + CP	Maize	Mi + BG
CAPS3	Strip	Intercrop	Maize	Mi + CP	Maize	Mi + BG

with maize; therefore, it was grown as the sole crop. However, alternative crops were available for the post-rainy season (July–November). Prior to our study period, in Nepal's hill region about 39% of the spring–summer maize area was planted in post-rainy season millet and about 33% in legumes in 2011. Therefore, to compare the performance of CAPS with the most dominant cropping pattern in the hill region of Nepal, spring–summer maize followed by post-rainy millet was selected as the traditional system (control). The tillage system for this traditional system was conventional tillage (CT). Among the three CAPS treatments, one of them had sole cropping of a legume and the remaining two had intercropping of millet and legume in the post-rainy season. Sole cropping a legume after maize increases soil cover for about 5 months, compared to leaving it fallow. Thus, a solitary post-rainy season legume crop was considered a conservation practice. Intercropping of millet and legume in the post-rainy season harnesses the benefits of legumes in terms of nitrogen fixation and also maintains the production of millet, which was required due to food security reasons. Hence, millet and legume intercropping was also considered as a conservation practice. Strip tillage (ST) was identified as a suitable option for reducing tillage, because it would reduce soil erosion in the sloping landscape. Therefore, it was selected as the conservation tillage practice. Combining legume intercropping and strip tillage practices, three new CAPS treatments were determined.

In the traditional cropping system tested, maize was followed by millet with CT. Second, CAPS1 was maize followed by cowpea or black gram with CT, which had already been adopted by some farmers in the study villages, but had lower acreage at the national level. Third, CAPS2 had maize followed by millet/legume intercropping with CT. The intercropping of millet and legumes was a new practice for farmers. Finally, CAPS3 had maize followed by millet/legume intercropping with ST. The ST system, which was also a completely new practice to the farmers, was done by digging a row to sow the crop, while leaving a strip of undisturbed land between rows. Since CAPS3 integrated most of the principles of conservation agriculture (i.e. legumes for nitrogen fixation, intercropping, and strip tillage), it was the regarded as the best-bet CAPS treatment in the on-farm trial.

3.2.3 On-farm trial

After finalizing four treatments for the on-farm trial, trial plots were established in the fields of 25 farmers in the aforementioned villages. The same set of treatments was established in all plots. Under the close supervision of agricultural technicians, the farmers themselves fertilized the land, planted the crops, carried out intercultural operations, and harvested the crops during the trials. The role of the agricultural technicians was to ensure the correct implementation of CAPS and to record data on agronomic variables such as plant density, yield, and labor requirements for different farm operations for evaluation. The evaluation and improvement of CAPS was a continuous process. In the first year, farmers selected cowpea as the legume crop for the CAPS treatments. In the second year, cowpea was replaced by black gram, because farmers felt that shading from the foliar growth of cowpea suppressed the growth and yield of the millet. Researchers also annually collected and analysed crops yields and soil benefits, the results of which were communicated to the farmers as feedback.

The impacts of crop yields under different CAPS treatment were analysed using a General Linear Model (GLM). The dependent variable for the model was crop yield (t/ha), and the explanatory variables were village, CAPS, and year. "Village" was treated as a block, due to the large degree of variation in geophysical characteristics and soil properties among villages. There were two contiguous years of crop yield data from different CAPS, so "year" was also considered as an explanatory variable. The variables of particular interest in the study were CAPS, which included the effects of four treatments, and the treatment by year interaction. The interaction shows how the yields of CAPS and the traditional system behaved in year 1 and year 2. Although 2 years of yield data is not enough to determine the full benefits of CAPS, it provides an analysis regarding crop yields in the initial years of transition from a conventional production system to CAPS. The possibility of yield decrease due to CAPS has been suggested as an obstacle to the adoption of CAPS (Hobbs, 2007); hence, analysis of the initial 2 years' yield is important for looking at the feasibility of adoption by smallholder farmers. Type III hypothesis testing was used to identify the significant independent variable.

3.3 Results and Discussion

3.3.1 Results of farm trials: crop yields under different CAPS

Due to high differentiation in microclimate and soil characteristics, differences in crop yields among villages were expected from the onset. As expected, there were significant differences in yields of maize ($P < 0.001$), millet ($P < 0.001$), and cowpea ($P < 0.001$) by villages. However, the difference in black gram yield between villages was not significant (Table 3.3). Average maize yield in Hyakrang (2.45 t/ha) was significantly higher than in Kholagaun (1.82 t/ha) and Thumka (1.87 t/ha). Millet yield in Hyakrang (1.04 t/ha) was also significantly higher than in Kholagaun (0.56 t/ha) and Thumka (0.46 t/ha). Cowpea yield followed a similar pattern, with the highest yield in Hyakrang (1 t/ha). However, the highest black gram yield was observed in Thumka (0.26 t/ha).

Despite huge variations by village and year, results indicated that maize and cowpea yields were not significantly different by treatments. However, millet and black gram yields were significantly different among treatments ($P < 0.001$ in both cases). The highest maize yield was recorded in CAPS1 (2.20 t/ha) and the lowest was recorded in CAPS3 (1.86 t/ha), but the mean comparison between those two was not statistically different. Similarly, cowpea yield was highest in CAPS1 (0.86 t/ha) and lowest in CAPS3 (0.59 t/ha), but again the statistical tests failed to prove significant differences. Instead, millet yield in CAPS2 (0.59 t/ha) and CAPS3 (0.54 t/ha) was significantly lower than the traditional system (0.93 t/ha) at the 95% confidence level. Black gram yield in CAPS2 (0.17 t/ha) and CAPS3 (0.16 t/ha) was also significantly lower than that of CAPS1 (0.35 t/ha). Thus, yields of millet and black gram in intercropping treatments (CAPS2 and CAPS3) were significantly lower than in sole cropping treatments (traditional system and

Table 3.3. Average yields of maize, millet, black gram, and cowpea by traditional system and CAPS in the three villages in the central part of the hill region of Nepal (2011–2012).

Village	On-farm trial treatments	Year	Maize (t/ha)	Millet (t/ha)	Black gram (t/ha)	Cowpea (t/ha)
Hyakrang			2.45 ± 0.19[a]	1.04 ± 0.12[a]	0.20 ± 0.08	1.01 ± 0.18[a]
Kholagaun			1.82 ± 0.19[b]	0.56 ± 0.13[b]	0.21 ± 0.08	0.30 ± 0.22[b]
Thumka			1.87 ± 0.19[b]	0.46 ± 0.13[b]	0.26 ± 0.07	0.78 ± 0.21[ab]
	Traditional		2.03 ± 0.23	0.93 ± 0.13[a]		
	CAPS1		2.20 ± 0.23		0.35 ± 0.07[a]	0.86 ± 0.21
	CAPS2		2.09 ± 0.23	0.59 ± 0.13[b]	0.17 ± 0.07[b]	0.64 ± 0.21
	CAPS3		1.86 ± 0.23	0.54 ± 0.13[b]	0.16 ± 0.07[b]	0.59 ± 0.21
		Year 1	2.14 ± 0.16	0.54 ± 0.10[a]		
		Year 2	1.95 ± 0.16	0.83 ± 0.11[b]		
ANOVA						
Source of variation		DF	SS	SS	SS	SS
Village (V)		2	16.58***	9.13***	0.05[NS]	5.45***
CAPS		3	2.9[NS]	4.08***	0.49***	0.84[NS]
V × CAPS		6	0.56[NS]	0.26[NS]	0.18[NS]	0.38[NS]
Year (Y)		1	1.69[NS]	3.02***		
Y × CAPS		3	0.92[NS]	0.43[NS]		
Fit statistics						
Pr > *F*			***	***	***	***
R^2 (%)			16.4	41.6	30.2	35.0

The same letter in the superscript of means indicates they are statistically not different in a Bonferroni adjusted simultaneous *t*-test. DF, degree of freedom; SS, sum of squares; NS, non-significant at 0.05 probability level; ***, significant at the 0.001 probability level. The values show least square means ±95% confidence interval.

CAPS1 for millet and black gram, respectively). Lower plant density in the intercropping treatments as compared to sole cropping was the main factor contributing to the lower crop yields from intercropping treatments. When millet and legume crops were grown under intercropping regime, half of the area was allotted for millet and the other half was allotted for legume crops. Therefore, both of the crops had almost half the number of plants compared to sole cropping treatments. Despite almost half the number of plants, cowpea yield was not reduced proportionally, due to its prolific growing nature.

Results also showed that millet and black gram yields under different intercropping regimes were not affected by types of tillage practices. Recalling that CAPS2 and CAPS3 had the same cropping pattern (i.e. maize–millet/legume for both) with different types of tillage (CAPS2 had CT whereas CAPS3 had ST), any difference in the crop yields, if it were significant, would have been caused by tillage practice. Previously, studies have also reported lower crop yields from conservation tillage systems during the initial years (Mando *et al.*, 2005; Iqbal *et al.*, 2008). However, our result did not support the previous claims about loss of crop yields in the initial years of practicing ST. Millet yields from both CAPS2 and CAPS3 were almost equal (0.59 t/ha for both). Similarly, black gram yield from both CAPS2 and CAPS3 were 0.17 t/ha and

0.16 t/ha, respectively. The cowpea yield in CAPS3 (0.54 t/ha) was slightly lower than in CAPS2 (0.64 t/ha), but not significantly so. Just as CAPS did not have an effect on maize and cowpea yield, ST did not reduce the yield of any of the crops in the system.

Given the importance of maize and millet as staple foods in the hill region of Nepal, it was important to examine the annual yield change of these crops more closely. Although CAPS treatments did not have significant effects on maize yield, there were a few noteworthy differences in year 1 and year 2. When the maize yield data from the year 2 were analyzed separately, maize yield on CAPS1 was highest among all. CAPS1 had a sole legume crop in the second season of year 1; hence, the effect of the preceding legume crop was observed in the maize yield in year 2. Although there was a general reduction in maize yield in 2012 (due to relatively unfavorable climatic conditions in 2012), the yield decrease was higher for CAPS3 compared to other treatments, indicating that bad weather caused higher yield reduction in strip tillage. Based on previous reports, it was expected that lower yield reduction would be observed in the ST system (Smith *et al.*, 2007). However, as higher yield reduction was observed in ST, this study failed to prove that ST systems reduce yield reduction during unfavorable weather. Reduction in yield variation during unfavorable environment condition is associated with the level of production risk. Hence, the study with 2 years of data could not confirm that CAPS with an ST system reduces production risks on maize cultivation.

Millet also showed different trends by year. Millet was intercropped with cowpea in the first year, while farmers opted to replace cowpea with black gram in the second year. The main reason for replacing cowpea was its higher vegetative growth, which caused a shading effect on the millet. Indeed, millet yield in the second year (0.83 t/ha) was significantly higher than millet yield in the first year (0.54 t/ha). Thus, it was clear that millet performed far better when it was intercropped with black gram as opposed to cowpea. When millet was taken as the primary interest crop for being a staple food, the decision to replace cowpea with black gram was sensible. However, the yield of black gram was not high enough to compensate the loss of profit derived from selling cowpea. Thus, replacing cowpea with black gram reduced the cash revenue of farmers. While maize and millet production are integral to farmers' food security status, the significance of legume crops on food security by either increasing income or by increasing the availability of a protein-rich diet should not be underestimated. Hence, to evaluate the effect of CAPS on different dimensions of food and nutritional security, further analysis was conducted.

3.3.2 Relating CAPS performances to food and nutritional security

Food and nutritional security have four interwoven dimensions (Gross *et al.*, 2000). First, *food availability* refers to physical availability, which can be increased by higher production. Second, *food access* refers to peoples' capacity to purchase food, which can be increased by increasing income. Third, *food quality and utilization* refers to the composition and consumption of different nutrients in food, which can be improved by eating diversified food with optimum intake of available nutrition. The fourth dimension is *stability*, which refers to long-term sustainability of food security. The adoptions of CAPS can potentially affect all four dimensions of food and nutritional security.

Since only less than 10% of household consumption is supplemented through purchased food in the villages (Chan-Halbrendt *et al.*, 2011, unpublished results), change in the yields of staple crops such as maize and millet directly affects food availability in villages. Although CAPS neither reduces nor increases the individual yields of major crops, it can affect food availability by introducing intercropping in which two crops are harvested from a given plot, as opposed to only one crop in sole cropping. Therefore, even if individual crop yields are lower in intercropping, it may still improve overall food availability by combining the production of two crops. Furthermore, while an increase in maize and millet yield directly affects food availability, an increase in legume yield improves household income. Cowpea and black gram have commercial market potential, and historically have had higher market prices than millet. In 2011, the prices of maize, millet, cowpea, and black gram were 35, 27, 72, and 102 Nepalese rupees per kilogram, respectively (based on rapid market appraisal collected by LI-BIRD). Therefore, a separate analysis was required to evaluate the effects of CAPS to the *access* and *quality* dimensions of food and nutritional security.

In fact, *maize equivalent yield* (MEY), *protein equivalent yield* (PEY) and *carbohydrate equivalent yield* (CEY) were calculated and compared among treatments to evaluate the general effect of CAPS to food access and food quality. MEY is the estimation of the quantity of maize that could be purchased if yields from all crops were sold and all income was used to purchase maize. The "yield equivalent" method is a popular tool to compare the performance of sole cropping and intercropping together (Haque *et al.*, 2013). Moreover, because of the less than required intake of both protein-rich food and carbohydrate-rich food, protein-energy deficiency is Nepal's main nutritional problem. Therefore, evaluation of the effect of CAPS on the availability of protein and carbohydrate-rich diet is important. To evaluate the effect of CAPS on the availability of carbohydrate-rich food and protein-rich food, the CEY and PEY were estimated and compared among CAPS. CEY is the accumulation of carbohydrate and PEY the accumulation of protein produced by all the crops in the system. These indices were similar to MEY, but instead of using market price as weight to estimate MEY, the percent of carbohydrate and protein in the grains was used as weights for CEY and PEY, respectively. All indices were compared among CAPS using GLMs, with village, treatments, years, village × treatment interaction and year × treatment interaction taken as explanatory variables.

The formulas used for MEY, CEY and PEY were as follows:

$$MEY = \sum_{i=1}^{4} \frac{Y_i \times P_i}{P_m}$$

where MEY = maize equivalent yield (t/ha), Y_i = yield of ith crop (t/ha), P_i = price of the ith crop (US$/t), P_m = price of maize (US$/t). The four crops included in the analysis were maize, millet, cowpea, and black gram.

$$PEY = \sum_{i=1}^{4} \frac{Y_i \times \%\ PR_i}{100}$$

where PEY = protein equivalent (t/ha), Y_i = yield of ith crop (t/ha), % PR_i = percent of crude protein in grain of ith crop.

$$CEY = \sum_{i=1}^{4} \frac{Y_i \times \% \ CA_i}{100}$$

where CEY = carbohydrate equivalent (t/ha), Y_i = yield of ith crop (t/ha), % CA_i = percent of carbohydrate in grain of ith crop.

The results showed a significant difference in MEY among different villages and different years ($P < 0.001$ for both; Table 3.4). Comparing villages, Hyakrang had a significantly higher MEY (4.14 t/ha) than Kholagaun (2.57 t/ha) and Thumka (3.02 t/ha). Comparing years, MEY was significantly higher in year 1 (3.55 t/ha) than in year 2 (2.93 t/ha), which indicated that total food availability decreased from year 1 to year 2. MEY was also significantly different by CAPS treatments ($P < 0.001$), such that MEY from CAPS1 (3.60 t/ha) and CAPS2 (3.47 t/ha) were significantly higher than the traditional system (2.76 t/ha). However, MEY of CAPS3 (3.13 t/ha) was in between these groups, since it was comparable with the MEY of CAPS1 and CAPS2, as well as that of the traditional system. Although the MEY of CAPS3 (strip tillage-based CAPS) was higher than the traditional system, the mean comparison test indicated that the difference was not significant. In other words, the most desirable conservation agriculture treatment, i.e. CAPS3, did not increase overall food access in the villages, but CAPS1 and CAPS2 both had the potential to improve overall food access in the villages.

Similar to MEY, CEY was also significantly different by villages ($P < 0.001$) and it was also highest in Hyakrang (2.5 t/ha), followed by Thumka (1.74 t/ha) and Kholagaun (1.65 t/ha) (Table 3.4). But unlike MEY, there was no significant difference of CEY by CAPS and by years. In essence, by failing to show any positive impact of CAPS for increasing CEY, the results indicated that the adoption of CAPS would not make any difference in the availability of energy-rich food. Maize contributed a large proportion of carbohydrate, and as reported in an earlier section, there was no significant difference in maize yield by treatments. Therefore, CAPS did not have an effect on CEY.

In contrast to the analysis of energy-rich food, significant differences were observed in the availability of protein-rich food among treatments. As with MEY, PEY was significantly different by villages, treatments, and years ($P < 0.001$, <0.05 and <0.001, respectively; Table 3.4). The PEY of CAPS1 (0.40 t crude protein, CP/ha) and CAPS2 (0.40 t CP/ha) were significantly higher than the traditional system (0.33 t CP/ha). Like MEY, the PEY of CAPS3 (0.36 t CP/ha) was also slightly higher than in the traditional system, but it was not significantly different in Bonferroni mean comparison. It was found that PEY depended on the choice of crop, because PEY in the first year (0.42 t CP/ha) was significantly higher than in second year (0.33 t/ha), indicating that farmers could produce a more protein-rich food if they grew cowpea in place of black gram.

However, before concluding any positive contribution of CAPS to increase protein-rich food, we must note that higher production of cowpea or black gram does not necessarily lead to higher intake of these grains. Farmers may prefer to sell these grains due to their high value in the market, thereby leaving less for

Table 3.4. Average of MEY, CEY, and PEY from on-farm trials by village, treatment, and year in three villages in the central part of the hill region of Nepal (2011–2012).

Village	On-farm trial treatments	Year	MEY (t/ha)	CEY (t/ha)	PEY (t/ha)
Hyakrang			4.14 ± 0.31^{a}	2.50 ± 0.18^{a}	0.48 ± 0.04^{a}
Kholagaun			2.57 ± 0.31^{b}	1.65 ± 0.18^{b}	0.30 ± 0.04^{b}
Thumka			3.02 ± 0.31	1.74 ± 0.18^{b}	0.34 ± 0.04^{b}
	Traditional		2.76 ± 0.36^{a}	2.05 ± 0.21	0.34 ± 0.04^{a}
	CAPS1		3.60 ± 0.36^{bc}	1.86 ± 0.21	0.40 ± 0.04^{bc}
	CAPS2		3.47 ± 0.36^{bc}	2.08 ± 0.21	0.40 ± 0.04^{bc}
	CAPS3		3.13 ± 0.36^{ac}	1.86 ± 0.21	0.36 ± 0.04^{ac}
		Year 1	3.55 ± 0.25^{a}	2.05 ± 0.14	0.42 ± 0.03^{a}
		Year 2	2.93 ± 0.26^{b}	1.87 ± 0.15	0.33 ± 0.03^{b}

ANOVA Sources of variation	DF	SS	SS	SS
Village	2	86.44^{***}	29.22^{***}	1.21^{***}
CAPS	3	20.97^{***}	2.1^{NS}	0.13^{*}
Village × CAPS	6	7.32^{NS}	0.68^{NS}	0.07^{NS}
Year	1	18.85^{***}	1.57^{NS}	0.32^{***}
Year × CAPS	3	4.6^{NS}	0.3^{NS}	0.08^{NS}
Fit statistics				
$Pr > F$		***	***	***
R^2 (%)		36.2	29.5	36.4

The same letter in the superscript of the means indicates they are statistically not different in Bonferroni adjusted simultaneous *t*-test. DF, degree of freedom; SS, sum of squares; NS, non-significant at 0.05 probability level. ***, *, significant at the 0.001 and 0.05 probability level, respectively. The values indicate least square means ± 95% confidence interval.

household consumption. However, even if this proves to be the case, it can be expected that when selling their produce at market, farmers will then have the resources to buy as well, thereby improving their food and nutritional security.

Taken on the whole, CAPS1 (maize–legume with CT) and CAPS2 (maize–millet/legume with CT) increased food availability and increased purchasing power compared to the traditional system. Furthermore, both of these CAPS could improve the nutritional imbalance of poor and subsistence households by increasing the supply of protein-rich diets. Although strip tillage-based CAPS (i.e. CAPS3; maize–millet/legume with ST) reduced food availability, purchasing power, and availability of protein-rich food as compared to CAPS1 and CAPS2, CAPS3 was still comparable or better than the traditional system.

3.3.3 CAPS and stability of food security

Long-term sustainability of food security is the foremost goal of CAPS; indeed, the principles of CAPS are built around sustainable utilization of resources. Although there is still debate about the potential of CAPS to increase yield in the short term, there is consensus regarding its long-term benefits on crop yields and soil health (Giller *et al.*, 2009; Kassam *et al.*, 2009). FAO (2013) has stated that CAPS can simultaneously increase yields and enhance soil conservation, which in the long run contributes to sustainability. However, different time frames have been reported to define the short-term or long-term benefits of CAPS. Hobbs (2007) has suggested that CAPS would take about 7 years to express the full advantages.

In fact, many long-term studies have shown that crop yields in CAPS decrease for the initial 3–4 years, and gradually start to increase after 4–5 years. For example, Thierfelder *et al.* (2012) studied the long-term yield trend of maize in CAPS on farmers' fields in Africa in which it took about 7 years of continuous CAPS to gain about 20% higher maize yield than conventional systems. Quinton and Catt (2004) also reported about 20% higher maize yield after 7 years of continuous CAPS in England, but it took about 6 years for CAPS to surpass the yield of the conventional system. Unlike these two studies, Liu *et al.* (2013) reported that maize yield started to increase after the fourth year of CAPS, but it was still lower than the conventional system after 7 years. In addition to these results from other parts of the world, a bioeconomic modeling exercise conducted by Das and Bauer (2012) in the hill region of Nepal suggested a slightly higher maize yield under a minimum tillage system (2.8 t/ha) compared to a conventional tillage system (2.75 t/ha). Since they assumed a linear growth trend, however, their results could not show any reduction in crop yield in the initial years of practicing CAPS.

In fact, no long-term experiments have been conducted in Nepal's hill region to evaluate the impacts of CAPS. Moreover, the 2 years of experimental data collected in this study were not sufficient to evaluate longer-term impacts. However, our results indicate that the adoption of CAPS would improve food security sustainably in the long run. A pessimistic projection indicates that farmers may not realize yield increase for 5–7 years after strip tillage adoption-based CAPS. In fact, if farmers have to wait so long for benefits, it does not provide an incentive for them to adopt CAPS, especially when they are facing a day-to-day food shortage. Therefore, government support may be required to encourage farmers to adopt sustainable food and nutritional security technologies such as CAPS, which may require short-term sacrifices in order for the long-term benefits to be realized.

Current national strategies for coping with food shortage in the hill and mountain regions are shortsighted, only treating the problem's symptoms rather than its root causes. Nepal's government regularly spends about US$4.5 million each year just to subsidize food transportation from the *terai* region to the food-deficient hill and mountain (MoF, 2013) regions. The United Nations World Food Programme (WFP), another main agency providing food assistance in Nepal, distributed food to about 1.6 million people in 2009 (WFP, 2010). Recently, WFP has allocated about US$215.3 million (US$43.0 million/year) for its WFP Nepal strategy: 2013–2017. Of this, about US$87 million is budgeted for purchasing food (WFP, 2013). If the amount allotted for food purchase is divided equally for 5 years, this will amount to about US$17.4 million/year. This is compared to the capital expenditure of the Ministry of Agriculture Development (MoAD) in 2011/12, which was about US$8.11 million (MoF, 2013). The comparison shows that the capital expenditure of MoAD is less than half of the budget used for food purchase. Without undermining the need for emergency food assistance in the food-deficient hill and mountain regions of the country, this investment scenario does not show that the country is planning for longer-term solutions of food and nutritional insecurity. Even if conservation technologies such as CAPS do not provide an immediate solution, the government should support farmers in adopting such practices in order to invest in long-term, sustainable solutions.

3.4 Conclusion

By all accounts, food and nutritional security of Nepal's poor and marginalized hill farmers cannot be improved without increasing food grain production. Food grain production can be increased either by increasing yield or by intensification. However, current practices of haphazard intensification in Nepal's hill region have several negative repercussions. While enough emphasis is required to reduce undernourishment and malnutrition in the short term, long-term strategies are needed to eradicate the problems permanently. Hence, there is an urgent need for sustainable agriculture technologies in Nepal's hill region. Considering the severity of immediate food insufficiency, however, farmers are unlikely to adopt any new technology that requires significant yield reduction in the short term.

The results of our on-farm trials verified that CAPS could be the suitable technology to cope with these challenges. Although CAPS did not increase the yields of individual crops, CAPS with an intercropping component increased overall food availability by allowing farmers to harvest two crops from the same land. From their first years, CAPS treatments with maize–legume and maize–millet/legume systems with conventional tillage increased economic access to food (by increasing the sales of cowpea and black gram), and improved diet quality (by increasing protein-rich food for consumption) compared to traditional systems. Although the most desirable CAPS (maize–millet/legume with strip tillage) did not have the advantage in terms of increasing food and nutritional security, it was still comparable to the traditional system, so farmers would not face yield reduction by adopting it. Unlike previous findings suggesting significant reduction of crop yields under reduced tillage practices, the reduction of yield from strip tillage was not significant in this study. Despite lack of local evidence of the long-term effects of CAPS on yield and environment in the mid-hills of Nepal, studies from other locations showed that CAPS with minimum tillage systems generally had positive effects on crop yield over time. Therefore, strip tillage-based CAPS could still be a feasible technology for long-term food security and sustainable agriculture development of Nepal's hill region. Although the adoption of CAPS appears to be the logical choice to achieve sustainable food and nutritional security in the hill region, the adoption of strip tillage-based CAPS is challenging at this time, due to the lack of short-term incentives for farmers to adopt this system. Hence, for higher adoption, the government, through the extension service, should educate farmers on the long-term benefits of CAPS, while informing them that the returns might be limited in the initial years of implementation. At the same time, the government should increase investment to encourage farmers to adopt sustainable technologies such as CAPS, especially in the initial years, when these technologies cannot generate sufficient gains to trigger autonomous adoption.

References

Acharya, G., McDonald, M., Tripathi, B., Gardner, R. and Mawdesley, K. (2007) Nutrient losses from rain-fed bench terraced cultivation systems in high rainfall areas of the mid-hills of Nepal. *Land Degradation and Development* 18(5), 486–499.

Adhikari, R.K. (2010) Food utilization practices, beliefs and taboos in Nepal – an overview. United States Agency for International Development – Global Health Technical Assistance Project. Available at: http://pdf.usaid.gov/pdf_docs/pnaeb773.pdf (accessed 28 October 2014).

Atreya, K., Sharma, S., Bajracharya, R.M. and Rajbhandari, N.P. (2006) Applications of reduced tillage in hills of central Nepal. *Soil and Tillage Research* 88(1), 16–29.

Atreya, K., Sharma, S., Bajracharya, R.M. and Rajbhandari, N.P. (2008) Developing a sustainable agro-system for central Nepal using reduced tillage and straw mulching. *Journal of Environmental Management* 88(3), 547–555, doi:http://dx.doi.org/10.1016/j.jenvman.2007.03.017 (accessed 27 October 2014).

Begum, F., Bajracharya, R.M., Sharma, S. and Staula, B.K. (2010) Influence of slope aspect on soil physic-chemical and biological properties in the mid hills of central Nepal. *International Journal of Sustainable Development and World Ecology* 17(5), 438–443.

Bharati, M.P., Whigham, D.K. and Voss, R.D. (1986) Soybean response to tillage and nitrogen, phosphorus, and potassium fertilization 1. *Agronomy Journal* 78(6), 947–950, doi:10.2134/agronj1986.00021962007800060002x (accessed 27 October 2014).

Bot, A. and Benites, J. (2005) The importance of soil organic matter: key to drought-resistant soil and sustained food and production (Vol 80). FAO, Rome. Available at: http://www.fao.org/docrep/009/a0100e/a0100e00.htm#Contents (accessed 28 October 2014).

CBS (Central Bureau of Statistics) (2002) Agriculture Census of Nepal, 2001/02. Central Bureau of Statistics, National Planning Commission, Government of Nepal, Kathmandu.

CBS (2005) Poverty Trends in Nepal (1995–96 and 2003–04). CBS, National Planning Commission, Government of Nepal, Kathmandu.

CBS (2008) Nepal Labour Force Survey 2008 (NLFS II). CBS, National Planning Commission Secretariat, Government of Nepal, Kathmandu.

CBS (2012) National Population Census (pp. 2–3). CBS, Nepal Planning Commission, Government of Nepal, Kathmandu.

CBS (2013) Nepal Thematic Report on Food Security and Nutrition 2013. CBS, National Planning Commission, Government of Nepal, Kathmandu.

CBS/WFP (2006) Small Area Estimation of Poverty, Caloric Intake and Malnutrition in Nepal. CBS, National Planning Commission, Government of Nepal, Kathmandu.

Das, R. and Bauer, S. (2012) Bio-economic analysis of soil conservation technologies in the mid-hill region of Nepal. *Soil and Tillage Research* 121, 38–48.

FAO (2010) Assessment of Food Security and Nutrition Situation in Nepal: An input for the preparation of NMTPF for FAO in Nepal. Food and Agriculture Organization of the United Nations, UN Complex, Pulchowk, Nepal.

FAO (2013) Conservation Agriculture. UN Food and Agriculture Organization. Available at: http://www.fao.org/ag/ca/1a.html (accessed 27 October 2014).

FAO/IFAD/WFP (2013) *The State of Food Insecurity in the World 2013. The Multiple Dimensions of Food Security.* FAO, Rome.

FAOSTAT (2013) Agri-Environmental Indicators: Soil. FAO. Available at: http://faostat3.fao.org/home/index.html (accessed 20 August 2013).

Floyd, C., Harding, A.H., Paudel, K.C., Rasali, D.P., Subedi, K., *et al.* (2003) Household adoption and the associated impact of multiple agricultural technologies in the western hills of Nepal. *Agricultural Systems* 76(2), 715–738, doi:http://dx.doi.org/10.1016/S0308-521X(02)00152-X (accessed 27 October 2014).

Gangwar, K., Singh, K. and Sharma, S. (2004) Effect of tillage on growth, yield and nutrient uptake in wheat after rice in the Indo-Gangetic Plains of India. *The Journal of Agricultural Science* 142(04), 453–459.

Gardner, R.A.M. and Gerrard, A.J. (2003) Runoff and soil erosion on cultivated rainfed terraces in the Middle Hills of Nepal. *Applied Geography* 23(1), 23–45, doi.org/10.1016/S0143-6228(02)00069-3 (accessed 27 October 20104).

Giller, K.E., Witter, E., Corbeels, M. and Tittonell, P. (2009) Conservation agriculture and smallholder farming in Africa: The heretics' view. *Field Crops Research* 114(1), 23–34, doi.org/10.1016/j.fcr.2009.06.017 (accessed 27 October 2014).

Grabowski, P.P. and Kerr, J.M. (2013) Resource constraints and partial adoption of conservation agriculture by hand-hoe farmers in Mozambique. *International Journal of Agricultural Sustainability* 1–17, doi:10.1080/14735903.2013.782703 (Accessed 27 October 2014).

Gross, R., Schoeneberger, H., Pfeifer, H. and Preuss, H. (2000) The four dimensions of food and nutrition security: definitions and concepts. *SCN News* 20, 20–25.

Hall, M., Dixon, J., Gullivar, A. and Gibbon, D. (2001) *Farming Systems and Poverty. Improving Farmers' Livelihoods in a Changing World.* FAO, Rome.

Haque, M., Kumar, B., Kalyani, S. and Kumar, R. (2013) Effect of maize (*Zea mays*) based intercropping system on maize yield and associated weeds under rainfed upland ecosystem. *Research Journal of Agricultural Sciences* 4, 416–419.

Hobbs, P.R. (2007) Conservation agriculture: What is it and why is it important for future sustainable food production? *Journal of Agricultural Science-Cambridge* 145(2), 127–137.

Hobbs, P.R., Gupta, R., Malik, R. and Dhillon, S. (2003) Conservation agriculture for the rice–wheat systems of the Indo-Gangetic Plains of South Asia: A case study from India. In: García-Torres, L., Benites, J., Martínez-Vilela, A. and Holgado-Cabrera, A. (eds) *Conservation Agriculture: Environment, Farmers' Experiences, Innovations, Socio-economy, Policy.* Kluwer Academic Publishers, Dordrecht, the Netherlands, pp. 169–174.

Iqbal, M., Ul Hassan, A. and Ibrahim, M. (2008) Effects of tillage systems and mulch on soil physical quality parameters and maize (*Zea mays* L.) yield in semi-arid Pakistan. *Biological Agriculture and Horticulture* 25, 311–325.

Kassam, A., Friedrich, T., Shaxson, F. and Pretty, J. (2009) The spread of conservation agriculture: justification, sustainability and uptake. *International Journal of Agricultural Sustainability* 7(4), 292–320.

Lienhard, P., Panyasiri, K., Sayphoummie, S., Leudphanane, B., Lestrelin, G., *et al.* (2013) Profitability and opportunity of conservation agriculture in acid savannah grasslands of Laos. *International Journal of Agricultural Sustainability* 12(4), 391–406.

Liu, S., Zhang, X.-Y., Yang, J. and Drury, C.F. (2013) Effect of conservation and conventional tillage on soil water storage, water use efficiency and productivity of corn and soybean in Northeast China. *Acta Agriculturae Scandinavica, Section B–Soil & Plant Science* 63(5), 383–394.

Manandhar, G., Adhikary, S. and Sah, G. (2009) *Sustainable Agricultural Practices and Technologies in Nepal.* United Nations Social and Economic Commission for Asia and the Pacific, Asian and Pacific Centre for Transfer of Technology Tech Monitor, New Delhi.

Mando, A., Ouattara, B., Sédogo, M., Stroosnijder, L., Ouattara, K., *et al.* (2005) Long-term effect of tillage and manure application on soil organic fractions and crop performance under Sudano-Sahelian conditions. *Soil and Tillage Research* 80(1), 95–101.

Maskey, R. and Joshi, D. (1991) Soil and nutrient losses under different soil management in the middle mountains of Nepal. In: Shah, P.B., Schreier, H., Brown, S.J. and Riley, K.W. (eds) *Soil Fertility and Erosion Issues in the Middle Mountains of Nepal, Jikhu Khola Watershed.* Integrated Survey Section, Government of Nepal, Kathmandu, pp. 105–120.

MoAC (Ministry of Agriculture and Cooperatives) (2011) Statistical information in Nepalese Agriculture, 2010–11. MoAC, Government of Nepal, Kathmandu.

MoAC/WFP/FAO (2011) *Crop Situation Update: A Joint Assessment of 2010 Summer Crops and Outlook for 2011 Winter Crops.* Ministry of Agriculture and Cooperatives, World Food Program and Food and Agriculture Organization, Kathmandu.

MoEST (Ministry of Environment, Scienct and Technology) (2006) Nepal: Third National Report on the Implementation of the UN Convention to Combat Desertification. MoEST, Nepal.

MoF (Ministry of Finance) (2012) Nepal Economic Survey Report, 2011–12. MoF, Government of Nepal, Kathmandu.

MoF (2013) Nepal Economic Survey Report 2012–13. MoF, Government of Nepal, Kathmandu.

Neupane, R.P., Sharma, K.R. and Thapa, G.B. (2002) Adoption of agroforestry in the hills of Nepal: A logistic regression analysis. *Agricultural Systems* 72(3), 177–196.

Nyamangara, J., Masvaya, E.N., Tirivavi, R. and Nyengerai, K. (2013) Effect of hand-hoe based conservation agriculture on soil fertility and maize yield in selected smallholder areas in Zimbabwe. *Soil and Tillage Research* 126, 19–25.

Pariyar, D. (2005) Country Pasture/Forage Resource Profiles. Pasture and Fodder Research Division, Khumaltar, Kathmandu.

Partap, T. (1999) Sustainable land management in marginal mountain areas of the Himalayan region. *Mountain Research and Development* 19(3), 251–260.

Partap, T. and Watson, H.R. (1994) Sloping agricultural land technology (SALT): A regenerative option for sustainable mountain farming. International Center for Integrated Mountain Development (ICIMOD) Occasional Paper No 23 (1994), Kathmandu.

Quinton, J.N. and Catt, J.A. (2004) The effects of minimal tillage and contour cultivation on surface runoff, soil loss and crop yield in the long-term Woburn Erosion Reference Experiment on sandy soil at Woburn, England. *Soil Use and Management* 20(3), 343–349.

Regmi, S., Pokharel, A., Ojha, S., Pradhan, S. and Chapagain, G. (2004) Nepal mental health country profile. *International Review of Psychiatry* 16(1–2), 142–149.

Rockström, J., Kaumbutho, P., Mwalley, J., Nzabi, A., Temesgen, M., *et al.* (2009) Conservation farming strategies in East and Southern Africa: Yields and rain water productivity from on-farm action research. *Soil and Tillage Research* 103(1), 23–32.

Rusinamhodzi, L., Corbeels, M., van Wijk, M.T., Rufino, M.C., Nyamangara, J., *et al.* (2011) A meta-analysis of long-term effects of conservation agriculture on maize grain yield under rain-fed conditions. *Agronomy for Sustainable Development* 31(4), 657–673.

Schreier, H., Shah, P.B., Lavkulich, L.M. and Brown, S. (1994) Maintaining soil fertility under increasing land use pressure in the Middle Mountains of Nepal. *Soil Use and Management* 10(3), 137–142. doi:10.1111/j.1475-2743.1994.tb00474.x (accessed 27 October 2014).

Shrestha, D.P. (1997) Assessment of soil erosion in the Nepalese Himalaya: A case study in Likhu Khola Valley, Middle Mountain Region. *Land Husbandry* 2(1), 59–80.

Smith, R.G., Menalled, F.D. and Robertson, G. (2007) Temporal yield variability under conventional and alternative management systems. *Agronomy Journal* 99, 1629–1634.

Thierfelder, C., Cheesman, S. and Rusinamhodzi, L. (2012) Benefits and challenges of crop rotations in maize-based conservation agriculture (CA) cropping systems of southern Africa. *International Journal of Agricultural Sustainability* 11(2), 108–124, doi:10.1080/14735903.2012.703894 (accessed 27 October 2014).

Thierfelder, C., Cheesman, S. and Rusinamhodzi, L. (2013) Benefits and challenges of crop rotations in maize-based conservation agriculture (CA) cropping systems of southern Africa. *International Journal of Agricultural Sustainability* 11(2), 108–124.

Tiwari, T., Brook, R. and Sinclair, F. (2004) Implications of hill farmers' agronomic practices in Nepal for crop improvement in maize. *Experimental Agriculture* 40, 397–417.

Upadhyaya, S.K. (2010) Upland poverty in nepal: The role of environment. Technical Report, Institute for Integrated Development Studies, Kathmandu. Available at: http://www.iids.org.np/Upland%20Poverty.pdf (accessed 28 October 2014).

WFP (World Food Program) (2010) Nepal Country Strategy: 3 Year Plan 2010–2013. World Food Program, United Nations, Kathmandu.

WFP (2013) Nepal Country Program Strategy: 2013–2017. World Food Program, United Nations, Kathmandu.

World Bank (2013) Agricultural irrigated land (% of total agricultural land). Available at: http://data.worldbank.org/indicator/AG.LND.IRIG.AG.ZS/countries (accessed 20 August 2013).

WORLDSTAT (2012) Nepal: Agriculture Land. Available at: http://en.worldstat.info/Asia/Nepal/Land (accessed 27 October 2014).

4 Effect of Tillage, Intercropping and Residue Cover on Crop Productivity, Profitability, and Soil Fertility under Tribal Farming Situations in Odisha, India

Aliza Pradhan,[1*] Travis Idol,[1] Pravat Kumar Roul,[2] Kshitendra Narayan Mishra,[2] Catherine Chan,[1] Jacqueline Halbrendt[1] and Chittaranjan Ray[1]

[1]*University of Hawai'i at Mānoa, Honolulu, Hawaii, USA;* [2]*Orissa University of Agriculture and Technology, Bhubaneswar, India*

4.1 Introduction

4.1.1 The need for resource-conserving agriculture

More than 50 years after the start of the Green Revolution, increasing population growth in many developing countries continues to make achieving food security a challenge. There appears to be no alternative but to increase agricultural productivity to meet future food demand and to alleviate poverty and hunger. However, agricultural intensification to increase crop production has had negative effects on natural resources such as surface- and groundwater pollution, sinking of groundwater levels, waterlogging and salinization of irrigated land, soil erosion, increasing pest resistance and resurgence, and loss of biodiversity and ecosystem services. These negative effects are especially pronounced in marginal crop production areas, where intensification practices have sometimes failed to increase crop yields sustainably. This realization has shifted the agricultural movement towards sustainable crop intensification that optimizes productivity while conserving, and even enhancing, natural resources. Even though an increased food supply is integral to eliminating hunger, malnutrition, and poverty, it alone cannot solve the problem. It is perhaps more important that we ensure people have access to the technology, knowledge, and purchasing power necessary to produce the food they need. Of the developing world's 5.5 billion people,

*Corresponding author; e-mail: apradhan@hawaii.edu

3 billion (nearly half of the entire world's population of 7 billion) live in rural areas. Of these rural inhabitants, an estimated 2.5 billion are in households involved in agriculture, and 1.5 billion of these are in smallholder households (World Bank, 2008; Herren, 2010). Poor farmers need low-cost and readily available technologies and practices to increase both local food production and their income.

In India specifically, there are several other key issues besides poverty and malnutrition. One of the most pressing is the limited future for expansion of irrigated areas. Even though India's agricultural growth has been sufficient to move the country from the severe food crises of the 1960s to aggregate food surpluses today, most of the increase in agricultural output over the years has taken place under irrigated conditions. However, opportunities for continued expansion of irrigated areas are limited, due to unsatisfactory performances of formal large canal systems, the massive public investment required in building the infrastructure of efficient irrigated systems, corruption in the construction process, and acknowledgement of the environmental impacts of irrigation projects (FAO, 2009). Thus, future productivity increases must come largely from rainfed agriculture. Despite a historic bias towards irrigated agriculture in terms of research and infrastructure development, rainfed agriculture has always been an important part of the agricultural sector (Kerr, 1996). Rainfed agriculture in India currently accounts for about two-thirds of the total cropped area and nearly half of the total value of agricultural output. Nearly half of all food grains are grown under rainfed conditions, and hundreds of millions of the rural poor depend on rainfed agriculture as the primary source of their livelihoods. However, farming without the aid of irrigation leaves farmers vulnerable to erratic or unpredictable rainfall, creating risk and yield uncertainty (Kerr, 1996).

Under India's current rainfall pattern, most rain occurs during a few months as part of the monsoonal cycle. These rain events tend to be intense, with strong erosive force. At the same time, agricultural land is often open to unrestricted grazing during the prolonged dry season. This leaves little plant cover, making fields highly vulnerable to accelerated erosion, especially on sloping lands. As a result, crops fail, yield becomes uncertain, and relative land productivity decreases, ultimately affecting the livelihoods of many poor and marginalized farmers.

This precarious food situation is especially dangerous in a country such as India, in which 260 million people live below the poverty line (Ansari and Akhtar, 2012). Of these, 75% live in rural areas, are totally dependent on agriculture, and produce much of what they eat. Small landholdings and their low productivity, along with uncertainties in rainfall patterns, increase economic and social risks for these farmers. This increased risk reduces their capacity and willingness to make investments in the farm system improvements associated with the Green Revolution, such as purchasing improved crop variety seeds, labor-saving equipment like tractors, and fertilizers. With degraded soils and unreliable weather patterns, return on investment is uncertain, and likely to be much lower overall than under irrigated conditions with better soils.

4.1.2 Tribal farming situations in Odisha, India

Though the region is rich in natural resources, the central Indian tribal belt (also known as the poverty belt) is a typical rainfed area facing challenges due to geography, traditional subsistence agriculture, poor irrigation infrastructure, dismal extension services, and lack of agricultural education within tribal communities (Phansalkar and Verma, 2004). India has as many as 427 tribal communities, second only to the continent of Africa. India's tribal communities, including the state of Odisha, are concentrated in the central and northeast regions, with 62 recognized tribal groups, totaling 8.15 million people. These tribal communities are spread among 30 administrative districts, six of which are considered fully tribal, with more than half of the area's total population comprised of tribal people. Keonjhar is one of Odisha's districts with a large proportion of tribal communities (56%). Keonjhar is one of the most poorly developed areas of the state, with a tribal literacy rate of approximately 25% (Sahoo, 2011). Tentuli, one of the study sites, is a typical tribal-dominated village in Keonjhar, representative of the plight of many smallholder farming communities in India. The village consists of 65 households, with more than 90% of the population dependent on agriculture. Average farm size is less than 1 ha. Upland agriculture in this area is predominantly rainfed, with low crop yields due to poor soil fertility and productivity, poor soil moisture retention, susceptibility to water erosion, and other external pressures of development and environmental conflicts of climate change (SMARTS, 2009). Most of the staple crops, such as local varieties of rice (*Oryza sativa* L.) and maize (*Zea mays* L.), are consumed by the farm household, with small surpluses and fruits and vegetables from backyard gardens sold in local markets.

The predominant upland crop in this tribal region is maize, grown during the rainy season (June–September), followed by mustard (*Brassica juncea* L.), an oilseed grown in the post-rainy season (October–January). Fields are cross-plowed two to three times prior to sowing. Seeds of local, open-pollinated maize varieties are broadcast sown on the fields at the anticipated start of the rainy season (i.e. mid-June). Hoes and hand weeding are used to control weeds.

Uncomposted farmyard manure is added in piles to fields once before the land preparation and is not actively incorporated into the soil. These levels and types of inputs are recognized as insufficient to meet crop demands and to replace nutrients lost in harvested yields (Wang *et al.*, 2009). No pesticides are generally used for maize, and the maize is harvested approximately 90–100 days later. The field is plowed again in preparation for sowing mustard. As with maize, seed from local varieties is broadcast sown on to the fields, with one weeding but no additional farmyard manure (FYM) application. The leaves of young mustard plants are harvested occasionally as a leafy vegetable for home consumption. The entire plant is harvested 90–100 days after sowing. After the mustard harvest, fields are generally left fallow for the remainder of the year, but are subject to unmanaged grazing by livestock. Deep tillage, along with high-intensity rainfall and no residue cover makes the land vulnerable to accelerated erosion during the rainy season. All crop residues are generally used for fuel, and prior to this study, there had been no legume cultivation in the fields for at least 40 years.

Under such conditions, intercropping systems such as introducing legumes into an existing single cropping system have been reported to have several benefits (Jaganathan *et al.*, 1974; Kalra and Gangwar, 1980), which include stability of production, insurance against crop failure, better resource use and income, and employment generation (Rao and Willey, 1980; Koshta and Karanjkar, 1986).

On the basis of crop growth morphology and duration, Herrera and Harwood (1973) suggested a variety of crop combinations, consisting of cereals and legumes such as maize and cowpea for intercropping. Cereals and legumes of varying maturity are used to ensure the efficient utilization of above- and underground resources, irrespective of rainfall conditions (Ofori and Stern, 1987). Maize seems to dominate as the cereal component, and can be combined with many different legumes like pigeon pea, mung bean, and cowpea (Remison, 1980). Besides these, other cultural factors such as plowing and residue management play an important role in sustainable crop management. Among many sustainable cropping systems, conservation agriculture is the one that helps in reversing soil degradation and improving crop production, as well as the socio-economic conditions of the smallholder farmers.

4.1.3 Definition and description of conservation agriculture

Conservation agriculture (CA) strategies and practices have been developed and promoted to reduce risk and improve natural resource conditions, as well as to address the combination of low yields, production risks, and poor natural resource conditions typically seen in developing tribal areas such as Odisha. According to the United Nations Food and Agriculture Organization (FAO), CA is an approach to managing agroecosystems for improved and sustained productivity, increased profits and food security, while preserving and enhancing the resource base and the environment. It is based on enhancing natural biological and ecological processes above and below ground. Interventions such as mechanical soil tillage are reduced to an absolute minimum, and the use of external inputs such as agrochemicals and nutrients of mineral or organic origin are applied at an optimum level and in a way and quantity that does not interfere with or disrupt biological processes (FAO, 2008). Conservation agriculture is characterized by three linked principles, namely:

1. Continuous, no or minimal mechanical soil disturbance (i.e. no-tillage and direct sowing or broadcasting of crop seeds, or direct placing of planting material in the soil; minimum soil disturbance from cultivation, harvest operation, or farm traffic).
2. Permanent organic soil cover, especially from crop residues and cover crops.
3. Diversified crop rotations in case of annual crops, or association of plants in case of perennial crops, including a balanced mix of legume and non-legume crops.

While CA principles are universal, implementation must obviously be adapted to local agroecological conditions and farmer capabilities and preferences. Currently, CA is being practiced on over 105 million ha (Mha) worldwide, with some farms having practiced it for 30 years and more (Friedrich *et al.*, 2012). CA has been reported to reduce production costs, increase water-use efficiency, and improve yields (FAO, 2002; Hobbs, 2007; Ernstein *et al.*, 2008, Govaerts *et al.*, 2009; Kassam *et al.*, 2009; Wall, 2009; Thierfelder and Wall, 2010). Since improved crop varieties are generally compatible with CA practices, long-term yields in CA systems are comparable with conventional intensive tillage systems (FAO, 2011). The loss of tillage to reduce weed presence and aerate the soil is offset by improved soil quality. With reduced tillage and higher organic soil cover, soil structure and structural integrity increase, resulting in higher water infiltration, and thus lower runoff and erosion (Tisdall and Oades, 1982; Six *et al.*, 2000; Bronick and Lal, 2005). Increased soil organic matter improves soil water- and nutrient-holding capacity, and water- and nutrient-use efficiency, increasing the capture and availability of these resources, and increasing the efficiency of crop fertilizer uptake (Ladd *et al.*, 1977; Carter, 1992; Graham *et al.*, 2002; Lal, 2002). Crop rotations and diversification reduce nutrient depletion and break up pest and disease cycles, improving overall crop and soil health (Sturz and Christie, 2003; Six *et al.*, 2004; Denef and Six, 2005). These improvements in various production factors result in long-term yield increases. In areas where the actual yield levels of tillage-based systems are low compared to the genetic and agroecological potential of the crops, the changeover to CA often results in yield increases even in the short term (Lal, 1986; Vogel, 1993). Thus, ideally, CA achieves sustainable crop intensification by improving natural resource quality and reducing farmers' risks.

Conservation agriculture systems also tend to be better suited for smallholder and resource-poor farmers. Labor requirements are reduced by about 50%, as the generally male-driven heavy work of soil tillage is reduced. This also reduces the need for heavy machinery and implements to turn the soil (Bishop-Sambrook, 2003), which is particularly appropriate for resource-poor farmers living on sloping hills. The labor savings can then be devoted to other tasks, including off-farm employment. Crop diversification with intercropping and rotation helps in improving the nutritional security of farm families, and reduces the risk of total crop failure in unfavorable or erratic weather. Increased water- and nutrient-use efficiencies reduce the need, and thus the costs, for irrigation and fertilizers. Better crop health reduces the need for pesticides. Thus, there should be long-term benefits to food security and agricultural income for farmers and rural communities.

Conservation agriculture is thus best conceptualized as an integrated production system that is universally applicable but locally adapted to achieve sustainable crop production. Careful consideration of farmer capabilities and preferences is just as important as understanding the production capabilities of the agroecological system. This study utilized this concept of conservation agriculture production systems (CAPS) to develop strategies and practices to address food production and livelihoods in a rural, tribal-dominated region of India. A baseline survey among the tribal farmers of the villages indicated no previous record of conservation agriculture. Therefore, with tribal famers' participation, a field research trial through the introduction of CAPS was implemented in the area as a solution to the posed problems of environmental degradation and food security and to enhance economic livelihood.

4.2 Methodology

4.2.1 Background

Through baseline surveys in Tentuli, a tribal village in Keonjhar, Odisha, along with expert advice from scientists, we investigated the study area's established practices relevant to each CA principle (tillage, cropping patterns, and residue cover), and identified appropriate CA alternatives. In Tentuli, conventional soil preparation involved plowing two to three times with an oxen-pulled, single-blade plow prior to sowing. Plowing in this manner breaks up the surface soil layer but does not turn it over as extensively as a moldboard plow. In consultation with farmers, a zero-tillage option for reduced soil disturbance was discarded in favor of a reduced-tillage approach in which there was one-time plowing to prepare the soil prior to sowing. At the time of sowing, a trench hoe was used to create a planting line for sowing maize seed. After maize harvest, the trench hoe was used again to create lines for sowing mustard seed in the next planting season.

With regards to cropping pattern, farmers grow only maize followed by mustard, and keep the fields fallow for the rest of the year. Because of the small farm size and dependence on maize as a staple food and cash crop, we decided to employ a legume for intercropping rather than crop rotation with legumes, as is common in larger-scale farms. Among all legume options, cowpea was chosen due to the availability of seed and a high market price in the study area's regional markets.

In terms of residue cover, farmers traditionally allow unmanaged livestock grazing during the dry season. Harvested mustard plants are typically threshed to release the seeds, leaving piles of stalks and leaves that are generally burned as fuel or simply as waste material. To apply the CA principles of retained residue cover, it was determined that crop residue management was preferable to cover cropping, due to the lack of water availability during the dry season. Instead, farmers were supplied with woven polyethylene tarpaulins to collect the mustard plant residue for redistribution back on the fields immediately after harvesting and threshing. As part of the experiment, we introduced improved varieties of maize, cowpea, and mustard supplied by the Orissa University of Agriculture and Technology. This was chosen partly for control of genetic variability in crop response, but also because these varieties were more suited for planting in rows, which was part of the reduced tillage treatment.

4.2.2 Experimental design

The experiment was conducted in June 2011 and 2012 in 20 adjacent farmer's fields, forming a single contiguous area. Fields were rectangular in shape and arranged perpendicular to the primary slope, which was generally <5%. Plots within fields were laid out in a randomized block design with four treatments: (CT-M) conventional tillage with sole cropped maize using an improved maize variety, 'Nilesh'; (CT-M+C) conventional tillage with maize intercropped with an improved variety of cowpea, 'Hariyalli Bush'; (MT-M) minimum tillage with sole cropped maize; (MT-M+C) minimum tillage with maize plus cowpea. Each plot was 5 m in width and as long as the farmer's field was wide, 10–15 m. Each field was considered as a blocked replicate of the treatments. In each farm field, conventional tillage was achieved by plowing plots twice, while in minimum tillage no part of the plot was plowed. Instead, seeds were sown along lines drawn with the help of a trench hoe. Spacing between maize rows was 60 × 30 cm. In the intercropped plots, cowpea was sown in between maize rows in a 1:1 ratio at a spacing of 60 × 15 cm. One gap filling (replanting of maize seedlings in gaps) and thinning (uprooting of extra seedlings for optimum use of space, light, nutrients, and water) was done 15 days after sowing. All other cultural operations, such as weeding and plant protection against diseases and pests, were carried out routinely and varied among treatments. Farmer reports of reduced 2011 cowpea yield due to shading from intense maize growth resulted in the 2012 use of an alternative improved variety, 'Utkal Manika', which was more shade tolerant, based on consulting with extension personnel and local scientists. After maize and cowpea harvest, the field was prepared for mustard planting in the post-rainy season. In the case of conventional tillage, the entire field was plowed. For minimum tillage, instead of entire field plowing, just the crop ridges were dismantled, leveled, and lines were demarcated. The mustard variety, 'Athagada Local', was sown as a cover crop in all the treatment plots. After harvesting the mustard seeds through threshing, all the plant residues were returned back to their respective plots as residue cover.

4.2.3 Soil sampling and analysis

As a baseline survey of cultural and socio-economic conditions is required to understand the study area, similarly baseline data of soil characteristics is required to understand initial soil status and the longer-term treatment effects on various soil properties. A study of initial soil status not only helps in developing and confirming better management strategies, but also gives a better platform for their improvement evaluation in consecutive years of CAPS practices. Therefore, initial soil samples were taken on 15 June 2011 from all the farming plots at both 0–5 cm and 5–10 cm soil depths, and analyzed for physical and chemical properties (Table 4.1) by the Orissa University of Agriculture and Technology (OUAT). Soil

Table 4.1. Initial soil characteristics (2011) of the farming plots of Tentuli, India.

Particular(s)	0–5 cm	Status	5–10 cm	Status	Method
pH	5.2	Slightly acidic	5.0	Slightly acidic	Digital electronic pH meter with 1:2.5 soil:water (Jackson, 1973)
Organic carbon (g/kg)	5.8	High	4.8	High	Walkley and Black method (Walkley and Black, 1934)
Available nitrogen (kg/ha)	259.41	High	237.15	High	Alkaline potassium permanganate method (Subbiah and Asija, 1956)
Available phosphorus (kg/ha)	5.8	Low	5.05	Low	Olsen's method (Olsen *et al.*, 1954)
Available potassium (kg/ha)	257.34	High	266.25	High	Ammonium acetate flame photometry method (Page *et al.*, 1982)
Bulk density (Mg/m^3)	1.48		1.49		Core sampling method (Black, 1965)
Soil texture	Silty loam		Silty loam		Bouyoucos hydrometer method (Piper, 1950)

sampling depths were decided based on literature reviews indicating that the short-term impacts of CA components were seen mostly in upper soil layers. Analysis of initial soil samples before CAPS implementation indicated that even though levels of organic carbon (C), available nitrogen, and potassium were relatively high, low availability of phosphorus might be a critical factor for crop productivity.

4.3 Results and Discussion

4.3.1 Maize and cowpea yield response

There was no significant difference in maize yield by tillage or intercropping over the 2 years of cropping seasons (Table 4.2). Maize yield under both conventional and minimum tillage were comparable. Even though results indicated that there was no yield gain from employing CAPS practices, there was also no yield penalty, indicating minimum tillage to be as good as conventional tillage. These 2-year results are important, as they dispel the idea of yield reduction due to CA practices, which might discourage the adoption of CA (Giller *et al.*, 2009). There are numerous other studies regarding the variability of short-term yield responses (positive, neutral or negative yield responses) to CA (Lal, 1986; Gill and Aulakh, 1990; Mbagwu, 1990; Mupangwa *et al.*, 2012).

Further, maize yield in intercropped plots was comparable to that of monocropping. Hence, the introduction of cowpea to maize fields did not have any negative influence on maize yield; rather, cowpea was an additional gain to the farmers from the same plot at the same time. Earlier studies have also reported no significant maize yield reduction when intercropped with legumes under conservation agriculture practices (Sakala, 1998; Myaka *et al.*, 2006; Kamanga *et al.*, 2010). However, maize yields achieved in both tillage types and intercropping

Table 4.2. Effect of treatments and year on maize and cowpea yield in on-farm trials in Tentuli, India.

Year	Treatments	Yield of maize (kg/ha)	Yield of cowpea (kg/ha)
2011	CT-M (control)	5,130	–
	CT-M+C	4,677	775
	MT-M	4,482	–
	MT-M+C	5,224	675
2012	CT-M (control)	5,200	–
	CT-M+C	5,125	1,500
	MT-M	4,685	–
	MT-M+C	5,352	1,350
Treatment (T)			
CT ~ MT		NS	–
M ~ M+C		NS	–
Year (Y)		NS	–
T × Y		NS	–

NS, significance at 5% levels.

systems over two seasons were greater than the national yield average of 2,285 kg/ha (Directorate of Maize Research, 2011/12) but comparable to other research findings (FAO, 2004; Jat *et al.*, 2010). The differences in maize yield between our findings and the national average can be attributed to varied rainfall patterns, varieties used, soil fertility of the sites, and general agronomic practices.

Similarly, the effect of cowpea on maize yield in an intercropping system seems to be variable. While there are reports claiming a yield reduction of maize in intercropped plots (Adeniyan *et al.*, 2007; Lemlem, 2013), others indicate an increase of maize yield (Nzabi *et al.*, 2000; Mpairwe *et al.*, 2002; Dapaah *et al.*, 2003). In our study, cowpea did not affect maize production, but represented an additional yield for farmers. Similar neutral responses of cowpea on maize yield were also reported by Watiki *et al.* (1993) and Thobatsi (2009). Cowpea yield was not analyzed statistically as varieties were not constant in both years. Even though cowpea yield was greater in 2012 than in 2011 (Table 4.2), this increase might be attributable to better performance of the variety, 'Utkal Manika'. However, the effect of intercropping on both the main and the intercrop is determined by cultivar maturity period, planting date, location, rainfall, and soil factors.

4.3.2 Mustard yield response

Average mustard seed yield under CAPS trials in 2011 and 2012 were 786 and 805 kg/ha, respectively, which were lower than the national average of 1,157 kg/ha but much higher than the state average of 416 kg/ha (Directorate of Economics and Statistics, 2012). Furthermore, both mustard seed and stover yields under intercropped plots were significantly higher by 30 and 40%, respectively, over sole cropping ($P < 0.01$) (Table 4.3). After threshing, the crop residues

Table 4.3. Effect of treatments and year on mustard seed and stover yield in on-farm trials in Tentuli, India.

Year	Treatments	Yield of mustard seed (kg/ha)	Yield of mustard stover (kg/ha)
2011	CT-M (control)	690	1,380
	CT-M+C	865	1,860
	MT-M	670	1,233
	MT-M+C	920	1,840
2012	CT-M (control)	695	1,400
	CT-M+C	900	1,980
	MT-M	682	1,350
	MT-M+C	945	2,000
Treatment (T)			
CT ~ MT		NS	NS
M ~ M+C		**	**
Year (Y)		NS	NS
T × Y		NS	NS

NS, significance at 5% level; **, significance at 1% level.

were used to cover the respective plots. As repeated addition of large amounts of crop residues lead to a greater soil C content in time (Erenstein, 2002), stover yield of the cover crop plays an important role in CA practice. Significant increase in mustard seed as well as stover yield in the case of intercropped plots seems to be advantageous for the farmers in terms of both economical and soil improvement. Furthermore, several studies have reported problems with retaining crop residues as surface mulch, largely due to competition with its use as livestock feed, particularly during the dry season (Mtambanengwe and Mapfumo, 2005; Umar *et al.*, 2011). An advantage of *B. juncea* is that the stover is non-edible, reducing the risk of loss when returned and applied as a surface mulch, thereby implying better prospects of CA adoption in the area.

4.3.3 Labor and profitability

Labor required for planting functions such as land preparation, sowing, weeding, and harvesting were affected by CAPS treatments. While minimum tillage affected plowing and weeding, intercropping had an impact on additional sowing and harvesting (Table 4.4). Minimum tillage reduced labor requirements by 31% during land preparation, as there was a single plowing instead of two. There was no change in labor requirements among treatments with respect to maize sowing. However, for intercropped plots, sowing and harvesting cowpea required additional labor. Minimum tillage increased labor by 39% for weeding and plant protection. This agrees with previous reports (Giller *et al.*, 2009; Mazvimavi and Twomlow, 2009) that found a higher number of days spent on weeding under CA compared to conventional tillage practices. The CT-M+C treatment required the least weeding and plant protection labor, a decrease by 49% over control (18.56 versus 37.13 labor days/ha). This decrease in labor might be attributed to cowpea protecting the space between maize rows by outcompeting weeds. The difference of labor days in mustard cultivation was due primarily to the higher crop yield in intercropped plots. The highest increase of total labor (71%) was reported under MT-M+C, followed by mustard over control (439.98 versus 257.16 labor days/ha), while the smallest increase was under MT-M (9%). Even though labor requirement was highest for minimum tillage with intercropping, higher maize and mustard yields along with additional cowpea yield increased profit (by 27% over control). Discussions with farmers revealed that mustard and cowpea were preferred crops, in addition to having attractive market prices. It is anticipated that the profit level will help in popularizing CA among smallholder farmers, as monetary gains act as a prime driver for adoption (Erenstein *et al.*, 2008).

4.3.4 Soil fertility

As soil is the basic resource of farmers for crop production, it needs to be nurtured in order to improve, conserve, and sustain its use. In general, the fertility of a soil depends on various climate and crop management factors, including the soil's physico-chemical properties. Most soil properties are relatively permanent and take years to decades to show any significant change towards management practices.

Table 4.4. Effect of tillage and intercropping on average labor requirement (labor days/ha) and profitability (US$/ha) in on-farm trials in Tentuli, India.

Treatment	Land prep	Sowing cowpea	Weeding	Harvest	Mustard cultivation	Total labor days/ha[a]	Total input costs (US$/ha)[b]	Total return (US$/ha)[c]	Profit (US$/ha)
CT-M (control)	39.09	–	37.13	49.13	95.12	257.16	635	1,153	518
CT-M+C	39.09	60.51	18.56	126.0	105.03	431.87 (67.94%)	854	1,370	516 (–0.34%)
MT-M	27.03	–	45.41	27.84	95.12	279.08 (8.52%)	583	1,038	455 (–12.12%)
MT-M+C	27.03	60.51	45.26	119.47	105.03	439.98 (71.01%)	797	1,457	660 (27.4%)

[a]Total labor days/ha also included labor required for sowing of maize (63.46 labor days/ha) and fertilization (19.22 labor days/ha).
[b]Total input cost was calculated by multiplying the labor days by wage rate (US$2.4/day).
[c]Crop yields were multiplied by their market price to get the total return under CAPS treatments (maize @ US$0.16/kg, cowpea @ US$0.27/kg, and mustard @ US$0.45/kg (market survey and feedback from farmers)). US$1 was equal to an average 55 Indian national rupees (during the crop harvest periods of 2011 and 2012).

In this study, even though CAPS had no significant effect on soil bulk density, organic C, and available nutrients after 2 years of crop cultivation, the trends were positive for minimum tillage with intercropping (Fig. 4.1). Furthermore, conventional tillage with sole cropping of maize was found to have a negative impact on soil properties, while the remaining two treatments (CT-M+C and MT-M) had variable effects. In general, the impacts of conservation practices on soil properties are site specific and there is no such definite time frame to have a significant effect. The study results also differ for specific soil properties. For example, some short-term studies (up to 5 years) reported that in the upper 30 cm soil layer, the bulk density of zero and minimum tillage was greater than that under tilled soil (Yang and Wander, 1999; Gal *et al.*, 2007), but other researchers indicated that soil bulk density was equal to or smaller under conservation systems relative to tilled plots (Angers *et al.*, 1997; Ussiri and Lal, 2009). Even though the addition of crop residue is one of the components of CA, roughly half of the experiments surveyed reported increased C over conventional practices, while 40% showed no change and 10% showed a reduction in soil C (Govaerts *et al.*, 2009). Therefore, even though the effect of conservation practices largely depends on soil type, climatic regime, and management practices, early adoption of the practices will check further soil degradation.

4.4 Conclusions and Future Outlook

After 2 years of CA implementation in this study, intercropping maize and cowpea under minimum tillage along with residue cover presents a win–win scenario due to improved crop yield, increased economic return, and trends of increasing soil fertility. Regardless of current results, requests from additional village farmers to implement CAPS during the third year of the study suggest that they recognize these benefits and are willing to modify conventional practices to achieve them. In fact, based on this study, a thorough assessment of the highlighted benefits of CA, and its future prospects, Keonjhar district Department of Agriculture and the Agricultural Technology Management Agency (ATMA) officials have approved a proposal to replicate minimum tillage with maize–cowpea intercropping in 500 ha of potential maize-grown area. Another neighboring district, Mayurbhanj, has approved adopting CAPS on 1,000 ha. This is indicative of the clear local benefits of the introduced CAPS, and hopefully points to increased dissemination in the near future. Especially in tribal regions, where farmer's socio-economic and cultural situation inhibits the adoption of mechanized and large-scale agricultural production, relatively small modifications of crop management such as intercropping with legumes have significant impact on livelihood improvements. However, community awareness of the roles of reduced tillage and crop residues in reversing soil degradation should also be stressed, as farmers may otherwise focus primarily on improved seed varieties and intercropping strategies to boost short-term yields and income.

As this experiment is ongoing, additional benefits should emerge after several cropping seasons. Significant improvements in soil quality are expected to emerge after several years of CA practices. In the meantime, however, a sustained

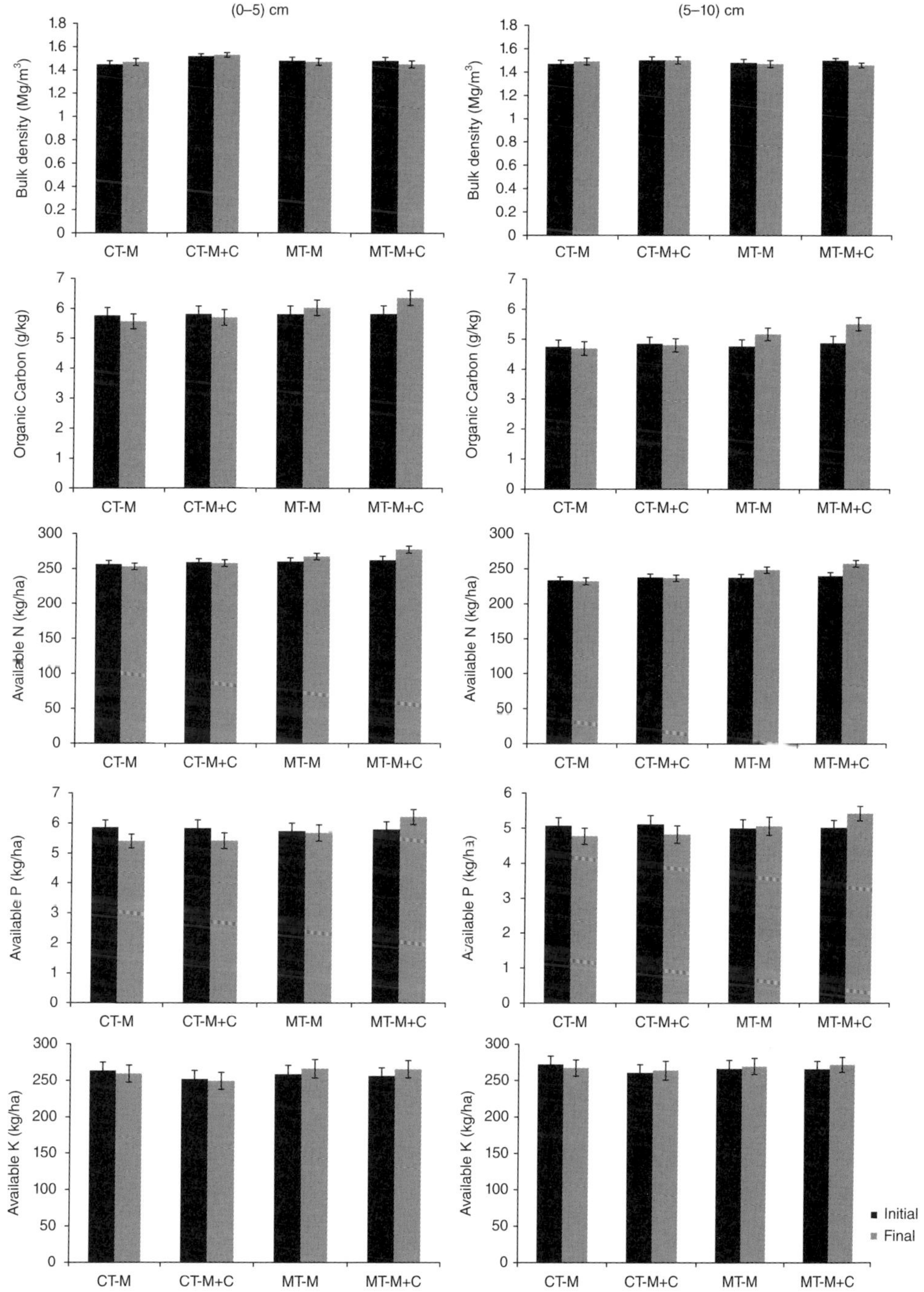

Fig. 4.1. Effects of CAPS on soil bulk density, organic carbon, and available nitrogen (N), phosphorus (P), and potassium (K) during initial (2011) and final (at end of 2012 cropping season) at (0–5) and (5–10) cm soil depths in on-farm trials of Tentuli. Vertical bars indicate standard error of means (*n* = 20).

policy and institutional support to provide incentives and required services to farmers would be greatly beneficial. Even though the present study demonstrated the technical performance of CAPS at the farm level through participatory action research, sustained improvement and further spread requires capacity building and networking among farmers, extension agents, researchers, and government officials. In the current state, even when the cost–benefit analysis at farm level indicates economic benefits, farmers may lack the opportunity to purchase inputs ahead of the cropping season, or lack the cash to invest. Therefore, timely availability of inputs and implements can be ensured through infrastructure development such as agricultural support via local agro-dealer shops and implement hiring services. The same applies to farmers' access to credit and to markets for agricultural inputs and produce.

Unfortunately, there is no one blueprint or rigid prescription for CA that will work across a range of socioecological circumstances. Hence, a better understanding of cultural, geographical, sociotechnical, and organizational issues is required to shape the idea of when and whether local adaptation of CA principles may be an appropriate way to address the needs of smallholder farmers. Conservation agriculture represents the core component of a new alternative paradigm for sustainable production intensification. But with the integrative nature of CA, adoption is unlikely to be immediate, but will be incremental, with farmers experimenting on small areas and only adopting on a large-scale when they are fully convinced of the technology. These CA innovations should be nurtured by frequent farmer training programs, workshops, and exposure visits, along with appropriate dissemination strategies such as publications, newsletters, leaflets, brochures, audio or video programs.

Acknowledgements

The financial support from USAID and Feed the Future Innovation Lab for Collaborative Research on Sustainable Agriculture and Natural Resource Management (SANREM), the efforts of scientists, students, and field technicians under Sustainable Management of Agroecological Resources for Tribal Societies (SMARTS) and the participating tribal farmers of Keonjhar are duly acknowledged.

References

Adeniyan, O.N., AKande, S.R., Balogun, M.O. and Saka, J.O. (2007) Evaluation of crop yield of African yam bean, cow pea and maize under intercropping systems. *America-Eurasian Journal of Agriculture and Environment Science* 2, 99–102.

Angers, D.A., Bolinder, M.A., Carter, M.R., Gregorich, E.G., Drury, C.F., *et al.* (1997) Impact of tillage practices on organic carbon and nitrogen storage in cool, humid soils of eastern Canada. *Soil and Tillage Research* 41, 191–201.

Ansari, I.Z. and Akhtar, S. (2012) Incidence of poverty in India: issues and challenges. *International Journal of Multidisciplinary Research* 2(3), 443–449.

Bishop-Sambrook, C. (2003) Labor saving technologies and practices for farming and household activities in eastern and southern Africa. IFAD/FAO, Rome.

Black, C.A. (1965) *Methods of Soil Analysis*. American Society of Agronomy, Inc, Madison, Wisconsin.

Bronick, C.J. and Lal, R. (2005) Soil structure and management: A review. *Geoderma* 124, 3–22.

Carter, M.R. (1992) Influence of reduced tillage systems on organic matter, microbial biomass, macro-aggregate distribution and structural stability of the surface soil in a humid climate. *Soil and Tillage Research* 23, 361–372.

Dapaah, H.K., Asafu-Agyei, J.N., Ennin, S.A. and Yamoah, C.Y. (2003) Yield stability of cassava, maize, soybean and cowpea intercrops. *Journal of Agricultural Science* 140, 73–82.

Denef, K. and Six, J. (2005) Clay mineralogy determines the importance of biological versus abiotic processes for macroaggregate formation and stabilization. *European Journal of Soil Science* 56, 469–479.

Directorate of Economics and Statistics (2012) *Agricultural Statistics at a Glance, 2012.* Compiled by Directorate of Economics and Statistics, Department of Agriculture Cooperation, Ministry of Agriculture, Government of India.

Directorate of Maize Research (2011/12) Indian Agricultural Research Institute, PUSA Campus, New Delhi.

Erenstein, O. (2002) Crop residue mulching in tropical and semi-tropical countries: an evaluation of residue availability and other technological implications. *Soil and Tillage Research* 67, 115–133.

Ernstein, O., Sayer, K., Wall, P., Dixon, J. and Hellin, J. (2008) Adapting no-tillage agriculture to the smallholder maize and wheat farmers in the tropics and sub-tropics. In: Goddard, T., Zoebisch, M.A., Gan, Y.T., Ellis, W., Watson, A. and Sombatpanit, S. (eds) *No-Till Farming Systems*. Special Publication 3. World Association of Soil and Water Conservation (WASWC), Bangkok, pp. 253–277.

FAO (Food and Agriculture Organization) (2002) Conservation agriculture: case studies in Latin America and Africa. *FAO Soils Bulletin 78*. FAO, Rome.

FAO (2004) Socio-economic Analysis and Policy Implications of the Roles of Agriculture in Developing Countries. Summary Report, Roles of Agriculture Project. FAO, Rome.

FAO (2008) What is Conservation Agriculture. Available at: http://www.fao.org/ag/ca/1a.html (accessed 28 October 2014).

FAO (2009) *The State of Food Security in the World: Economic Crises – Impacts and Lessons Learned*. FAO, Rome

FAO (2011) *Save and Grow, A Policymaker's Guide to Sustainable Intensification of Smallholder Crop Production*. FAO, Rome.

Friedrich, T., Derpsch, R. and Kassam, A. (2012) Overview of the global spread of Conservation Agriculture. *Field Actions Science Reports*, Special Issue 6.

Gal, A., Vyn, T.J., Micheli, E., Kladivko, E.J. and McFee, W.W. (2007) Soil carbon and nitrogen accumulation with long-term no-till versus moldboard plowing overestimated with tilled-zone sampling depths. *Soil and Tillage Research* 96, 42–51.

Gill, K.S. and Aulakh, B.S. (1990) Wheat yield and soil bulk-density response to some tillage systems on an Oxisol. *Soil and Tillage Research* 18, 37–45.

Giller, K.E., Witter, E., Corbeels, M. and Tittonell, P. (2009) Conservation agriculture and smallholder farming in Africa: the heretics' view. *Field Crops Research* 124, 468–472.

Govaerts, B., Sayre, K.D., Goudeseune, B., De Corte, P., Lichter, K., *et al.* (2009) Conservation agriculture as a sustainable option for central Mexican highlands. *Soil and Tillage Research* 103, 222–230.

Graham, M.H., Haynes, R.J. and Meyer, J.H. (2002) Soil organic matter content and quality: effects of fertilizer applications, burning and trash retention on a long-term sugarcane experiment in South Africa. *Soil Biology and Biochemistry* 34, 93–102.

Herren, H.R. (2010) Moving in the right direction. In: *Food Security: Creating an Affordable, Sustainable, Economically Sound and Socially Acceptable Food Future for Everyone*. London, p. 88.

Herrera, W.T. and Harwood, R.R. (1973) Crop inter-relationship in intensive cropping system (mimeo). IRRI, Los Banos, Philippines.

Hobbs, P.R. (2007) Conservation agriculture: What it is and why it is important for future sustainable food production? *Journal of Agricultural Science* 145, 127–137.

Jackson, M.L. (1973) *Soil Chemical Analysis*. Prentice Hall of India, Pvt. Ltd., New Delhi.

Jaganathan, N.T., Morachan, Y.B. and Ramiah, S. (1974) Studies on the effect of maize and soybean association in different proportions and spacing on yield. *Journal of Potassium Research* 10(2), 386–391.

Jat, M.L., Saharawat, Y.S. and Gupta, R. (2010) Conservation agriculture: Improving resource productivity in cereal systems of South Asia. Lead Paper. In: *Proceedings of the National Symposium on Resource Management Approaches Towards Livelihood Security*. University of Agricultural Sciences, Bengaluru, Karnataka, India, pp. 389–393.

Kalra, G.S. and Gangwar, B. (1980) Economics of intercropping of different legumes with maize at different levels of nitrogen under rainfed conditions. *Indian Journal of Agronomy* 25,181–185.

Kamanga, B.C.G., Wadddington, S.R., Robertson, M.J. and Giller, K.E. (2010) Risk analysis of maize–legume crop combinations with smallholder farmers in resource endowment in central and southern Malawi. *Experimental Agriculture* 46, 1–21.

Kassam, A., Friedrich, T., Shaxson, F. and Pretty, J. (2009) The spread of conservation agriculture: justification, sustainability and uptake. *International Journal of Agricultural Sustainability* 7, 292–320.

Kerr, J.M. (1996) Sustainable development of rainfed agriculture in India. EPTD discussed paper. International Food and Policy Research Institute, Washington, DC.

Koshta, A.K. and Karanjkar, S.V. (1986) Economics of intercropping system under rainfed farming. *Financing Agriculture* 18(22), 33–36.

Ladd, J.N., Parsons, J.W. and Amato, M. (1977) Studies of nitrogen immmobilization and mineralization in calcareous soils. 1. Distribution of immobolized nitrogen amongst soil fractions of different particle-size and density. *Soil Biology and Biochemistry* 9, 309–318.

Lal, R. (1986) Soil surface management in the tropics for intensive land use and high and sustained production. *Advances in Soil Science* 5, 1–109.

Lal, R. (2002) Soil carbon dynamics in cropland and rangeland. *Environmental Pollution* 116, 353–362.

Lemlem, A. (2013) The effect of intercropping maize with cowpea and lablab on crop yield. *Herald Journal of Agricultural and Food Science Research* 2(5), 156–170.

Mazvimavi, K. and Twomlow, S. (2009) Socioeconomic and institutional factors influencing adoption of conservation agriculture by vulnerable households in Zimbabwe. *Agricultural Systems* 101, 20–29.

Mbagwu, J. (1990) Mulch and tillage effects on water transmission characteristics of an Ultisol and maize grain yield in SE Nigeria. *Pedologie* 40, 155–168.

Mpairwe, D.R., Sabiiti, E.N., Ummuna, N.N., Tegegne. A. and Osuji, P. (2002) Effect of intercropping cereal crops with forage legumes and source of nutrients on cereal grain yield and fodder dry matter yields. *African Crop Science Journal* 10, 81–97.

Mtambanengwe, F. and Mapfumo, P. (2005) Organic matter management as an underlying cause for soil fertility gradients on smallholder farms in Zimbabwe. *Nutrient Cycling in Afroecosystems* 73, 227–243.

Mupangwa, W., Twomlow, S. and Walker, S. (2012) Reduced tillage, mulching and rotational effects on maize (*Zea mays* L.), cowpea (*Vigna unguiculata* (Walp) L.) and sorghum (*Sorghum bicolor* L. (Moench)) yields under semi-arid conditions. *Field Crops Research* 132, 139–148.

Myaka, F.M., Sakala, W.D., Adu-Gyamfi, J.J., Kamalongo, D., Ngwira, A., *et al.* (2006) Yields and accumulations of N and P in farmer-managed intercrops of maize-pigeonpea in semi-arid Africa. *Plant and Soil* 285, 207–220.

Nzabi, A.W., Makini, F., Onyango, M., Kidula, N., Muyonga, C.K., *et al.* (2000) Effect of intercropping legume with maize on soil fertility and maize yield. In: Mureithi, J.G., Gachene, C.K.K., Muyekho, F.N., Onyango, M.N., Mose, L. and Magenya, O. (eds) *Proceedings of the 2nd Scientific Conference of the Soil Management and Legume Research Network Projects*, held in Kenya on 26–30

June 2000, pp. 193–197. Available at: http://www.kari.org/fileadmin/publications/Legume_Project/Legume2Conf_2000/26.pdf (accessed 18 November 2014).

Ofori, F. and Stern, W.R. (1987) Cereal–legume intercropping systems. *Advances in Agronomy* 41, 41–90.

Olsen, S.R., Cole, C.V., Watanabe, F.S. and Dean, L.A. (1954) Estimation of available phosphorus in soils by extraction with sodium bicarbonate. USDA Circular No 939. United States Department of Agriculture, Washington, DC.

Page, A.L., Miller, R.H. and Keeny, D.R. (1982) *Methods of Soil Analysis*, 2nd edn. American Society of Agronomy, Madison, Wisconsin.

Phansalkar, S. and Verma, S. (2004) Improved water control as strategy for enhancing tribal livelihoods. *Economic and Political Weekly, Review of Agriculture* 31, 3469–3476.

Piper, C.S. (1950) *Soil and Plant Analysis*. University of Adelaide, Australia, pp. 368.

Rao, M.R. and Willey, R.W. (1980) Evaluation of yield stability in intercropping studies on sorghum + pigeon pea. *Experimental Agriculture* 16, 105–106.

Remison, S.U. (1980) Interaction between maize and cowpea at various frequencies. *The Journal of Agricultural Science* 94(3), 617–621.

Sahoo, L.K. (2011) Socio-economic profile of tribal populations in Mayurbhanj and Keonjhar districts. *Orissa Review* 63–68. Available at: http://odisha.gov.in/e-magazine/Orissareview/orissareview-2011.htm (accessed 13 November 2014).

Sakala, W.D. (1998) Nitrogen dynamics in maize/pigeonpea intercropping in Malawi. 1998 PhD thesis, Wye College, University of London.

Six, J., Paustian, K., Elliott, E.T. and Combrink, C. (2000) Soil structure and soil organic matter. I. Distribution of aggregate size classes and aggregate associated carbon. *Soil Science Society of America Journal* 64, 681–689.

Six, J., Bossuyt, H., Degryze, S. and Denef, K. (2004) A history of research on the link between (micro) aggregates, soil biota, and soil organic matter dynamics. *Soil and Tillage Research* 79, 7–31.

Sturz, A.V. and Christie, B.R. (2003) Beneficial microbial allelopathies in the root zone: the management of soil quality and plant disease with rhizobacteria. *Soil and Tillage Research* 72, 107–123.

Subbiah, B.V. and Asija, G.L. (1956) A rapid procedure for estimation of available nitrogen in soils. *Current Science* 25, 259–260.

SMARTS (Sustainable Management of Agroecological Resources for Tribal Societies) (2009) SMARTS Proposal: Response to EEP Comments. University of Hawai'i at Mānoa, Honolulu, Hawaii.

Thierfelder, C. and Wall, P. (2010) Rotation in conservation agriculture systems of Zambia: Effects on soil quality and water relations. *Journal of Experimental Agriculture* 46, 1–17.

Thobatsi, T. (2009) Growth and yield responses of maize (*Zea mays L.*) and cowpea (*Vigna unguiculata* L.) in an intercropping system. Masters dissertation submitted to Department of Plant Production and Soil Science, University of Pretoria, South Africa.

Tisdall, J.M. and Oades, J.M. (1982) Organic matter and water stable aggregates in soils. *Journal of Soil Science* 33, 141–163.

Umar, B.B., Aune, J.B., Johnsen, F.H. and Lungu, O.I. (2011) Options for improving smallholder conservation agriculture in Zambia. *Journal of Agricultural Science* 3, 50–62.

Ussiri, D.A.N. and Lal, R. (2009) Long-term tillage effects on soil carbon storage and carbon dioxide emissions in continuous corn cropping systems from an Alfisol in Ohio. *Soil and Tillage Research* 104, 39–47.

Vogel, H. (1993) Tillage effects on maize yield, rooting depth and soil-water content on sandy soils in Zimbabwe. *Field Crops Research* 33, 367–384.

Walkley, A.J. and Black, I.A. (1934) Estimation of organic carbon by chromic acid titration method. *Soil Science* 37, 29–38.

Wall, P. (2009) Strategies to overcome the competition for crop residues in southern Africa: Some light at the end of the tunnel. In: *4th World Congress on Conservation Agriculture*. M/S Print Process, New Delhi, pp. 65–70.

Wang, E., Cresswell, H., Xu, J. and Jiang, Q. (2009) Capacity of soils to buffer impact of climate variability and value of seasonal forecasts. *Agricultural and Forest Meteorology* 149, 38–50.

Watiki, J.M., Fukai, S., Banda, J.A. and Keating, B.A. (1993) Radiation interception and growth of maize/cowpea intercrop as affected by maize plant density and cowpea cultivar. *Field Crops Research* 35, 123–133.

World Bank (2008) World Bank Development Report 2008: Agriculture for Development. World Bank, Washington, DC.

Yang, X.M. and Wander, M.W. (1999) Tillage effects on soil organic carbon distribution and storage in a silt loam soil in Illinois. *Soil and Tillage Research* 52, 1–9.

5 Assessment of Maize-based Conservation Agriculture Production Systems (CAPS) in Rainfed Uplands of Odisha, India

Pravat Kumar Roul,[1*] Aliza Pradhan,[2] Kshitendra Narayan Mishra,[1] Plabita Ray,[1] Travis Idol,[2] Satya Narayan Dash,[1] Catherine Chan[2] and Chittaranjan Ray[2]

[1]*Orissa University of Agriculture and Technology, Bhubaneswar, India;*
[2]*University of Hawai'i at Mānoa, Honolulu, Hawaii, USA*

5.1 Introduction

Rainfed agroecosystems, the purported gray patches untouched by the Green Revolution or most technological advances, occupy a prominent position in Indian agriculture. However, since productivity of the country's irrigated areas has almost reached a plateau, future growth in farm productivity will likely come from rainfed agroecosystems. The rainfed zones of India, with annual rainfall ranging from 500 to 1,500 mm, constitute 60% of the country's net cultivated area. Calculations based on rainfall distribution pattern and soil type showed that even if the full irrigation potential of the country was realized, 50% of the net sown area would remain rainfed.

Cropping intensities and crop yields are low and unstable in rainfed areas, due to unpredictable patterns of rainfall, a host of biotic and abiotic stresses such as disease pest infestation, drought, flood, extreme temperature, and poor soil conditions, and adherence to age old traditional farm practices. Despite these challenges, rainfed agriculture supports 40% of India's population, and its production contributes 44% to the national food basket (CRIDA, 2003). Due to an increasing population, per capita land availability in rainfed areas is expected to decrease from 0.28 ha in 1990 to less than 0.10 ha by 2025 (CRIDA,

*Corresponding author; e-mail: roulpvt@yahoo.com.sg

 Conservation Agriculture in Subsistence Farming: Case Studies from South Asia and Beyond (eds C. Chan and J. Fantle-Lepczyk)

2003). The demand for food will continue to rise, however, necessitating higher productivity from rainfed regions. This emphasizes the critical importance of rainfed agriculture for food security and the Indian economy to meet the needs of a growing population (Katyal *et al.*, 1997).

Of the 85 million hectares (Mha) of rainfed cultivated area in India, about 78% is located in eastern India. Low and unstable yields are due mainly to the erratic and not always well-timed onset of southwest monsoon rains. A delay or lack of these rains can result in moisture stress during critical crop growth periods, leading to drastic yield reductions (Kar *et al.*, 2004; Jat *et al.*, 2005). Under these conditions, crop and varietal diversification with crops tolerant of drier conditions, like maize (*Zea mays*), black gram (*Phaseolus mungo*), cowpea (*Vigna unguiculata* L.), and sesame (*Sesamum indicum*), may be the best option for farmers to manage the risk of drought and ensure a reasonable yield and income (Vittal *et al.*, 2002). Among these options, maize is one of the most versatile emerging crops, having wider adaptability under varied agroclimatic conditions.

The state of Odisha, located in the eastern part of India, has 2.62 Mha of maize with a production of 6.08 million tonnes (Mt), resulting in an average productivity of 2,321 kg/ha (Government of Odisha, 2012) annually. Nationally, maize is cultivated on 8.78 Mha, with a production of 21.76 Mt and an average annual yield of 2,478 kg/ha (Directorate of Economics and Statistics, 2012). In Odisha, maize is cultivated primarily in interior districts such as Keonjhar (27,580 ha), Koraput (18,650 ha), Kandhamal (16,900 ha), Kalahandi (16,640 ha), Mayurbhanj (13,120 ha), and Gajapati (10,770 ha), predominantly populated by tribal ethnic groups. Maize is a staple tribal food crop and amounts to 7.4% of the state's total food consumption. These tribal communities constitute 23% of total state population (Census-2011, www.censusindia.gov.in) and are mainly comprised of resource-poor smallholder (<1 ha) farmers, with more than 90% of the population depending on agriculture. Upland agriculture in these areas is predominantly characterized by maize monocropping. Continuous cultivation of local, low-yielding maize varieties has resulted in significant declines in soil fertility due to: excessive mining of plant nutrients from soil; rapid loss of fertile topsoil due to water erosion in the beginning of the monsoonal rainy season; traditional cultivation practices, such as intensive tillage (plowing land three times before sowing); unbalanced nutrient management (application of farmyard manure only); and no residue management practice. The majority of soils in this tropical agroecological zone are weathered, acidic, course textured, and erodible, and are low in organic matter and chemical fertility. Aggregates of such soils are weak; they lose productivity quickly and do not sustain adequate water and nutrients for sustainable production under current management practices (Roy *et al.*, 1981). In order to minimize soil and other natural resource degradation, and to achieve a sustainable crop production, a set of crop–nutrient–water–land system management practices, such as conservation agriculture (CA), needs to be developed for the area.

Conservation agriculture can be defined as "a concept for resource-saving agricultural crop production that strives to achieve acceptable profit together with high and sustained production levels while concurrently conserving the environment" (FAO, 2010). The three key tenets of CA, as defined by the Food and Agricultural Organization of the United Nations (FAO) are:

- continuous, no or minimal mechanical soil disturbance;
- permanent organic matter soil cover, especially by crop residues and cover crops; and
- diversified rotations of annual crops or perennial plant associations, as relevant.

The first key principle, practicing minimum mechanical soil disturbance, is essential to bring stability of soil organic matter, enhance soil aggregation, and reduce soil erosion. There are reports of minimum tillage having positive influence on soil bulk density, soil compaction, infiltration, and water retention capacity (Bishop-Sambrook, 2003; Besconca *et al.*, 2006; Bhattacharya *et al.*, 2006). Other benefits include improved infiltration of rainwater into the soil, increased water availability in plants, reduced surface runoff, and improved groundwater recharge (Bhale and Wanjeri, 2009). Besides these, reduced surface cultivation reduces farming energy requirement and overall farming costs, as well as saves time, an important aspect for resource-poor tribal farmers as saved time can be allotted towards other farm activities (Joshi, 2011).

The second tenet of CA, permanent organic soil cover (especially with crop residues and cover crops), protects soil against the harmful effects of rain, such as the splashing effect of raindrops leading to soil erosion. Crop residues can be used as mulches to cover the soil, which in turn increases water-filled pore space (Karlen *et al.*, 1994). Besides these, soil cover also creates an ideal environment for microbes to grow, which further improves overall soil health, as well as provides humus, mitigates temperature, and reduces surface runoff (FAO, 2010), thereby maintaining the soil quality for sustainable crop production.

The third main facet of CA is the practice of crop rotation with more than two species. Crop rotation helps to build up a soil's infrastructure, in that establishing crops in a rotation allows for an extensive growth of roots to deeper soil depths, which in turn will allow for better nutrient cycling and water infiltration (Hobbs *et al.*, 2007). Crop rotation can also be used as a plant protection measure. Inclusion of more than one crop in a cropping cycle rather than monocropping prevents further spread of the disease pest in the cropping system (Hobbs *et al.*, 2007). In fact, selected crop species, when combined in a crop rotation system, will act as a natural insecticide and herbicide, thereby eliminating problems with yield reduction (FAO, 2010).

However, given the obvious variability of agroecological environments, cropping systems, and farmer capacities and preferences, there is not a single set of CA practices that applies worldwide. Therefore, the successful introduction of CA depends on adapting and tailoring the basic principles to the local context. The conservation agriculture production system (CAPS) concept was introduced by the Feed the Future Innovation Lab for Collaborative Research on Sustainable Agriculture and Natural Resource Management (SANREM). This approach is intended to develop locally relevant and preferred CA systems to ensure they are effective in meeting production and conservation goals, as well as being acceptable to producers and communities. As maize is the staple crop in tribal areas of

Odisha, a maize-based CAPS is needed to improve agronomic, environmental, and socio-economic sustainability in these areas.

In order to enhance and promote CAPS adoption among tribal farmers, we sought to understand and assess crop yield and system productivity, profitability, and soil fertility under the influence of different maize-based CAPS. In order to achieve this, an on-station field experiment was designed by the scientists from the University of Hawai'i at Mānoa, USA, and the Orissa University of Agriculture and Technology (OUAT), India, in consultation with extension personnel and local farmers.

5.2 Methodology

5.2.1 Description of site and experimental design

Our field experiment was conducted in rainfed uplands at the Regional Research and Technology Transfer Station (RRTTS) of OUAT in the Keonjhar district during the rainy and post-rainy seasons of 2011/12 and 2012/13. The experimental site's soil was a silty loam with pH 7.47. From 0 to 20 cm depth, soil organic carbon was 6.58 g/kg, available nitrogen (N) 267.1 kg/ha, available phosphorus (P) 15.8 kg/ha, and available potassium (K) 341.8 kg/ha. Keonjhar was chosen as the research site as it has the highest tribal population in the state, has the greatest area under maize cultivation (27,580 ha), and currently faces problems of poor yield, soil erosion, and land degradation. The district's usual cropping system is maize during the rainy season (mid-June–September), followed by mustard (*Brassica campestris* L.) as a post-rainy season crop (October–January).

A set of CAPS practices was selected based on the discussions with farmers, researchers, and extension personnel regarding their tillage and crop preference, past cropping history of the area, market demand, and other threats and challenges. In order to reduce soil erosion, a minimum tillage method of plowing once before planting was proposed as an alternative to the conventional practice of plowing three times. Because of the central importance of maize as a food staple and the limitation of land for rotation, an intercrop rather than crop rotation option was selected. Cowpea (*V. unguiculata* L.) was considered suitable as an intercrop, as it was a legume and would help in biological nitrogen fixation, had a high market value (double that of maize), and local farmers had some previous experience with growing and selling it. In order to address the cover crop and residue management principle, horse gram (*Macrotyloma uniflorum*) was selected as an alternative post-rainy season monocrop in addition to mustard. Horse gram was the preferred cover crop option, as it provided economic yield as well as acting as a legume soil cover. Both crops grow reasonably well on residual soil moisture, and mature better with dry weather during the late vegetative and reproductive stages (i.e. during January).

Given the selected tillage, cropping system, and cover crop factors, 12 experimental CAPS systems were identified for further investigation. Prior to the start of the rainy season, the trial experiment was laid out in a randomized block design (RBD) in three adjacent fields at the Keonjhar RRTTS (Fig. 5.1a). Treatments of

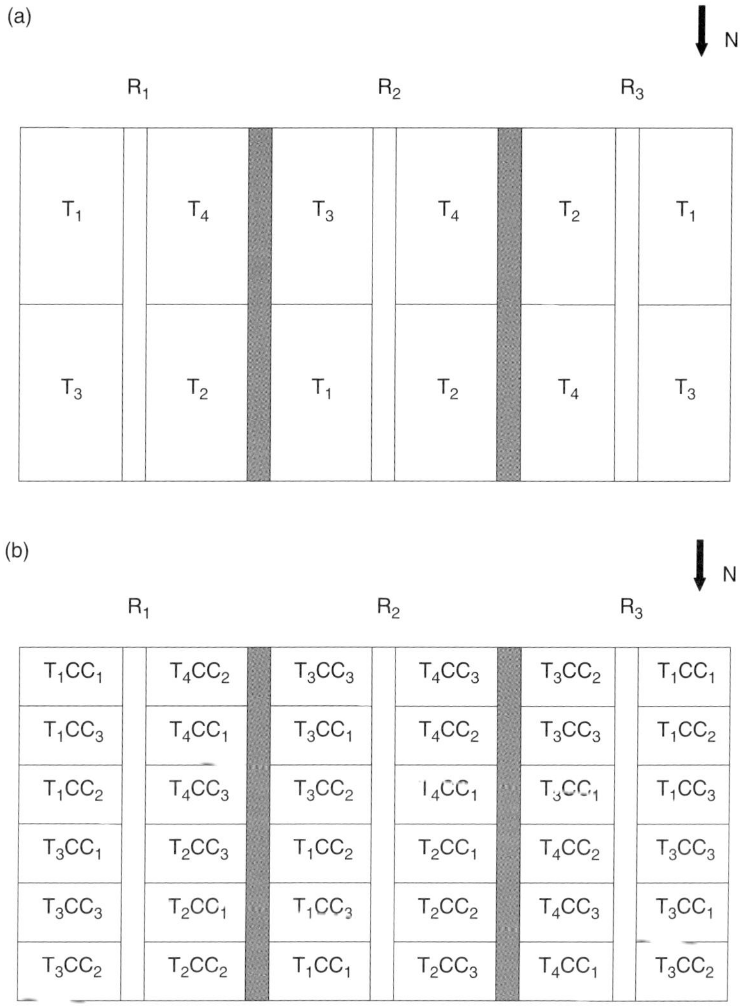

Fig. 5.1. (a) Rainy season experimental layout. Experimental design: randomized block design (RBD); number of treatments (T): 4; number of replications (R): 3; individual plot size: 10.2 m × 7.2 m. T_1, conventional tillage only maize; T_2, conventional tillage with maize + cowpea; T_3, minimum tillage only maize; T_4, minimum tillage with maize + cowpea. (b) Post-rainy season experimental layout. Experimental design: split plot; number of treatments (T): 12; number of replications (R): 3; individual plot size: 7.2 m × 3.2 m. Main plot: T_1, conventional tillage only maize; T_2, conventional tillage with maize + cowpea; T_3, minimum tillage only maize; T_4, minimum tillage with maize + cowpea. Subplot (cover crop): CC_1, fallow; CC_2, mustard; CC_3, horse gram.

tillage and cropping system were tested, each with two levels. The tillage treatment levels were conventional and minimum tillage, while the cropping treatment levels were sole maize and intercropped maize–cowpea. Treatments by level were fully crossed in each of the three 10.2 × 7.2 m plots, designated as blocks. Each treatment combination was replicated once per block. Conventional tillage consisted of a single pass of an oxen-drawn plow during the pre-monsoonal rains, a few days to weeks prior to the expected heavy monsoonal rains. After the onset of monsoonal rains, the field was cross-plowed, meaning two complete passes were made, with the second pass perpendicular to the first. Though farmers' normal practice is to broadcast maize seed throughout the plot, this study used line sowing of maize seed to maintain consistency with the minimum tillage treatment. Minimum tillage consisted of a single plowing prior to sowing, followed by strip-tilling rows with hand-held hoes to sow maize seed. In both tillage systems, plots were hand-weeded with hoes several times during the growth phase.

For sole cropping of maize, seed of the 'Nilesh' variety was hand-sown in rows at a density of three seeds/m. Row spacing was 60 cm, resulting in 17 rows of maize/plot and an expected plant density of 408 maize plants/plot (55,555 plants/ha). Ten days after maize sowing, plots were hand-weeded with hoes in the inter-row spaces. In the intercropped scenario, cowpea seeds of the variety, 'Hariyalli Bush', were planted in the disturbed soil at a density of six seeds/m. With 16 rows/plot, this resulted in an expected density of 768 cowpea plants/plot (104,575 plants/ha). Maize was harvested approximately 90 days after sowing. Because cowpea was an indeterminately flowering and fruiting crop, harvesting of mature seedpods began approximately 40 days after sowing and continued until 60 days after sowing.

After final harvest of both maize and cowpea, crop residues were left as such in the fields, and the plot was prepared by strip-tilling rows with hand-held hoes for planting of cover crops. Each plot was split into thirds and assigned to one of three cover crop treatments: no cover crop, mustard as the cover crop, and horse gram as the cover crop (Fig. 5.1b). The mustard variety, 'Parvati', and the horse gram variety, 'Athagada Local', were line sown at a density of 10 seeds/m, with 24 rows/plot, which resulted in an approximate density of 768 plants/plot (333,333 plants/ha). The crops were harvested approximately 75 days after sowing, and after threshing all residues were returned back to their respective plots.

5.2.2 Measurements and data analysis

Yield measurements of maize seeds, cowpea pods, mustard seed, and horse gram seeds were recorded after harvesting of the crops. To estimate the treatment effects on system productivity, the yield per crop was weighted by average yearly market price and normalized to that of maize, to estimate a "maize equivalent yield" (MEY) (Eqn 5.1). The market prices were collected from local farmer markets during 2011/12 and 2012/13.

$$\text{MEY (kg/ha)} = \frac{\text{Yield of other crop produce (kg/ha)} \times \text{price of that produce (US\$/kg)}}{\text{Price of maize grain (US\$/kg)}} \quad (5.1)$$

The total cost of cultivation was calculated by taking input costs such as seed and fertilizer and labor requirements during various intercultural operations and harvesting. The total return under a system was calculated by multiplying MEY with the average market price of maize for 2011/12 and 2012/13 cropping seasons. Furthermore, the net return was calculated by subtracting the total cost of cultivation from the total return.

As the whole field experiment was designed as a split plot, data were analyzed by taking tillage and intercropping as the main plot factor and the cover crop effect in the subplot. Similarly, soil samples were collected from 0–10 cm soil depth, processed, and analyzed for the CAPS effect on the various soil properties.

5.3 Results and Discussion

5.3.1 Effect of CAPS on maize yield and system productivity

Average maize yields under CAPS systems for year 1 (2011/12) and year 2 (2012/13) were 4,890 and 5,411 kg/ha, respectively, much higher than the state average (Table 5.1). This may be attributed to the improved variety and cultivation practices such as line sowing, and better nutrients and plant protection measures. However, after 2 years of growing under CAPS, there was no significant difference in maize yield by tillage or intercropping. Even though minimum tillage was not associated with any significant increase in maize yield over conventional tillage, neither was there any yield reduction for shifting the tillage practice. Similarly, for intercropping, cowpea was an additional gain without affecting maize growth and yield.

The total system productivity for both cropping seasons was considered by analyzing the yield contributions of cowpea, mustard, and horse gram towards MEY under different CAPS (Table 5.1). While tillage had a significant impact on the MEY of cowpea, intercropping had a significant effect on the MEY of mustard (Table 5.1) over 2 cropping years. The MEY of cowpea was significantly lower ($P > 0.05$) by 27% and 14% in year 1 and year 2, respectively, under minimum tillage. Similarly, the MEY of mustard grown after maize–cowpea intercropping was significantly higher ($P > 0.05$) by 28% and 49% in year 1 and year 2, respectively, towards MEY. The weed-suppressive benefits of cowpea for maize may have carried over for mustard, but a more likely explanation is the addition of organic matter and associated nutrients, especially nitrogen, from the incorporation of cowpea residues. That is, it likely acted as green manure for the subsequent crops (Mantiamely, 2005). Neither treatment nor year had any significant effect on MEY of horse gram.

5.3.2 Effect of CAPS on total cost of production

The cost of cultivation, averaged over 2011/12 and 2012/13, varied with respect to intercropping and cover crop treatments (Table 5.2). In the rainy season, conventional tillage along with intercropping had the highest cultivation cost, an increase of 14% over the traditional practice, conventional tillage only with maize (US\$550/ha), while cultivation cost was lowest under minimum tillage with monocropped maize (US\$482/ha). This US\$68/ha cost increase for conventional tillage with intercropping might be due to the cost incurred by tilling

Table 5.1. Effect of CAPS on maize and maize equivalent yield (MEY), 2011/12 and 2012/13.

				Effect on MEY (kg/ha)		
Year	Treatments	State average maize yield (kg/ha)	Maize yield (kg/ha)	Cowpea as intercrop	Mustard as cover crop	Horse gram as cover crop
2011/12	Conventional tillage – maize	2,321	5,210	–	2,344	1,235
	Conventional tillage – maize + cowpea		4,440	2,895	2,798	1,045
	Minimum tillage – maize		4,300	–	2,017	903
	Minimum tillage – maize + cowpea		5,610	2,114	2,784	1,474
2012/13	Conventional tillage – maize	2,411	4,657	–	2,578	1,282
	Conventional tillage – maize + cowpea		4,455	3,553	3,498	1,120
	Minimum tillage – maize		4,237	–	2,100	1,054
	Minimum tillage – maize + cowpea		5,489	3,050	3,500	1,485
Tillage			NS	*	NS	NS
Intercropping			NS	–	*	NS
Year			NS	NS	NS	NS
Interaction			NS	–	NS	NS

Average yearly local market price of maize seed US$0.16/kg, cowpea green pod US$0.27/kg, mustard seed US$0.45/kg, and horse gram seed US$0.27/kg (personal communication) were taken into consideration for calculating maize equivalent yield (MEY) during the cropping season of 2011/12 and 2012/13. NS, $P > 0.05$; *, $P < 0.05$.

more frequently during land preparation along with cowpea cultivation. In the overall average post-rainy season, fallow plots had no additional cultivation cost, whereas mustard cultivation costs (US$232/ha) were approximately 18% greater than those for horse gram (US$191/ha). Considering the total cost of cultivation, conventional tillage with intercropping followed by mustard had the highest value of US$847/ha.

Table 5.2. Effect of CAPS on the total cost of production (averaged over 2011/12 and 2012/13).

Rainy season treatment	Cultivation costs (US$/ha)	Cost of cultivation in post-rainy season (US$/ha)			Total cost of cultivation (US$/ha)	Total system yield (MEY, kg/ha)	Total return (US$/ha)	Net return (US$/ha)
		Fallow	Mustard	Horse gram				
CT-M	550	–			550	4,934	789	239
	550		231		781	7,395	1,183	402
	550			197	747	6,193	991	244
CT-M+C	654	–			654	7,672	1,228	574
	654		245		899	10,820	1,731	832
	654			185	839	8,755	1,401	562
MT-M	482	–			482	4,269	683	201
	482		200		682	6,328	1,012	330
	482			160	642	5,248	840	198
MT-M+C	597	–			597	8,132	1,301	704
	597		250		847	11,274	1,804	957
	597			220	817	9,612	1,538	721

CT-M, conventional tillage, maize; CT-M+C, conventional tillage, maize + cowpea; MT-M, minimum tillage, maize; MT-M+C, minimum tillage, maize + cowpea. Cost of cultivation includes input costs such as: (i) seeds: maize at US$0.02/kg, cowpea at US$0.07/kg, mustard at US$0.09/kg, and horse gram at US$0.06/kg; (ii) fertilizer: urea at US$0.1/kg, diammonium phosphate (DAP) at US$0.34/kg, and muriate of potash (MOP) at US$0.22/kg; (iii) labor wage at US$2/day (personal communication).

Both total return and net return were highest under minimum tillage with intercropping followed by mustard (34% and 65% higher, respectively) over the farmers' standard practice of conventional tillage with maize followed by mustard. Increased yield of maize and mustard along with the high market price of both cowpea and mustard resulted in such large net gains.

5.3.3 Effect of CAPS on soil properties

As soil is one of the basic resources for crop production, the impact of CA on various soil properties over time needs to be assessed. In general, CA establishes new dynamics in the soil through interactions among soil fauna, plant roots, water, air, soil temperature, and recycling of nutrients. Though some soil properties are relatively permanent and change very slowly with time, others may change even during the initial years of treatment implementation. In this study, after 2 years of cropping, the impact of CAPS on two soil physical properties, bulk density (BD) and water-stable aggregate (WSA), were taken into consideration. Bulk density is a basic physical property of soil that regulates soil water-holding capacity, pore space, nutrient and water movement, etc., in soil. The lower the value of BD, the better is the soil quality. Similarly, WSA in the surface soil of sloping land acts as an important predictor of soil erosion through water. With regards to soil chemical property, soil organic carbon was analyzed to assess the treatment impact, as it affects various other soil properties such as soil bulk density, porosity, aggregate stability, nutrient availability, water-holding capacity, and soil microbial activity.

Soil bulk density

Tillage practices with different cropping systems (sole versus intercrop) affected the soil bulk density (BD) significantly at the end of the second cropping cycle (Fig. 5.2). The practice of conventional tillage with sole maize significantly increased BD by 5% over the initial value of 1.24 Mg/m^3 ($P < 0.05$), whereas it remained unchanged under minimum tillage (MT). Loss of finer soil particles due to water erosion and low soil organic matter (SOM) contents, leading to less aggregation, might be the reason for significantly higher soil BD in the conventional tillage (CT) system (Lafond *et al.*, 2011). The lack of significant change of soil BD under MT is also likely due to improved SOM, resulting in more aggregation and enhanced biological activity (Jemai *et al.*, 2013). The impact of cover crops (mustard and horse gram) on soil BD was comparable to that under no crop cover (1.27 Mg/m^3). The lowest BD of 1.22 Mg/m^3 was recorded in the soils under MT-M+C followed by horse gram.

Water-stable aggregates

Soil organic matter plays the most vital role in aggregate formation. Significant changes in the percentage of water-stable macroaggregates (>0.250 mm) and microaggregates (0.053–0.250 mm) were observed in the soil under different CAPS (Fig. 5.3). Continuous practice of CT significantly reduced the macroaggregates by 2.4% ($P < 0.05$) over the initial status (73.7%) as the maximum soil disturbances might have resulted in rapid decomposition of SOM. Minimum tillage, on the other hand, significantly increased the macroaggregates by 8.7% ($P < 0.05$), which might be due to the accumulation and preservation of SOM. The M+C intercrop showed significantly ($P < 0.05$) higher aggregates (+6.2%) over sole maize (74.2%).

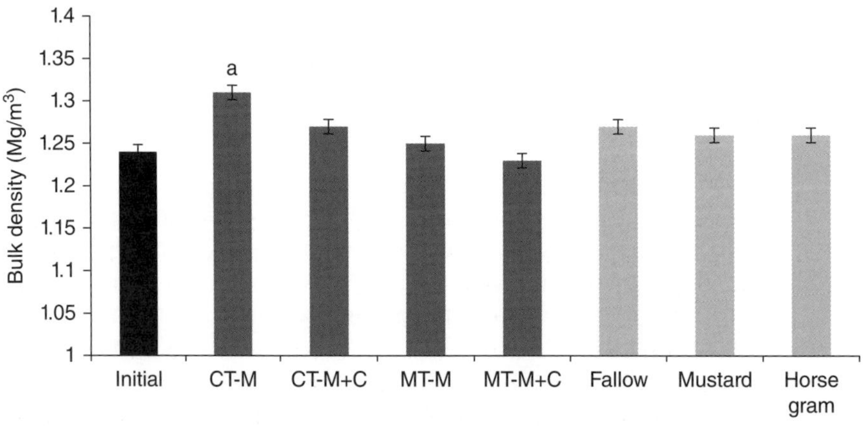

Fig. 5.2. Effect of CAPS on soil bulk density (Mg/m^3) over 2 years of cropping cycle at 0–10 cm soil depth; treatment with a lower case letter was significant at $P = 0.05$. CT-M, conventional tillage, maize; CT-M+C, conventional tillage, maize + cowpea; MT-M, minimum tillage, maize; MT-M+C, minimum tillage, maize + cowpea.

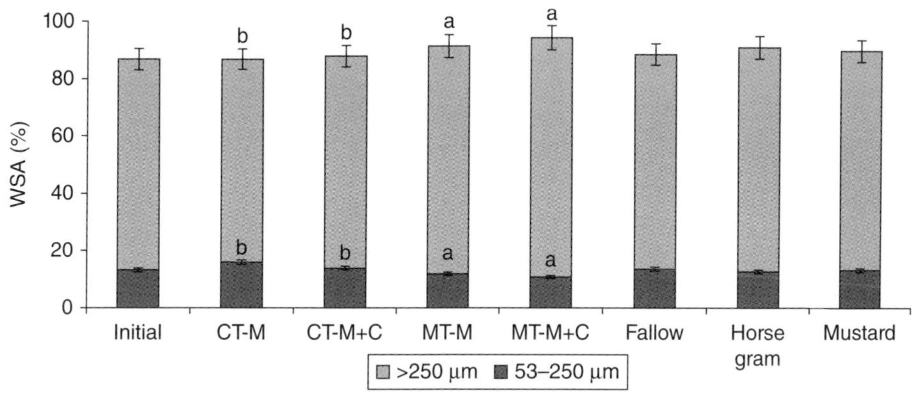

Fig. 5.3 Effect of CAPS on soil macro- (>0.250 mm) and microaggregates (0.053–0.250 mm) over 2 years of cropping cycle at 0–10 cm soil depth; treatments with different lower case letter were significant at $P = 0.05$. CT-M, conventional tillage, maize; CT-M+C, conventional tillage, maize + cowpea; MT-M, minimum tillage, maize; MT-M+C, minimum tillage, maize + cowpea.

Similarly, extensive soil disturbances in CT resulted in significantly higher ($P < 0.05$) contents of microaggregates (0.053–0.250 mm) (+13.3%) over the initial value of 13.12%, which could be due to physical turnover of macroaggregates (Fig. 5.3). The practice of MT, on the other hand, significantly lowered ($P < 0.05$) the contents of microaggregates by 13.4% over initial values, due mostly to turnover of microaggregates into macroaggregates in the presence of higher organic binding agents. However, soils associated with intercropping and cover crops did not show any pronounced changes in the contents of water-stable macroaggregates. Higher macroaggregate contents in MT (81.1%) are related to a higher stock of fresh organic matter, hence increased microbial activity and production of microbial binding agents (Mikha and Rice, 2004). Conventional tillage (CT), in contrast, disrupts macroaggregates, thereby enhancing its turnover to microaggregates (Balesdent *et al.*, 2000; Six *et al.*, 2000; Zotarelli *et al.*, 2007).

Soil organic carbon

There were no significant effects of CAPS on soil organic carbon (SOC) at the end of the second cropping cycle (Fig. 5.4). However, the minimized soil disturbance in MT increased SOC contents by 14% over the initial level of 6.62 g/kg. The practice of CT, on the other hand, reduced the organic carbon contents of the soils by 2.4% over the initial status. Similarly, neither fallow nor cover crops had any significant effect on SOC over the initial level (Fig. 5.4). In general, fields that receive large inputs of organic matter in the form of crop residues are generally rich in carbon, while fields that receive no or little organic matter have small soil carbon contents (Samake *et al.*, 2005; Tittonell *et al.*, 2007; Zingore *et al.*, 2007). Thus, even though the effect of CAPS on SOC was not significant over the 2 years of cropping, the benefits of enhanced SOC might be recovered in the long run.

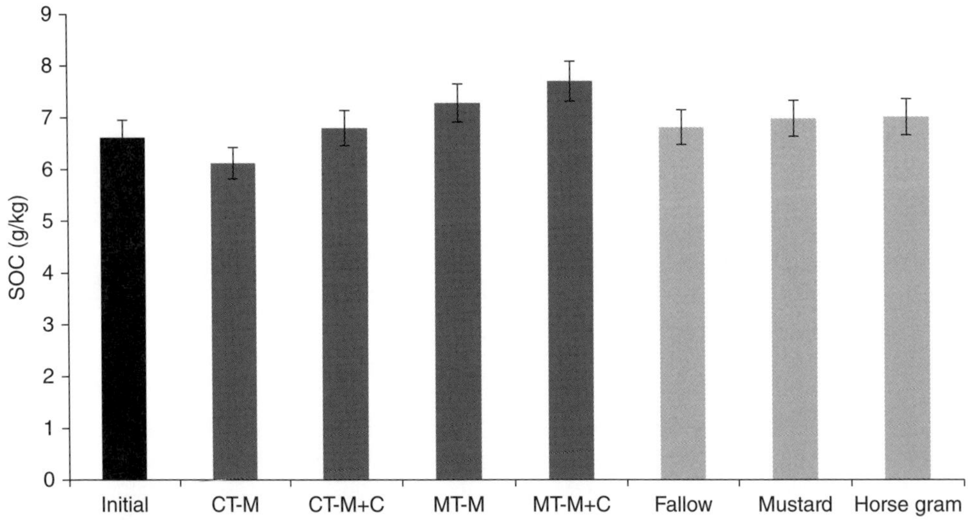

Fig. 5.4. Effect of CAPS on soil organic carbon (g/kg) over 2 years of cropping cycle at 0–10 cm soil depth. CT-M, conventional tillage, maize; CT-M+C, conventional tillage, maize + cowpea; MT-M, minimum tillage, maize; MT-M+C, minimum tillage, maize + cowpea.

5.4 Conclusion

Agriculture in developing countries focuses primarily on finding a sustainable agricultural technology that meets the demands of smallholder farmers while maintaining or improving soil fertility. Though there is no single universal clear-cut strategy to end food insecurity and rural poverty issues, the results from this study show that with location-specific and low-input conservation agriculture practice such as CAPS, farmers can intensify crop production in marginal lands without any environmental issue. As economic gain is an important factor of a technology adoption, we suggest that CAPS with maximum system productivity and highest net return (profit) will be attractive to smallholder tribal farmers. At the same time, technology promotion through appropriate agronomic practice and crop varieties should be done in a package, rather than one element at a time, to achieve maximum impact. Furthermore, appropriate technology and techniques should be disseminated to farmers through adaptive research, extension agents, policy supports, and public investments. Assuming that farmers prioritize immediate economic benefits, proper education should be offered and training programs held to assist these key players in understanding the long-term goals of CAPS such as soil fertility and sustainability. In the long run, government support is integral in building confidence among marginal smallholder farmers to try new varieties and practices. Governmental support provides insurance to farmers against unforeseeable risks and uncertainties; the assurance of a reliable safety net to fall back on can only increase their willingness to assume the risk. Furthermore, local institutions that enable farmers to connect with their peers and other actors along the value chain should be supported and encouraged by government whenever needed. This support can take the form of training sessions, exchange visits, subsidy programs, and workshops. International organizations should support CAPS adoption among poor smallholder farmers through supportive policies, research, and funding arrangements. Similarly, specific conservation technologies suitable for particular locations and groups of farmers should be researched, disseminated, and supported.

Acknowledgements

The authors thank USAID-SANREM Feed the Future Innovation Lab for funding this research through the LTRA 11 project: Sustainable Management of Agroecological Resources for Tribal Societies (SMARTS). We are also grateful to field and laboratory staff for assisting in experimental management, data collection, and soil analysis.

References

Balesdent, J., Chenu, C. and Balabane, M. (2000) Relationship of soil organic matter dynamics to physical protection and tillage. *Soil and Tillage Research* 53, 215–230.

Besconsa, P., Imaz, M.I., Virto, I., Enrique, A. and Hoogmoed, W.B. (2006) Soil and water retention as affected by tillage and residue management in semi arid Spain. *Soil and Tillage Research* 87, 19–27.

Bhale, V.M. and Wanjeri, S.S. (2009) Conservation agriculture: A new paradigm to increase resource use efficiency. *Indian Journal of Agronomy* 54(2), 149–158.

Bhattacharya, R., Prakash, V. and Kundu, S. (2006) Effect of tillage and crop rotation on pore size distribution and soil hydraulic conductivity in sandy loam soil of Indian Himalayas. *Soil and Tillage Research* 86, 129–140.

Bishop-Sambrook, C. (2003) Labor saving technologies and practices for farming and household activities in eastern and southern Africa. IFAD/FAO, Rome.

CRIDA (Central Research Institute for Dryland Agriculture) (2003) *Annual Report 2003–2004*. All India Co-ordinated Research Project for Dryland Agriculture, Central Research Institute for Dryland Agriculture, Hyderabad, India, pp. 291–301.

Directorate of Economics and Statistics (2012) *Agricultural Statistics at a Glance, 2012*. Directorate of Economics and Statistics, Department of Agriculture and Cooperation, Ministry of Agriculture, Government of India.

FAO (Food and Agriculture Organization) (2010) What is Conservation Agriculture. Available at: http://www.fao.org/ag/ca/1a.html (accessed 29 October 2014).

Government of Odisha (2012) Odisha Agriculture Statistics, 2011–12. Directorate of Agriculture and Food Production, Government of Odisha, India. Available at: www.agriodisha.nic.in (accessed 13 November 2014).

Hobbs, P.R., Sayre, K. and Gupta, R. (2007) The role of conservation agriculture in sustainable agriculture. *Philosophical Transactions of the Royal Society B* 363, 543–555.

Jat, M.L., Singh, R.V., Solanki, N.S., Kumpawat, B.S., Balyan, J.K., *et al.* (2005) Rainfall characteristics and crop planning in Bhilwara region. *Research on Crops* 6(1), 170–172.

Jemai, I., Aissa, N.B., Guirat, S.B., Ben-Hammouda, M. and Gallali, T. (2013) Impact of three and seven years of no-tillage on the soil water storage, in the plant root zone, under a dry subhumid Tunisian climate. *Soil and Tillage Research* 126, 26–33.

Joshi, P.K. (2011) Conservation agriculture: An overview. *Indian Journal of Agricultural Economics* 66(1), 53–63.

Kar, G., Singh, R. and Verma, H.N. (2004) Alternative cropping systems for assured and efficient crop production in upland rainfed areas of Eastern India based on rainfall analysis. *Agricultural Water Management* 37, 47–62.

Karlen, D.L., Wollenaupt, N.C., Erbach, D.C., Berry, E.C., Swan, J.B., *et al.* (1994) Crop residue effect on soil quality following 10 years of no-till corn. *Soil and Tillage Research* 31, 149–167.

Katyal, J.C., Ramachandran, K., Reddy, M. and. Rao, C. (1997) Indian agricultural profile of land resources, crop performance and prospects. In: Ravinchandran, V. (ed.) *Regional Land Cover Changes and Sustainable Agriculture in South Asia.* Proceedings of the COSTED/EU International Workshop on Regional Land Cover Changes, Sustainable Agriculture and their Interactions with Global Change, Chiang Mai, Thailand, December 1996. COSTED, Chennai, India, pp. 16–34.

Lafond, G.P., Walley, F., May, W.E. and Holzapfel, C.B. (2011) Long-term impact of no-till on soil properties and crop productivity on the Canadian prairies. *Soil and Tillage Research* 117, 110–123.

Mantiamely, B., Machado, P.L.O.A., Torres, E., Andred, A.G. and Valencia, L.I.O. (2005) No tillage and crop rotation effects on soil aggregation and organic carbon in a rhodferralsol from southern Brazil. *Soil and Tillage Research* 80, 185–200.

Mikha, M.M. and Rice, C.W. (2004) Tillage and manure effects on soil and aggregate associated carbon and nitrogen. *Soil Science Society of America Journal* 68, 809–816.

Roy Shanma, R.P., Sharma, H.M. and Thakur, H.C. (1981) Studies on intercropping in long duration pigeon pea, on sandy loam soil of North Bihar. *Indian Journal of Agronomy* 26, 77–82.

Samake, O., Smaling, E.M.A., Kropff, M.J., Stomph, T.J. and Kodio, A. (2005) Effects of cultivation practices on spatial variation of soil fertility and millet yields in the Sahel of Mali. *Agriculture, Ecosystems and Environment* 109, 335–345.

Six, J., Elliott, E.T. and Paustian, K. (2000) Soil macroaggregate turnover and microaggregate formation: A mechanism for C sequestration under no-tillage agriculture. *Soil Biology and Biochemistry* 32, 2099–2103.

Tittonell, P., Vanlauwe, B., de Ridder, N. and Giller, K.E. (2007) Heterogeneity of crop productivity and resource use efficiency within smallholder Kenyan farms: Soil fertility gradients or management intensity gradients? *Agricultural Systems* 94, 376–390.

Vittal, K.P.R., Maruthi Sankar, G.R., Singh, H.P. and Sharma, J.S. (2002) Sustainability of practices in dryland agriculture – Methodology and assessment. Research Bulletin, Central Research Institute for Dryland Agriculture, Hyderabad, India.

Zingore, S., Murwira, H.K., Delve, R.J. and Giller, K.E. (2007) Influence of nutrient management strategies on variability of soil fertility, crop yields and nutrient balances on smallholder farms in Zimbabwe. *Agriculture, Ecosystems and Environment* 119, 112–126.

Zotarelli, L., Alves, B.J.R., Urquiaga, S., Boddey, R.M. and Six, J. (2007) Impact of tillage and crop rotation on light fraction and intra-aggregate soil organic matter in two Oxisols. *Soil and Tillage Research* 95, 196–206.

6 Risk as a Determinant of Adoption of Conservation Agriculture by Smallholder Farmers in Malawi

Jessica Rust-Smith*

Hove, East Sussex, UK

6.1 Introduction

For many farmers throughout Africa, their traditional methods of farming are no longer sufficient. Agricultural growth in sub-Saharan Africa is slow and based mainly on expansion of the area farmed (World Bank, 2013). Maintaining the status quo and sticking with conventional farming methods will not ensure the food security of some farmer families, who are faced with the effects of climate change and marginal lands. Instead, sustainable farming methods are needed (FAO, 2014). An increasing number of development actors, including the World Bank, the Food and Agriculture Organization (FAO) of the United Nations, and numerous non-governmental organizations (NGOs) – ActionAid, Christian Aid, Concern Universal, and the Malawi-based, Total LandCare (TLC) – are advocating for and implementing conservation agriculture (CA) programs. Conservation agriculture is a form of agricultural practice that is intended to increase yields and be more sustainable than conventional forms of production techniques because of its application of ecological concepts and principles. Through the use of these programs, development aides hope to improve smallholder productivity in developing countries, to mitigate the effects of climate change, and to reduce environmental degradation.

*E-mail: j.rust.smith@gmail.com

What is Conservation Agriculture?

The three core principles of conservation agriculture are: minimum soil tillage, maximum soil cover, and crop rotations. These are alternative methods to conventional agriculture, which relies on tilling the land, hand weeding, mono-cropping, and burning off crop residue. Conservation agriculture can increase yields, saves labor, reduces a household's vulnerability to climate change, reduces erosion, and helps to control weeds, diseases and pests. Conservation agriculture also complements the use of fertilizer and herbicides, and over time reduces the need for these external inputs (TLC, 2012).

The potential yield increases in smallholder farmers' crops as a result of CA would be of particular significance in Malawi, a least developed country that is largely rural and highly dependent on rainfed agriculture. Malawi has a growing population, increasing land scarcity, and high levels of household food insecurity (TLC, 2012). An increase in crop resilience in the likely event of drought, and an increase in food production (both of which CA can provide), could significantly change the lives of these farmers for the better. Unfortunately, however, while CA could be quite advantageous in the region, smallholders face significant obstacles to transitioning from conventional farming methods to CA methods. These include farmers' perceptions that they do not have adequate knowledge to implement CA, or that the risks associated with experimenting with a new kind of technology are too high (Concern Universal, 2011). This research investigates this seeming paradox: why are the farmers (often vulnerable farmers with little education and very small landholdings) who would benefit the most from adopting CA often the least able to make the switch?

In order to understand the basis of getting farmers to employ CA techniques, it is necessary to understand the motivations for CA adoption – why do some farmers adopt and others do not? In order to best promote what international development organizations believe to be a solution to the looming food scarcity, it is essential to understand how best to promote its adoption by increasing farmers' incentives to adopt and decreasing obstacles to adoption. It is important to understand whether farmers consider CA a technology to mitigate risk *or* as a risky technology for which additional risk-mitigating measures must be offered to encourage adoption. If the former, promoters must stress the risks of conventional farming; and if the latter, promoters must work to buy down the risk of trying CA.

To identify gaps in our understanding of why farmers adopt CA, we have reviewed the existing literature. This review is divided into three parts. The first examines the literature that addresses constraints on farmers adopting new technology generally (for the purposes of this analysis, CA will be treated as a technology). Next, we look at constraints on farmers adopting CA specifically, followed by an examination of risk aversion as one of these constraints. We then theoretically assess the barriers that exist to CA adoption (lack of knowledge and/or capital, perceptions of difficulties in carrying out CA, and risk aversion that prevents experimentation), with a focus on risk. We use primary research involving a comparative analysis of adopters and non-adopters of CA in Malawi, conducted at the invitation of the NGO, Total LandCare. Finally, we will also offer ideas for future research and potential areas where adoption constraints can be reduced or overcome.

6.1.1 Constraints on technology adoption: An overview

In order to understand what affects a smallholder farmer's decision to adopt CA, first we must understand some of the factors that affect a smallholder farmer's decision to adopt *any* new technology. Generally, research on the adoption of new farming techniques has found that it is those farmers who are older, more educated, wealthier, and who know more farmers who have already adopted, who are most willing to adopt new technologies (Bandiera and Rasul, 2006; SADC, 2007). According to Bandiera and Rasul (2006), literate, older, and less vulnerable farmers are more likely to adopt a new crop – in this case, sunflowers in Mozambique. Adoption is related to the number of adopters in a farmer's social network. Adoption rates increase until the number of adopters in a social network reaches a certain threshold of members of that social network, at which point adoption rates decrease. Farmers are also more influenced by the adoption rate within their group of family and friends, less so by the adoption rate within their religious group, and not at all by the adoption behavior of people outside of these two groups. Furthermore, adoption rates might be lower if there is risk sharing within the social network (Bandiera and Rasul, 2006). In Malawi specifically, the Southern African Development Community (SADC, 2007) also cites poverty and illiteracy as limiting factors in farmers adopting new technologies.

6.1.2 Constraints on the adoption of conservation agriculture: An overview

Constraints on the adoption of CA are similar to constraints on the adoption of new agricultural technologies in general. Again, research suggests that older, more educated farmers are more likely to adopt. Furthermore, farmers belonging to farmer groups and those with larger landholdings are also more likely to adopt CA (FAO, 2011). For instance, farmers in Lesotho are more likely to become full-time CA farmers if they have more education and greater crop diversity (Wilcox *et al.*, 2012). In Malawi, older farmers are more likely to adopt CA, as are farmers who are members of farming groups (Ngwira *et al.*, 2014). Farmers with more access to credit are also more likely to adopt (FAO, 2011). Some constraints to adoption include farmers' insecurity of land tenure and the small size of landholdings (FAO, 2011). Additional constraints are farmers' inadequate technological know-how and a general lack of understanding of CA, and severe problems with access to markets (Concern Universal, 2011). Further adoption barriers include time discounting by farmers – this is when farmers consume in the present at the expense of future consumption – so these farmers face liquidity constraints (Concern Universal, 2011). Therefore, these farmers do not have cash when they need it to buy inputs such as fertilizer and herbicide, inputs that some organizations (including TLC) recommend for practicing CA. Thus, a farmer's lack of capital to purchase inputs is another barrier to adoption. This barrier might be overcome by giving farmers opportunities to diversify their sources of income (FAO, 2011): relying on multiple sources of income can mean a more regular cash flow for a farmer. Finally, risk aversion among farmers is also considered an adoption constraint (Concern Universal, 2011). A farmer who is generally unwilling to take risks will be unwilling to try his or her luck on a new kind of farming technology such as CA.

6.1.3 Risk as a technology adoption constraint

Risk is a key factor to investigate because if it does affect a smallholder farmer's decision-making process, those promoting CA can look at ways to "buy down" a farmer's risk to encourage adoption. Households with lower levels of wealth allocate assets in such a way so as to reduce their exposure to risk, and so they trade-off potential gains from a risky endeavor for lower returns (Carter and Barrett, 2007). The question of interest is: are these vulnerable households *less* willing to adopt conservation agriculture, because they are risk averse and therefore avoid potentially risky new technologies, or are they *more* willing to adopt CA because of the risk-mitigating potential of this new technology?

If a poor household's assets are eroded to a level below a certain threshold, referred to as the Micawber threshold, recovery to normal levels of production can be impossible, and the household may become stuck in a poverty trap (Carter and Barrett, 2007). An indication that a household is close to the Micawber threshold is if it engages in asset smoothing rather than consumption smoothing – that is, saving assets and reducing consumption rather than selling assets to maintain the same level of consumption (Carter and Barrett, 2007). In theory, vulnerable households could benefit from a productive social safety net to prevent them from slipping below an asset threshold (Carter and Barrett, 2007). Conservation agriculture has the potential to act as such a productive social safety net, to reduce a household's risks associated with climate change and other negative shocks, including illness and food price fluctuations.

However, despite the fact that the most vulnerable households might indeed benefit most from the risk-mitigating possibilities of CA, these households may be the least likely to adopt this new technology. Morduch (1995, cited in Cole *et al.*, 2010) writes that households self-insure against weather risk by reducing inputs, to avoid losses in the event of a poor harvest. Therefore, vulnerable households would be unlikely to want to try a new technology such as CA, which might involve the use of inputs including pesticides and herbicides.

Ngwira *et al.* (2013) analyze the risk level of economic returns to CA technologies, examining CA as a risk-mitigating technology. They hypothesize that, based on their findings, risk-averse farmers in Malawi would prefer CA. In Kenya, CA is preferred by risk-averse farmers (Guto *et al.*, 2011, cited in Ngwira *et al.*, 2013). While this gives credence to the idea that CA might be considered a form of insurance to reduce farming production risks in Malawi, detailed household surveys to assess levels of risk aversion and CA adoption tendencies among Malawian farmers would be required to confirm this hypothesis.

Despite existing research on the topic, there is not yet a definitive answer as to whether the adoption of CA increases or decreases with an increase in a farmer's level of risk aversion. If the perception of risk does indeed decrease CA adoption, one way to ameliorate this may be to employ risk-mitigating financial products such as microinsurance. In Section 6.3.3, we will explore whether these mitigating products can be used to increase the adoption of CA, or if, instead, the adoption of CA involves the same obstacles (such as a lack of knowledge) as the uptake of these products (e.g. microinsurance and indemnified loans). If so, these products will *not* be useful in encouraging CA adoption.

6.2 Methodology

We wanted to further our understanding of farmers' perceptions of CA vis-à-vis its riskiness, farmers' levels of risk aversion, and farmers' current knowledge of CA. To do this, we administered semi-structured interviews to 24 farmers in rural Malawi, 12 of whom were CA adopters, and 12 of whom were non-adopters. The relatively small sample size was due to focusing on more in-depth discussions with fewer farmers within the limited time allotted to this research. This small sample size does not allow us to extrapolate the causes of adoption or non-adoption in the population as a whole, but may be illustrative of farmers' motivations. An equal number of farmers were chosen from adopting and non-adopting groups to give equal weight to both in this analysis. The farmers selected to be interviewed were either adopters participating in a TLC study, or selected by TLC field coordinators, or non-adopters chosen by adopting neighbors. Non-adopters were chosen based simply on their proximity to their neighbor and their availability. To the extent possible, we tried to attain gender balance when selecting interviewees.

Interviews were conducted in three separate trips to two different districts: the environs of Nkhotakota town, in Nkhotakota district, on 11 and 12 July 2012 (seven adopters and seven non-adopters interviewed); the environs of Dowa town, in Dowa district, on 17 July 2012 (two adopters and two non-adopters interviewed); and the environs of Mvera town, also in Dowa district, on 20 July 2012 (three adopters and three non-adopters interviewed). Nkhotakota district has a population of about 301,000, the town has a population of about 33,150; Dowa district has a population of about 411,387, Dowa town has a population of about 6,176; and the greater Mvera area has a population of about 16,300. The NGO, Total LandCare, is already operating in these locations, and so the farmers are familiar with its work and with their staff members.

The interview questions were formulated to test the various theories put forward in the existing literature on the risks and constraints to the adoption of new agricultural technologies. The questions drew heavily from the work of Cole *et al.* (2008). The questions were intended to assess the levels of risk aversion that farmers experienced, their perceptions of the risk, and the actions they took to mitigate the risk. The questions were designed to collect basic statistics about farmers, such as adoption status, gender, education, and income levels. Additional questions about their perceptions of risk and their methods to mitigate risk began with short answer questions, which then precipitated follow-up questions based on the nature of their answer. Farmers were asked which type of farming – conventional or CA – they considered more risky and why. "Risk" was defined as a problem or obstacle, as there was not an exact equivalent to the English word "risk" in Chichewa. Questions were re-phrased if farmers were inclined to list only the advantages and disadvantages rather than the risks of the two different kinds of agriculture. Farmers were also asked what they thought the likelihood was of drought and flooding occurring in the upcoming growing season. They were given five options: 0% (no drought or flood); 25% (not likely); 50% (equally likely or unlikely); 75% (very likely); 100% (there will be drought or flood).

To determine a farmer's level of risk aversion, he or she was asked to choose one of two options if he or she were given 1,000 kwacha to invest (approximately US$4 at the time of asking). The first option (1) was a safe investment in which the farmer would always earn a 5,000 kwacha payout. Choosing this option indicated a farmer was *more* risk averse. The second option (2) was an investment in which the farmer had a 50% chance of getting 10,000 kwacha and a 50% chance of getting 500 kwacha. Choosing this option indicated a farmer was *less* risk averse.

Farmers were also asked what they did to prepare for potential negative agricultural shocks (*ex ante* risk mitigation techniques); and what they did to cope with a negative shock (*ex post* techniques). In both instances, farmers were free to give multiple answers, and responses were coded. In order to test levels of trust and if there was any relationship between a farmer's most trusted person or organization and a farmer's decision to adopt CA, farmers were asked who or what they trusted most to help them in times of need. There were also follow-up questions posed to ask how or why a farmer behaved in a certain way, resulting in more in-depth answers. However, the interview script was kept as short as possible, in recognition of the fact that farmers' time was valuable and that the need for translation would make the entire interviewing process lengthier. Interviews ran for approximately 30 min.

In addition, from 9 to 18 July 2012, interviews were conducted in the capital city of Lilongwe with numerous professionals involved in CA, including representatives from the World Bank, TLC, ActionAid, Christian Aid, National Smallholder Farmers' Association of Malawi (NASFAM), and the Chitedze Agricultural Research Station. The purpose of these interviews was to gain insight into the various methods being used by these organizations to promote CA.

6.3 Results and Discussion

Our results are derived from a very small sample size, and are not representative of the whole farming population of Malawi. Based on the statistic that 80% of Malawians are smallholders (USAID, 2014) and the populations of the districts visited for this research, as stated earlier, we would have needed to conduct between 535 and 599 interviews in *each* of the three regions in order for the findings to be representative of the whole population – to do so was infeasible.

In terms of sociodemographics for the sample group as compared to the wider population in the districts, we compare statistics from larger studies to the findings from our survey. In Malawi as a whole, the gross national income

(GNI) per capita is estimated at US$320 (World Bank, 2012). The poverty line is about 37,000 kwacha/person (approximately US$140 at the time of the survey), and 50.7% of Malawians fall below the poverty line. In Nkhotakota district, 32.1% of the population fall below the poverty line; and in Dowa district, 45.6% of the population fall below the poverty line (NSO, 2012). In our study, 28.6% of Nkhotakota farmers interviewed fell below the poverty line (3.5% less than in the Integrated Household Survey), and 40% of Dowa farmers interviewed fell below the poverty line (5.6% less than in the Integrated Household Survey).

In terms of education levels, in Malawi as a whole, 65.4% of the population is literate, and 21.1% have never attended school. In Nkhotakota district, 71.5% of the population is literate, and 15.1% have never attended school. In Dowa district, 70.3% of the population is literate, and 16.4% have never attended school (NSO, 2012). In our study, 14.3% of Nkhotakota farmers interviewed had never attended school (0.8% less than in the Integrated Household Survey), and 10% of Dowa farmers interviewed had never attended school (6.4% less than in the Integrated Household Survey).

6.3.1 Characteristics of non-adopters and adopters

Though extrapolation to the population at large is somewhat limited due to our small sample size, within the sampled group we were able to identify several characteristics of farmers who had adopted CA techniques. Specifically, adopters were significantly older on average than non-adopters (47.3 versus 36.1 years; $P < 0.05$), and they had a higher level of education than non-adopters (7.4 versus 5.2 years of schooling; $P = 0.1$). Previous studies have also identified more years of education as a covariate with adopting CA (Wilcox *et al.*, 2012), while illiteracy and poverty have been found to limit technology adoption in Malawi (SADC, 2008). While a greater percentage of CA adopters were male (75% versus 67%), this difference was not significant ($P > 0.1$), and this finding might be biased as we attempted to have a balance between the genders.

Related to years of education, results showed adopters were significantly more experienced farmers than non-adopters (Fig. 6.1). Adopters had an average of 20.9 years of farming experience, while non-adopters had an average of 15 years of experience ($P = 0.1$). One possible theory is that farmers with more experience are better able to assess the merits of CA, because they have spent more years learning the difference between bad and good farming practices: they recognize a good thing when they see it, after having been introduced to CA by a neighbor, or government or NGO extension agent. In addition, adopters tend to know significantly more fellow adopters than do non-adopters (23 versus 4 adopters on average; $P < 0.1$). Being exposed to more adopters may instil a belief in farmers that CA can be trusted. That is, because it is familiar to them they consider it less risky. In support of the above findings, TLC horticulture expert, Brand Mbale, said in an interview (2012) that in his experience adopters were more educated, experienced, and had traveled more than non-adopters. It is important to note, however, that the number of lead farmers with TLC ($n = 4$) who were interviewed as adopters in the survey may present a selection bias in some

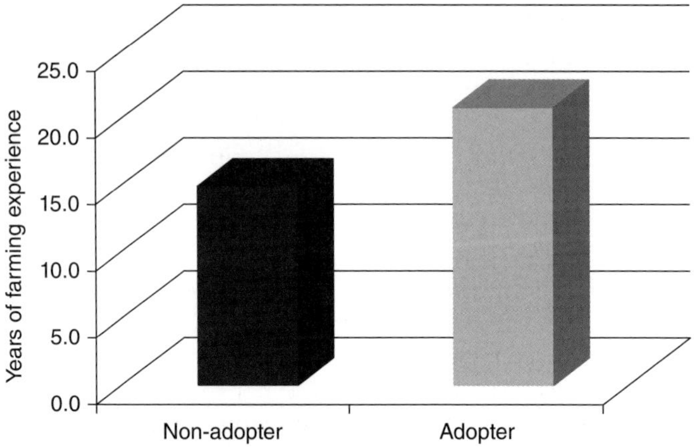

Fig. 6.1. Average number of years of farming experience of CA adopters and non-adopters.

of our results. Lead farmers know more adopters by virtue of their role, as they are responsible for contacting farmers and informing them about CA. This relationship also introduces the possibility of reverse causality, as those who adopt CA may subsequently come to know more fellow adopters, rather than farmers being influenced to adopt by knowing other adopters.

As an example of the differences between adopting and non-adopting farmers' farming practices, CA farmers we interviewed were significantly more likely to grow groundnuts (83% versus 50%; $P < 0.1$), a notably resilient crop, although there was no significant difference between the growing of maize or cassava between adopters and non-adopters. Traditional rainfed crops popular across the three locations in this study include maize, groundnuts, and soybeans. Other crops grown in Nkhotakota include cassava, rice, and some cotton (TLC is promoting growing cowpeas using CA), while traditional crops in Dowa and Mvera include tobacco. Furthermore, TLC is currently promoting CA-grown pigeon peas in Dowa.

We did not find significant differences, however, between adopters and non-adopters when looking at asset possession. Within the sampled group, adopters had larger landholdings (13.2 acres) than non-adopters (8.2 acres). Although this difference was not significant, it did seem to support the findings of the FAO (2011). It is possible that those with larger landholdings were better able to experiment with new farming methods such as CA. In fact, adopters were growing 39.5% of their total crops using CA techniques, suggesting that they were experimenting with this new technology, with few devoting more than 50% of their landholdings to CA. However, this 39% was still large, considering that the organization, Christian Aid, was encouraging adopters to start experimenting with CA on much smaller percentages of their land (S. Makoloma, Lilongwe, 2012, interview).

In addition, the findings that adopters had higher incomes and experienced less hunger were not statistically significant, and could be a result of "noise" in the data and the small sample size. Adopters earned an average of 85,000 kwacha/year, while non-adopters earned only 76,000 kwacha/year and were less likely to experience hunger than their non-adopting counterparts (0.8 months for adopters versus 1.3 months for non-adopters).

6.3.2 Income diversification and conservation agriculture

Based on anecdotal observation, it seems a farmer with different and diverse sources of income may be better able to spread risk, and therefore be in a better position to adopt a new farming technology. Participating in cash-generating activities such as small businesses can also increase their liquidity. This can enable adoption by allowing them to purchase inputs, particularly herbicides. Although adopters tended to have a greater variety of income sources (2.2) than non-adopters (1.7), the difference between the two groups was not statistically significant in our small study. There may be reverse causality at play, such that adoption leads to diversification. In fact, TLC (2012) states that the adoption of CA increases yields, which decreases labor requirements, allowing for farming diversification and an increase in non-farm activities. While the present data were insufficient to explore the direction of causality rigorously, the data did allow us to make some logical observations on the differences in diversification and risk aversion characteristics between adopters and non-adopters.

Consider, for example, the difference between adopter Gumbwa and non-adopter Kapanga. Gumbwa listed four different sources of income and was among the highest income earners interviewed (more than 40,000 kwacha/year). His most important source of income was his government pension, then his hardware store, and then earnings from agricultural trading and sales. Compare this to non-adopter M. Kapanga, a subsistence farmer and therefore in the lowest income bracket of interviewees. Earning no cash, she farmed only maize and cassava. It is possible that Gumbwa was in a position to adopt CA because he had a diverse "portfolio" of earnings and so felt confident to experiment and would be relatively unscathed if the experiment failed. The reverse, that adopting CA allowed him to diversify to other sources of income, was less likely – he had been practicing CA for 3 years only, and had a government job prior to this. In the case of Kapanga, she was typical of the most vulnerable farmers interviewed – female, with very few years of schooling (2), and very small landholdings (0.4 ha). If she were to have *any* source of cash income, let alone multiple sources including non-farm activities, she might have been able to attempt CA. Adopters have a greater reliance on non-farm activities (NFAs) – such as running a restaurant, as a primary source of income, when compared to non-adopters – none of whom gave a non-farm activity as a primary source of income (Fig. 6.2).

Forty two per cent (41.67%) of adopters and 58.33% of non-adopters interviewed relied on cash crops as their most important source of income, but the difference was not statistically significant. Theoretically, growing cash crops can inject much-needed liquidity into a household for the purchase of farm inputs, some of which, such as herbicides, are associated very strongly with CA. However, getting a good harvest from a cash crop is still weather dependent, and so in many ways is still riskier than reliance on a non-farm activity such as a restaurant. This diffusion of risk through NFAs may be very important for adoption to occur.

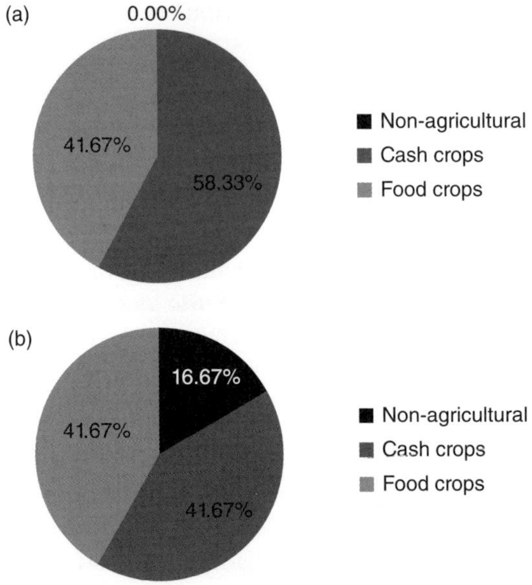

Fig. 6.2. Sources of primary income with (a) non-adopters and (b) adopters of CA.

Cash from an NFA can enable a household to buy inputs needed for CA. Only two out of the 12 adopters (16.67%) said their most important source of income was a non-farm activity (one of them was Gumbwa), but *none* of the non-adopters stated that a non-farm activity provided their most important source of income ($P = 0.16$).

Other studies have found that households with higher income find it cheaper to produce using CA, because it is labor saving (Ngwira *et al.*, 2014). In an interview, Ngwira (2012) noted that what adopting farmers saved on labor could be spent on herbicides. Adopters and non-adopters alike stated the problem of the expense of farm inputs and a lack of funds to buy these as a risk when engaging in both conventional *and* conservation agriculture. Liquidity was also mentioned as an obstacle to adoption by Ken Matekenya of ActionAid, which is part of a coalition promoting a no-/low-inputs form of CA (Ken Matekenya, Coordinator for Food, Nutrition and Human Security, ActionAid Malawi, 10 July 2012, interview, Lilongwe). In fact, Christian Aid (also a member of the aforementioned coalition) does not provide any inputs when promoting CA, based on the belief that the eventual withdrawal of inputs will lead to dis-adoption (Sophie Makoloma, Programme Manager, Christian Aid, 18 July 2012, interview, Lilongwe). Serving as a case in point, among the non-adopters there was one farmer who considered herself a dis-adopter, M. Kapanga (mentioned above). She had made preparations to adopt CA by laying down crop residue on her fields as mulch, but she could not proceed due to a lack of funds to purchase inputs, specifically chemical herbicides.

6.3.3 Risks and conservation agriculture

We found no correlation between adoption and risk aversion among the farmers interviewed. Our findings indicate adopters interviewed for this study are, on average (though not statistically significantly so), less risk averse than non-adopters (75% of adopters less risk averse versus 50% of non-adopters; Table 6.1). It is, of course, possible with such a small sample size that this finding is a result of data noise.

However, others *have* found a correlation between risk aversion and uptake of a risk-mitigating technology/product. The take-up of microinsurance in India was *lower* for risk-averse households (Giné *et al.*, 2010), and that seemed to be the case here – non-adopters were *more* risk averse. When considering CA, it is possible that risk-averse households may not adopt, even though the insurance CA could provide would *lower* their risk profile. Conversely, less risk-averse households seem more willing to "take a gamble" on a new kind of farming technology. This seems contrary to the preceding discussion of adopting farmers diversifying to reduce their risk, and therefore presumably being risk averse. However, perhaps because of the greater security that comes with diversified sources of income, adopters feel more comfortable with making risky decisions. Alternatively, there may be something inherent in these farmers who adopt CA that makes them willing to take a gamble on a new technology. The idea that farmers who adopt may be less risk averse is also contrary to the hypothesis of Ngwira *et al.* (2013), who suggest that CA will appeal to farmers who have greater risk aversion.

While 83% of all farmers regarded conventional agriculture as more risky than conservation, this was more evident for adopters (92%) than for non-adopters (75%) (Table 6.2). However, the difference between adopters' and non-adopters' perceptions was not statistically significant. This shared perception of increased risks associated with conventional agriculture is supported by the findings of Ngwira *et al.* (2013), which show that overall, CA yields are more reliable.

Because a majority of non-adopters recognized that conventional agriculture was more risky, this led to the question of why non-adopters, who were overall more risk averse (though not significantly), would not therefore adopt a set of farming technologies that they considered less risky. There may have been other obstacles, such as their lower levels of income diversification, and therefore less

Table 6.1. Risk aversion among non-adopters and adopters of CA. (From author's own research, 2012.)

	Non-adopter (12 respondents)	Adopter (12 respondents)	*P* value
Respondents who chose Option 2: More risk averse	50%	25%	0.22^{NS}
Respondents who chose Option 1: Less risk averse	50%	75%	0.22^{NS}
Total	100%	100%	

NS, not significant.

Table 6.2. Perceived riskiness of conventional agriculture versus conservation agriculture. (From author's own research, 2012.)

	Non-adopter (12 respondents)	Adopter (12 respondents)	All (24 respondents)	*P* value
Conventional is more risky than CA	75%	92%	83%	0.3^{NS}
CA is more risky than conventional	25%	8%	17%	0.3^{NS}
Total	100%	100%	100%	

NS, not significant.

liquidity for the purchase of fertilizer and herbicides, which may have prevented them from adopting CA.

Having established *which* of CA and conventional farming the farmers considered more risky, we next asked *what* the farmers found to be the biggest risks associated with CA and conventional farming. Adopters considered conventional agriculture more risky because of the lower yields and the amount of labor required (where there is a risk of not being able to fill the requirements in the event of illness or competing interests for a farmer's time). Non-adopters considered other risks more salient in the practice of conventional farming, specifically a lack of access to fertilizer, and poor rains (Fig. 6.3a). The adopters' concerns reflected the manner in which CA was promoted as a set of technologies that would increase yields and reduce the amount of labor required in comparison to conventional agriculture. The responses from adopting farmers suggest they may have internalized the advantages of CA as promoted by its proponents; or they may suggest that adopters really do value what CA can offer in contrast to conventional farming, and that CA proponents have hit on these advantages successfully when extolling the virtues of CA.

When asked about what risks they associated with CA, non-adopters cited erosion and the amount of labor required (they also cited labor constraints as a risk of conventional agriculture) in greater numbers than adopters (Fig. 6.3b). This is interesting because CA is promoted as a means to *reduce* the risks associated with losses in soil quality and labor required. In addition, more non-adopters than adopters said there were *no risks* associated with CA. This is contrary to what is expected. It may indicate an unrealistic perception of CA by non-adopters, or it may indicate that perceptions of risk have no bearing on adoption decisions and some other factor is at play. In addition, significantly far fewer non-adopters said there was a risk of lower rainfall, compared to adopters (8% versus 67%; $P < 0.05$), who were evidently more aware that, while CA might make them *more* drought resilient, it would not make them immune to the effects of drought.

Adopters and non-adopters were equally concerned with the expense and availability of inputs – herbicides, fertilizer, and seeds – for practicing CA. One input supply problem cited in interview responses from farmers was concerned with the government's Farm Input Subsidy Programme (FISP). Specifically, this included fears that farmers might not receive coupons for the subsidy every year, or inputs might arrive late; retailers might not have a certain brand of, or any,

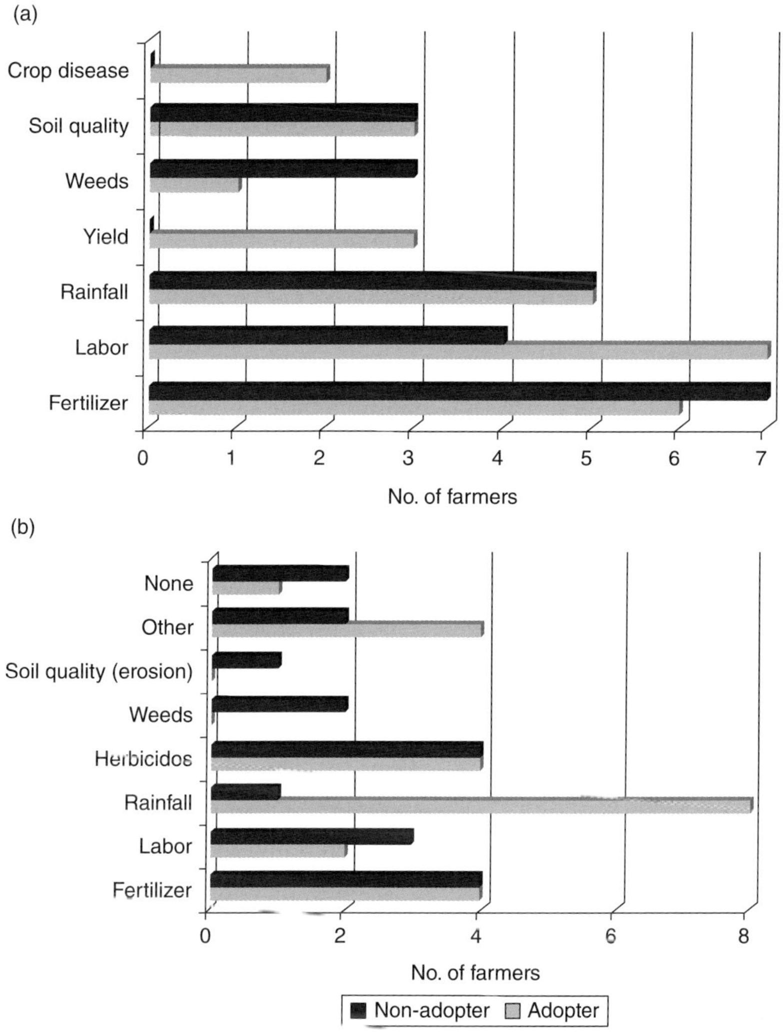

Fig. 6.3. Perceived risks of (a) conventional agriculture; (b) conservation agriculture.

herbicide available, or those that they did have might be expired or fake, or only available in small quantities. These responses are in keeping with the findings of Ngwira *et al.* (2014) – limited access to farm inputs is a constraint to adoption. This is interesting because while proponents of CA, including TLC, stress that it can be practiced with or without these inputs, farmers clearly associate CA with their usage. This may be because some CA proponents such as TLC tend to practice a form of CA that uses fertilizer and herbicides.

Farmers were also asked what they thought the likelihood was of drought and flooding occurring in the upcoming growing season. Adopters were more pessimistic about the likelihood of drought than non-adopters (45.8% versus 39.6%; Table 6.3) – though the difference between the two groups was not statistically significant ($P = 0.7$). When it came to flooding, there *was* a statistically significant difference

Table 6.3. Perceived probability of drought and flooding among non-adopters and adopters. (From author's own research, 2012.)

	Non-adopter (12 respondents)	Adopter (12 respondents)	All (24 respondents)	*P* value
Average of farmer responses to probability of drought	39.6%	45.8%	42.7%	0.7[NS]
Average of farmer responses to probability of flooding	50.0%	12.5%	31.25%	0.03*

NS, not significant; *, significant.

between adopters, who thought flooding was, on average, much less likely than did non-adopters (12.5% versus 50%; $P < 0.05$). Therefore, there was no correlation between adoption and the expectation of drought, but there was a negative correlation between adoption and the expectation of flooding. Conservation agriculture is promoted as a method of drought resilience by TLC (2012); and the FAO (2011) hopes to promote CA by doing the same. But if these farmers do not see CA as a way of ensuring against drought, these promotion attempts may be misguided. Alternatively, adopters may have adopted for other reasons, but have become more aware of the probability of drought and the effects of climate change because of the rhetoric used by CA promoters. The three farmers who mentioned climate change in reference to these questions were all adopters.

It should be noted that many farmers were unwilling to make a guess at the probability of drought, suggesting that they might not give consideration to how seasons would differ. Many farmers also reasoned that because the rains had been bad this season, they would be good in the coming season. Therefore, in subsequent interviews in Mvera, farmers were asked if they ever listened to and took heed of the government's meteorological forecast for the coming season. All but one farmer, a non-adopter, said that they did listen to the forecast and also planned accordingly. However, this does not fit with the seemingly superstitious phenomena of farmers believing that poor rains could not strike 2 years in a row.

In Malawi, there is one growing season – the rainy season between November and April – on which farmers depend for watering their rainfed crops. Rainfall risk is endemic in Malawi, and the 2004–2005 drought left 40% of smallholder farmers reliant on food aid (World Bank, 2008). An improper understanding of probabilities can prevent the uptake of a risk-mitigating device (Cole *et al.*, 2008). Therefore, it would seem that those farmers who fail to recognize persistent rainfall risk are also those less likely to adopt conservation agriculture.

6.3.4 *Ex ante* risk-mitigation techniques – is CA one of them?

Farmers use a variety of methods to prepare themselves in the event of a negative agricultural production shock such as drought or an illness that affects their ability to work. These methods can be divided into formal approaches, including the usage of financial products and services, and informal approaches, including the use of traditional institutions such as lending groups in faith-based organizations (FBOs) to cushion themselves against a possible financial blow.

In terms of formal mitigation measures, there is a suggestion that adopters are more "plugged-in" to financial services including savings accounts, formalized loans, and insurance than non-adopters (Fig. 6.4a); however, none of these findings are statistically significant. Two adopters had insurance versus zero non-adopters ($P = 0.16$); and though not statistically significant, five adopters had loans from an official source versus four non-adopters ($P = 0.68$); and seven adopters had savings accounts versus four non-adopters ($P = 0.24$). In the case of NGOs (excluding TLC), adopters were more involved than non-adopters, but not significantly so (three adopters versus two non-adopters; $P = 0.63$). While others suggest that farmers with better access to these formal services are better placed to adopt new farming methods (Ngwira *et al.*, 2014), it may also be the case that farmers who are more knowledgeable about such formal services also self-select into NGO programs such as TLC's CA program.

However, the finding that adopters were significantly more likely to be members of a farmers' group (five adopters versus one non-adopter; $P < 0.1$) matched the findings of Ngwira *et al.* (2014). Belonging to a TLC farmers' group, for example, enables adopters to borrow from a revolving fund that provides cash to purchase inputs. TLC offers one-time loans to adopters in which farmers pay 2,500 kwacha prior to harvest for inputs and an additional 2,500 kwacha after harvest (B. Mbale, Lilongwe, 2012, interview). In this case, though, adoption enables borrowing, not the reverse.

Of the 24 farmers interviewed, two CA-practicing tobacco farmers benefited from financial services provided by the tobacco industry, including home and life insurance and insured loans. The ties these farmers have to the tobacco sector are significant in determining their access to financial products, because the sector is conscientious about maintaining a strong tobacco supply chain. This access might have increased the farmers' interest in trying CA, because they were guaranteed more financial stability than most of the farmers interviewed. The loans were not contingent on the farmers practicing CA, but with the insured loan farmers theoretically could afford to take a chance on a new farming technique. Only one of the two farmers was growing his tobacco using CA techniques.

According to a NASFAM representative, banks and microfinance institutions are only willing to lend to farmers in the tobacco sector, which generates more reliable cash returns (K. Makiyoni, Lilongwe, 2012, interview). NASFAM piloted an index-based weather insurance program for farmers growing a food crop (groundnuts). Farmers were given insurance embedded in a loan for inputs, and this provided financial protection if a lack of rainfall affected their groundnut crop. However, the program did not proceed beyond the pilot phase, because other problems in the supply chain affected loan repayment rates and resulted in very little profit. The World Bank (2008) also considers tobacco as the only insurable agricultural sector in Malawi, because the supply chain is well established. CA promoters might learn from the tobacco industry by encouraging

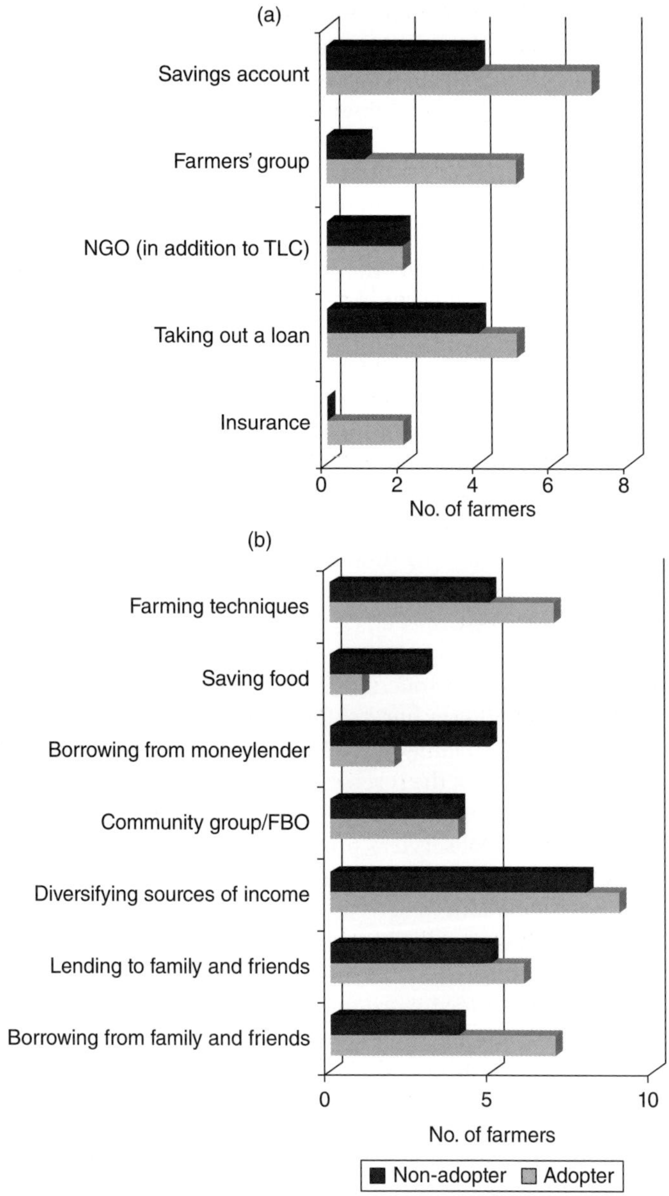

Fig. 6.4. Risk mitigation techniques. (a) Formal; (b) informal.

CA adoption using insured loans in well-established commodity sectors first, and then moving to food crops.

Moving from formal to informal methods to reduce risk, there were differences between the two groups, but most were not statistically significant. Non-adopters were more likely to store food (three non-adopters versus one adopter; $P = 0.3$;

Fig. 6.4b), which could be an informal risk-mitigation technique, although this difference was not statistically significant. Non-adopters were also more likely to borrow from a moneylender, but not significantly (five non-adopters versus two adopters; $P = 0.2$). The findings that adopters are more likely both to lend to family and friends (six adopters versus five non-adopters; $P = 0.2$) *and* borrow from family and friends (seven adopters versus four non-adopters; $P = 0.2$) were not significant. If significant, this might have suggested adopters had a stronger safety net within their communities. That is, adopters might have more and stronger connections with immediate and extended family members, neighbors, and friends whom they could rely on, and who in turn could rely on them. This might place adopters in a better position to try something new, such as CA, which might potentially be risky. However, this would contradict the prediction that risk sharing within a community could decrease the likelihood of adoption (Bandiera and Rasul, 2006).

Note: "Farming techniques" include planting cassava and other crops considered "safe", using more irrigation, planting more food in *dimba* (low-lying) gardens, and planting crops earlier.

6.3.5 *Ex post* risk-mitigation techniques

Farmers were asked what methods they used *ex post* to cope with a negative agricultural production shock such as drought or crop failure, to assess how adopters and non-adopters dealt with such shocks. Non-adopters were significantly more likely to plant cassava – a crop considered "safe" (42% versus 25% of adopters; $P < 0.1$) (Fig. 6.5). Non-adopters and adopters were equally likely to plant crops earlier. Only *one* adopter said that their response to a shock was to increase the amount of land under CA cultivation. However, adopters were significantly more likely to change the amount they were investing in their farms than non-adopters – 33% of adopters said they would invest more, as opposed to 0% of non-adopters ($P < 0.05$); 42% of adopters said they would invest less as opposed to 8% of non-adopters ($P < 0.1$). This suggests that adopters may be in a better position to use funds to compensate for a lack of farm productivity, whereas non-adopters are more reliant on changes in production that do not require money. However, it is a concern that more adopters would *reduce* their investment in their farms as compared to the number of adopters who would *increase* investment. Reducing investment is likely a form of self-insurance, reducing inputs to avoid losses (Morduch, 1995, cited in Cole *et al.*, 2010). This would seem to indicate that even though adopters have greater access to formal methods of risk mitigation (though not significantly more), they are still lacking some services such as weather or crop insurance that might prevent them from having to reduce their farm investments in times of negative production shocks.

Note: "Changes to farming methods" include planting cassava and other crops considered "safe", irrigating, planting more food in *dimba* (low-lying) gardens, planting earlier, and planting trees.

As for other informal methods of coping with shocks, non-adopters were more likely to consume less (usually meaning eating less) than adopters (75%

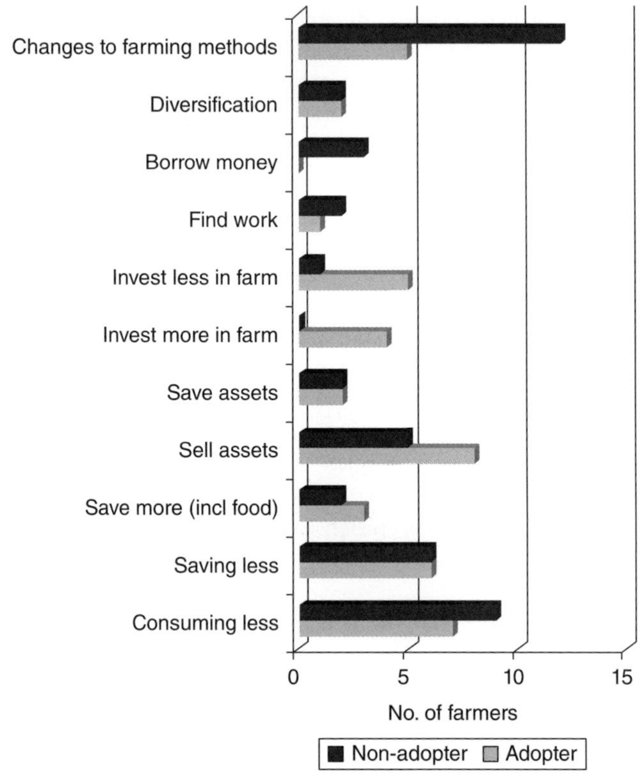

Fig. 6.5. Coping mechanisms.

versus 58%); and adopters were more likely to sell their assets than non-adopters when faced with a shock such as drought or flooding (67% versus 42%); however, neither of these findings were statistically significant. Questions about savings and consumption were intended to determine if farmers were above the Micawber threshold. Because these results were not significant, we could not conclude that non-adopters appeared to be asset smoothing: saving assets and reducing consumption rather than selling assets to maintain the same level of consumption (Carter and Barrett, 2007).

6.3.6 Individuals and entities farmers trust most

Adopters were significantly more likely to trust themselves most (25% versus 0% of non-adopters; $P < 0.1$; Fig. 6.6) to protect themselves against negative agricultural production shocks like drought. A greater trust in oneself might indicate that the farmers most likely to adopt have an inherent confidence in themselves

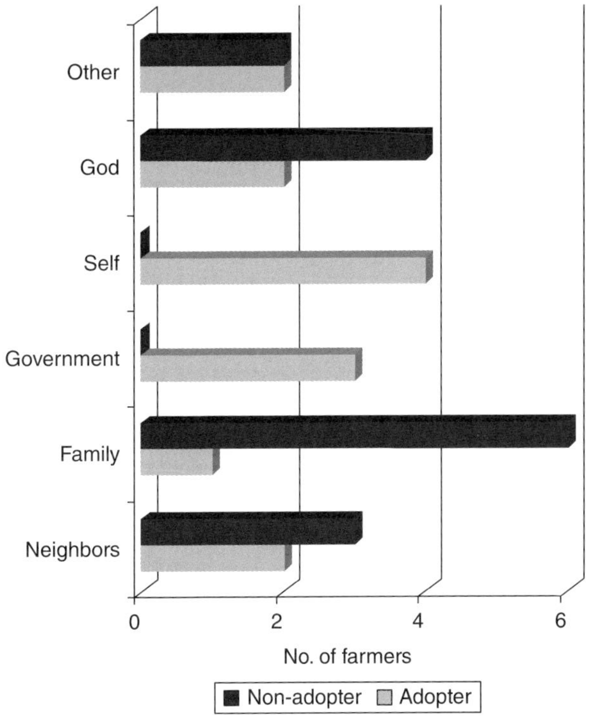

Fig. 6.6. Individuals/entities farmers trust most for protection against shocks (sum of farmers who chose each individual/entity as most trusted for protection).

and their skills, which non-adopters lack. This confidence is a characteristic that may make them more willing to try a new approach to farming. Adopters were also significantly more likely to trust the government most to protect them (25%, versus 0% of non-adopters; P <0.1). Farmers did not specify if “the government” meant extension officers or another branch or representative of government.

Non-adopters were significantly more likely to trust family most to protect them against shocks (50%, versus 8% of adopters; $P < 0.05$). The second most trusted entity among non-adopters was God (33%, versus 17% of adopters; $P = 0.4$); however, this difference was not statistically significant. Therefore, this finding does not necessarily mean that a greater trust in God precludes adoption. Christian Aid develops trust with communities, and encourages the adoption of CA through faith-based organizations (S. Makoloma, Lilongwe, 2012, interview); and in Kenya, CA is promoted through church groups (ASSS, 2008).

An overarching theme that emerges is that those farmers who would most benefit from CA – farmers with less income diversity, less liquidity, and more exposure to shocks – are also those who are less likely to have adopted CA. This suggests there is room for greater efforts to communicate the benefits of CA and a need for complementary activities to support adoption and build trust, such as insurance and/or agricultural microfinancing.

6.4 Conclusions

It is clear there are a myriad of obstacles to the adoption of CA, of which risk, and the perception of risk, are very significant constraints. Risk compromises farmers' abilities to adopt CA, because they may see it as a new and unfamiliar technology and so consider an attempt to try something new as too big a risk. They may see adoption as risky because they do not trust their own farming abilities, or they do not trust the proponents of this new technology. Because many farmers face serious liquidity constraints, they may see spending on inputs for CA as too big a risk to take because of the household's competing needs for that cash. Alternatively, farmers may not be willing to adopt CA because they lack awareness of the risks they face by not adopting it, in terms of protecting themselves against the effects of climate change – especially more unpredictable rainfall patterns, and the risks associated with labor (illness, etc.).

While risk may seem like a major obstacle to CA adoption, two important suggestions can be made to address this. The first is to target vulnerable groups by providing more information and by providing more opportunities to increase their liquidity. The second is to make adoption more attractive to *all* farmers, by reducing the risks associated with securing inputs and offering microinsurance packages in conjunction with CA start-up loans. In addition, there are steps CA proponents can take to promote trust between themselves and farmers, as a means of encouraging adoption.

6.4.1 Targeting more vulnerable farmers

The following suggestions are applicable to all farmers, but especially to "more vulnerable" farmers, who are defined as those with less education, lower earnings, and very significant liquidity constraints. Vulnerable farmers can be encouraged to adopt CA by providing them with targeted, salient information and education, and by providing means for them to increase their liquidity. One method of educating these vulnerable farmers would be an information campaign that raises awareness of the effects of climate change and informs farmers of CA facts and its benefits, as well as its downsides, and risks. Recommendations include the use of print media and simple technical guidelines (FAO, 2011), as well as intensification of extension services (Ngwira *et al.*, 2014). There is an opportunity to educate smallholder farmers about CA and its merits due to the fact that Malawian farmers are becoming more aware of the effects of climate change and the probability of drought, and so can be made to realize the mitigating effects of CA (FAO, 2011). Literature may be useful, but it is important to realize that many Malawian farmers – a great many of them women – are not literate, and so must be targeted through other means and methods. Other means of outreach could include: radio programs, which are used by NASFAM; visits by extension officers and lead farmers, also common practice for many NGOs and the government of Malawi; and community meetings, such as those used by Christian Aid.

Another method for getting information to farmers might be through agrodealers – retailers who sell agricultural inputs to farmers. If agrodealers were to be familiarized with CA, they could promote the approach to customers, as well as sell the appropriate inputs to farmers. CA-adopting farmers would then get an embedded service akin to extension services with their purchase. Offering embedded services can also build agrodealers' customer loyalty, an incentive for agrodealers to buy into this "market facilitation" approach. Finally, this approach would be more sustainable than an externally donor-funded information campaign.

Additional adoption barriers faced by more vulnerable farmers include liquidity constraints, such as an absence of cash on hand to buy essential inputs for CA (herbicides and fertilizers). Furthermore, it appears that the less liquidity a household has, the less willing it is to risk fluctuations in its income by changing its methods of production. Some organizations, like the FAO (2011), recommend breaking away from the inputs "dependency syndrome". They suggest providing cheaper options for CA inputs instead, with less emphasis on the use of chemical fertilizers and herbicides. While this would make switching to CA less costly in terms of cash, the risk in switching production methods is still high for a risk-averse household.

Another option to overcome CA adoption barriers is to offer grants for inputs. However, some CA proponents such as the FAO are concerned that this could translate into farmers adopting CA purely because they are chasing a grant. Cases of adverse selection might be reduced by offering one-time grants or subsidies to increase their liquidity in the long term, such as providing farmers with a small subsidy for CA inputs postharvest, when they have the most cash at their disposal. At this time of year, they are already most able to buy inputs, but might prioritize other household purchases because the next planting season is so far in the future (a "present bias"). Offering a time-sensitive subsidy might encourage farmers to buy inputs while they can. Such a subsidy ought to be withdrawn over time, so that behavior change lasts even in the absence of a financial incentive, however small (as in Duflo *et al.*, 2011).

There are also alternative livelihood options that enable farmers to raise the capital to purchase inputs. One farmer interviewed was a participant in the Farmer Income Diversification Programme (FIDP), implemented by the Government of Malawi. One possibility to introduce CA through a diversification scheme like the FIDP might be a partnership between organizations promoting CA and those organizations carrying out income diversification projects. As a result, a household willing to adopt CA can also benefit automatically from a diversification program. This could increase adoption rates, because it might attract farmers who were interested in income diversification projects but had not previously considered adopting CA. Such a partnership is also complementary to the idea of income diversification, because adopting CA also allows farmers to diversify within their fields, between crops grown using conventional methods and others grown using CA.

6.4.2 Targeting all farmers

All farmers, even those not deemed "vulnerable", face obstacles to adoption that involve risk, including difficulties with accessing inputs and limited access to financial services. A large percentage of farmers in the study cited difficulties in accessing farming inputs for purchase, including fertilizer and herbicides. In fact, 54% considered access to fertilizer a problem with conventional agriculture. If farmers are to adopt CA that involves herbicides, they must have confidence in the markets, and feel assured that herbicides of good quality and adequate quantity will be available when needed, which is not currently the case in Malawi. This problem might be overcome by partnerships between CA-promoting organizations and input producers and/or retailers. Perhaps in return for guaranteed availability of supplies of inputs for farmers, producers and retailers can benefit if their particular product is promoted among CA adopters, providing retailers with a solid consumer base.

Lastly, more farmers might be encouraged to adopt CA if there was the added benefit of increased access to financial services including credit and insurance. Ngwira *et al.* (2014) state that there needs to be enhanced availability of credit and loan facilities because these can increase a farmer's cash flow. In terms of loans, there are a number of microcredit options available; for example, TLC and government extension workers have assisted farmers in the formation of groups and the creation of a revolving fund. Farmers pay a deposit to the fund and receive a payout that enables them to pay for herbicides and maize seed. Currently in Malawi, a smallholder is likely only able to get insurance if he or she is involved in the tobacco industry. However, there could be the potential for a partnership between CA promoters and microfinance institutes and banks. Organizations promoting CA do not have the same resources as a financial institution. TLC, for example, does offer a loan to first-time adopters, but this loan is not insured. Thus, if a farmer defaults due to crop failure, he or she has lost the down payment. However, if a for-profit financial institution were to offer insured loans for CA adopters, more farmers might be inclined to make the move to CA. The insurance would offer the opportunity to access credit and try a new technology at a much lower level of risk. A national insurer, and an international reinsurer (an international financial body that insures a national insurance institute) might be attracted to the idea of partnering with a CA-promoting organization because clients (CA-adopting farmers) will already be engaged in a less risky method of farming, unlike non-adopting farmers practicing conventional agriculture. It is a much-coveted win–win situation.

6.4.3 Building trust

A final consideration for increasing the adoption rates of CA is to improve the trust between promoters of CA and would-be adopters. Christian Aid builds trust by partnering with both faith-based and secular organizations in communities, and involves the community in decision making. TLC uses lead farmers to foster peer-to-peer learning within communities, which involves the use of CA demonstration plots. Another method for ensuring CA information reaches farmers is to use trusted information channels, be they radio stations, politicians, etc. It is in the interest of all promoters of CA to investigate the diversity of approaches to building trust within communities, and therefore reduce the risks that farmers associate with this new technology.

The above suggestions for encouraging adoption are based on findings from Malawi only, and from a very small sample size, and require more research and subsequent implementation projects. Regardless, CA has huge potential for smallholders in Malawi and elsewhere in the developing world. Risk-related obstacles to adoption are present but not insurmountable. Careful research and incremental experimentation are called for to work towards increasing CA adoption rates.

References

ASSS (Africa Soil Science Society) (2008) Conservation agriculture in Lesotho: Current practices and prospects. A paper presented to the Africa Soil Science Society on the Promotion of Conservation Agriculture in Africa. Nakuru, Kenya, February 2008.

Bandiera, O. and Rasul, I. (2006) Social networks and technology adoption in northern Mozambique. *Economic Journal* 116, 862–902.

Carter, M.R. and Barrett, C.B. (2007) Asset thresholds and social protection: A 'think-piece'. *Institute of Development Studies Bulletin* 38.3, 34–38.

Cole, S., Tobacman, J. and Topalova, P. (2008) Weather insurance: Managing risk through an innovative retail derivative. Working Paper, Harvard Business School, Cambridge, Massachusetts; Oxford University, Oxford, UK; and the International Monetary Fund, Washington, DC.

Cole, S., Bastian, G., Giné, X., Oliver, J., Vyas, S., *et al.* (2010) The effectiveness of micro-insurance in helping small-holders manage weather-related risks. Department for International Development, UK.

Concern Universal (2011) Conservation agriculture research study 2011. Concern Universal Malawi, Blantyre, Malawi; Concern Universal UK, Hereford, UK.

Duflo, E., Kremer, M. and Robinson, J. (2011) Nudging farmers to use fertilizer: Theory and experimental evidence from Kenya. *American Economic Review* 101, 2350–2390.

FAO (Food and Agriculture Organization) (2011) Why is adoption of conservation agriculture still low among smallholder farmers in Malawi? The case for inhibiting food security and developing sustainable rural livelihoods projects. United Nations Food and Agriculture Organisation, Royal Norwegian Embassy, and Republic of Malawi Ministry of Agriculture and Food Security, Lilongwe, Malawi.

FAO (2014) A Statement by FAO Director-Geenral José Graziano da Silva, "2014 African Year of Agriculture and Food Security: from subsistence to success." Available at: http://www.fao.org/about/who-we-are/director-gen/faodg-statements/detail/en/c/213162/ (accessed 28 April 2014).

Giné, X., Menand, L., Townsend, R. and Vickery, J. (2010) Microinsurance: A case study of the Indian rainfall index insurance market. Policy Research Working Paper Series 5459, The World Bank.

NSO (National Statistical Office of Malawi) (2012) Third integrated household survey 2010–2011: Household socio-economics characteristics report. Commissioner of Statistics, National Statistical Office, Zomba, Malawi.

Ngwira, A., Thierfelder, C., Eash, N. and Lambert, D.M. (2013) Risk and maize-based cropping systems for smallholder Malawi farmers using conservation agriculture technologies. *Experimental Agriculture* 49(4), 483–503.

Ngwira, A., Johnsen, F.H., Aune, J.B., Mekuria, M. and Thierfelder, C. (2014) Adoption and extent of conservation agriculture practices among smallholder farmers in Malawi. *Journal of Soil and Water Conservation* 69(2), 107–119.

SADC (Southern African Development Community) (2008) Multi-country Agricultural Productivity Programme. 2007. Analysis of the Agricultural Technologies and Dissemination Situation in Malawi. SADC, Lilongwe, Malawi.

TLC (Total LandCare) (2012) TLC Booklet No 4: Guidelines for conservation agriculture in east and southern Africa. Total Land Care, Lilongwe, Malawi.

USAID (2014) Feed the future: Malawi country profile. Available at: http://www.feedthefuture.gov/country/malawi (accessed 16 April 2014).

Wilcox, M.D., Bisangwa, E., Lambert, D.M., Eash, N., Walker, F.R., *et al.* (2012) Smallholders, sustainability and food security: Conservation agriculture in a developing country context. Paper presented at the International Conference on Climate Change, Recycling of Agricultural Resources, Technology Improvement and Agriculture Management. Wuhan, China, 7–10 January 2012.

World Bank (2008) *The International Task Force on Commodity Risk Management in Developing Countries: Activities, Findings and the Way Forward.* World Bank, Washington, DC.

World Bank (2012) Malawi country data: World development indicators. Available at: http://data.worldbank.org/country/malawi (accessed 17 February 2014).

World Bank (2013) Fact sheet: The World Bank and agriculture in Africa. Available at: http://web.worldbank.org/WBSITE/EXTERNAL/COUNTRIES/AFRICAEXT/0,,contentMDK:21935583~pagePK:146736~piPK:146830~theSitePK:258644,00.html (accessed 28 April 2014).

7 Economic Potential of Conservation Agriculture Production Systems (CAPS) for Tribal Farmers in the Hill Region of Nepal

BIKASH PAUDEL,* CATHERINE CHAN, ALIZA PRADHAN AND BRINTON FOY REED

University of Hawai'i at Mānoa, Honolulu, Hawaii, USA

7.1 Introduction

7.1.1 Background

Hill farming systems, characterized by crop cultivation on sloping agricultural lands, provide food for millions of people worldwide. However, in recent years, conventional farming practices in Nepal's hilly areas, which provide food for about 43% of Nepal's population, have been forced to weather challenges such as population growth, deforestation, and climate change (Craswell *et al.*, 1997; Templeton and Scherr, 1999). At the same time, the region is facing increasing food demands and declining crop productivity. Unfortunately, expansion of agricultural lands is not generally feasible in Nepal's hill farming systems, where arable land is extremely scarce. Therefore, farmers have intensified production per unit area rather than expanding it (Hall *et al.*, 2001). In fact, although Nepal's total arable land area has increased by only 2% (WORLDSTAT, 2014), staple crops have increased by 16% for the 1991–2012 period. Maize and millet, the main crops from Nepal's hill region, have increased in area by 15% and 40%, respectively (FAOSTAT, 2014). Although agriculture intensification has contributed to increased food supply, at the same time, it has also increased soil degradation and the non-sustainability of Nepal's hill farming systems (Raut *et al.*, 2011).

*Corresponding author; e-mail: paudel@hawaii.edu

The rate of soil degradation on agricultural lands is very high in Nepal (MEST, 2006). The lack of sustainable soil management practices and haphazard intensification of farming systems are the major causes of high soil degradation (Gardner and Gerrard, 2003; Acharya *et al.*, 2007). High levels of soil degradation have not only reduced crop yields significantly (Tiwari *et al.*, 2004); they have also decreased soil nutrient stock, to the degree that it affects the stability of the hill agroecosystems (Manandhar *et al.*, 2009), ruining any chance for the region's sustainable development (Sharma *et al.*, 2007). Despite these problems, food demand is increasing steadily in the hill region, causing a high regional food deficit (MoAC/WFP/FAO, 2011). Purchasing food from outside is not a feasible option, due to physical challenges and affordability. Hence, achieving sustainable production from Nepal's hill farming systems must become a high priority for policy makers.

7.1.2 Overview of agriculture in Nepal

Nepal is a small country in the southern part of the great Himalayas, with an area of 147,181 km^2 and a population of about 30.4 million people (CBS, 2012). Nepal's land mass is divided into three geographic regions: mountain, hill, and *terai*. The mountainous region in the north covers about 33% of Nepal's total area (Fig. 7.1). The wide middle region, covering approximately 53% of the country's area, is referred to as the hill region (MoAC, 2011). The *terai* region, in the south, covers about 14% of Nepal (Chaudhary, 2000). In both mountain and hill regions, farming systems are characterized by crop cultivation on sloping lands (Fig. 7.1). With a few exceptions in and around valleys and river belts, sloping land agriculture is the only available option in these regions. Thus, it is estimated that over 63% of Nepal's agricultural lands are sloped (Pratap, 1999).

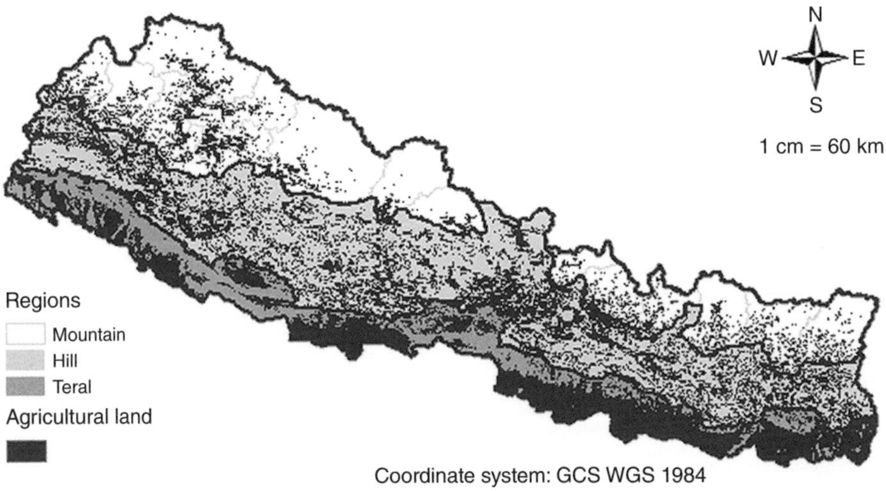

Fig. 7.1. Map of Nepal showing geophysical regions and distribution of agricultural lands. (Prepared by author; data sources GLC, 2000; GADM, 2012.)

Agricultural land in Nepal is generally categorized into three types: *khet*, *bari*, and *khoria*. *Khet* is the highest-quality land and is often irrigated. It is flat land and used for rice cultivation. *Bari* is upland, which slopes gently to a level plain, generally terraced and usually not irrigated. *Khoria* is the lowest-quality land, is often on a slope, and is never irrigated. Available agricultural lands in the mountain and hill region are less fertile than in the *terai*. Currently, per capita arable land in the hill region of Nepal is small, about 0.011 ha (CBS, 2011), which is half of the world's per capita arable land availability (0.02 ha) (WORLDSTAT, 2014). Given current low crop yields, the land in the hill region is insufficient to feed the people living in that region.

Disappointingly, previous efforts to increase crop yields in the hill region have been ineffective, as evidenced by persistently low crop yields in the region. There are many reasons for lower crop yields in the hill region, including high rates of soil degradation, use of traditional farming practices, lack of irrigation facilities, lack of access to high-yielding varieties, poor fertilization, and lack of proper plant protection, among others. Growth rates of crop yield in the hill region are slower than those of the *terai* region. In Nepal, from 1992 to 2013, rice and wheat (major crops in *terai*) yields increased by 54% and 85%, respectively, while maize yields (the hills' main crop) increased by about 41%, and millet yields (another main hills' crop) declined by about 5% (Fig. 7.2) (FAOSTAT, 2014). Despite low crop yields, people do not have alternatives to agriculture for

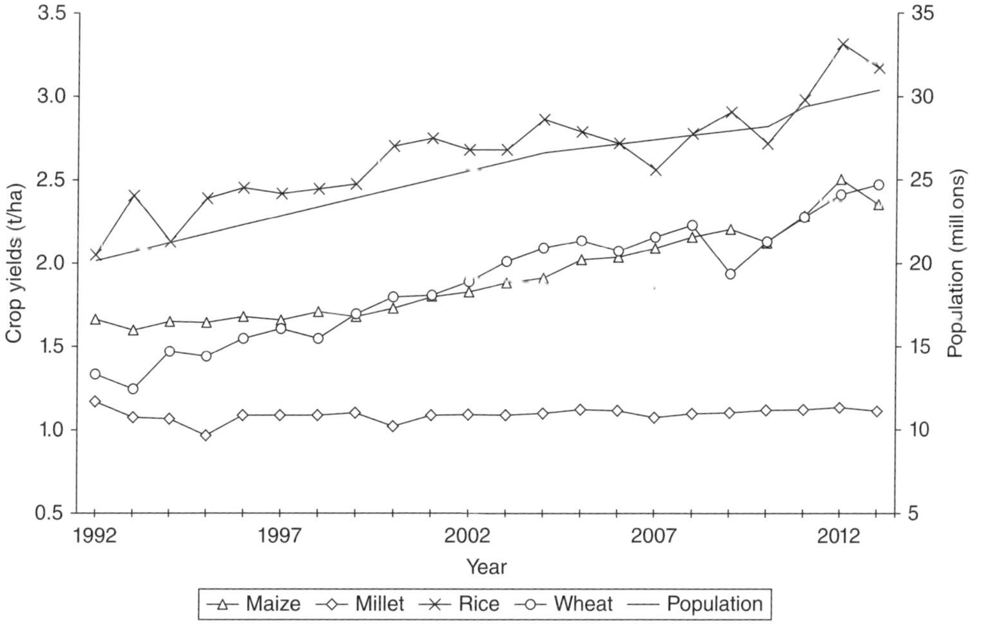

Fig. 7.2. Trend of crop yields and population growth in Nepal, 1992–2013. The crop yields has been shown in the primary *y*-axis (left), while population growth has been shown in the secondary *y*-axis (right). Both axes are in multiples of 7 from lower to higher value; hence, the scale of change is comparable. (From IMF, 2011; FAOSTAT, 2014.)

their livelihood. Therefore, lower crop yields have contributed to greater poverty, food insecurity, and malnutrition in the region (UNDP, 2009; MoAC/WFP/FAO, 2011). Thus, there is an immediate need to increase regional yields, while recognizing that this cannot be accomplished without addressing the issues directly related to the sustainable management of soil fertility. Studies have predicted that, if current soil management is continued, crop yields in the hill region of Nepal will decline in the future, due to decreasing soil fertility (Das and Beaur, 2012) and/or the effects of climate change (Malla, 2009). So, to achieve sustainable growth of crop yields, it is crucial to address the problem of rapid soil fertility decline (Tiwari *et al.*, 2004).

7.1.3 Causes of soil fertility decline in the hill region of Nepal

There are two main causes of soil fertility decline in the region, the first of which is farming system intensification, due to the increased frequency of cultivation. In Nepal's hill region, intensification has taken the form of increasing from one to two growing seasons per year. When the same piece of land is cultivated twice instead of once, the amount of tillage increases, making the soil susceptible to erosive forces that, due to the bare soil surface, destroy soil aggregates, and loosen soil particles. In addition, farming intensification has also reduced the soil's nutrient balance by increasing the cultivation of nutrient-exhaustive crops such as maize and millet. More than 40% of the summer-season maize crop is now followed by millet in the post-rainy season (MoAC, 2011), while most of these lands used to be fallow. When two nutrient-exhaustive crops are grown in sequence, this doubles nutrient uptake and depletes the soil nutrient balance. A second cause for decline of soil fertility is that farmers have reduced practices such as terracing, agroforestry, and shifting cultivation, all of which reduce soil erosion and help to maintain soil nutrition. Specifically, terraces can reduce the intensity of surface runoff drastically, which reduces the sheet and rill erosion in sloping land (Neupane and Thapa, 2001). Similarly, the traditional agroforestry system practiced throughout the hilly region of Nepal reduces soil erosion by stabilizing the terraces, and enhances soil nutrient balance by providing organic matter to the soil when the trees shed leaves, which then decompose in the winter (Amatya and Newman, 1993; Neupane *et al.*, 2002). Shifting cultivation (leaving land fallow under grass cover without physical disturbance for a few years following sequential seasons of cultivation) also reduces soil erosion, increases the biological activities in soil, reduces plant nutritional uptake, and increases the accumulation of soil organic matter (Kerkhoff and Sharma, 2006). Farmers used this practice to rehabilitate soil degraded due to continuous cultivation. However, because of higher food demand and limited access to fertile land, farmers can no longer practice shifting cultivation (Rasul and Thapa, 2003). With declining practice of traditional farming methods supporting sustainability of the hill production system, new methods and technologies are needed urgently to increase productivity under increasing intensification and higher food demand.

7.1.4 History of sustainable agricultural development in the hill region

Marginal, low productive areas such as those of the hill region's maize-based farming system do not get much agricultural development attention. Green Revolution technologies (such as growing high-yielding varieties and use of external inputs like fertilizers, irrigation, insecticides, and pesticides) promoted in the *terai* regions are simply not suitable for the hill region, due to the lack of irrigation, accessibility, and commercial inputs. Therefore, alternative development strategies have been suggested for hill region farming. Since 1995, the Agriculture Perspective Plan (APP) has tried to develop the hill region through promoting orchards and fruit gardens (APP, 1997). Improved agroforestry is another technology being promoted to enhance livelihoods and soil health. The International Centre for Integrated Mountain Development (ICIMOD) has piloted and promoted Sloping Agricultural Land Technology (SALT), a combination of practices such as terracing and hedgerow management to manage sloping agricultural lands for sustainable development of the hill region's agriculture (Maskey *et al.*, 2003). All these new technologies have elements of soil and water conservation, or sustainable management of soil nutrients, in common. However, farmer adoption of these technologies has been limited, partly because all would require farmers to replace field crops with orchard, forest, or animal husbandry. As an alternative that would support the continued cultivation of field crops, conservation agriculture production systems (CAPS) have been suggested (Atreya *et al.*, 2006; Begum *et al.*, 2010).

CAPS is a simultaneous practice of three conservation agricultural principles: minimizing soil disturbance by reduced tillage, managing year-round soil cover, and practicing optimum crop rotation (Kassam *et al.*, 2009). Worldwide, CAPS have shown the potential to increase crop yields (in the range of 15–30%) and improve soil quality simultaneously (Quinton and Catt, 2004; Thierfelder *et al.*, 2013; FAO, 2014). More locally, a few studies conducted in Nepal's hill region supported CAPS as a viable alternative, because under these systems, crop yields were maintained while reducing soil erosion (Atreya *et al.*, 2006). However, previous attempts, which have focused on crop yields only, or sometimes on soil characteristics, have not been sufficient, because either they have taken a single conservation agriculture practice or they have evaluated CAPS over a single year only, which is not long enough to realize the full benefits of CAPS. Additionally, previous studies have not paid due attention to the importance of the economic aspects of adoption.

7.1.5 Economic factors and adoption of CAPS

Despite the importance of geophysical and social factors, economic factors play a predominant role in the adoption of conservation technologies in poor and marginalized production environments such as in Nepal. There are many economic-related factors that might lead to non-adoption, including size of farm and availability of labor (McNamara *et al.*, 1991; Marenya and Barrett, 2007), timing and substitutability of the needed inputs (Neill and Lee, 2001), land tenureship (Soule *et al.*, 2000; Lee, 2005), ability to restore inputs (Smale *et al.*, 1994;

Uaiene *et al.*, 2009), accessibility of capital or credit (Makokha *et al.*, 1999; Jara-Rojas *et al.*, 2013), provisioning of government subsidies and support (Jara-Rojas *et al.*, 2013), and degree of the risk of adoption (Marra *et al.*, 2003). Most of these factors exist in Nepal and are thus very pertinent to understanding CAPS adoption. In fact, land and labor availability were more important than non-economic constraints such as technological know-how in Nepal's hill region (Floyd *et al.*, 2003). Similarly, decisions regarding the adoption of new varieties and fertilizers were affected significantly by credit availability, labor markets, agricultural extension services, and household labor endowments (Thapa, 2009).

Nepal's hill farmers will possibly face several economic constraints to CAPS adoption. Agricultural credit services are almost non-existent in Nepal. In addition, labor supply is unreliable, as there has been a huge out-migration of laborers in recent years, mainly from rural areas to find non-farm jobs (CBS, 2012). Furthermore, since the country's agriculture extension service is underfunded, it has limited capability to disseminate new technology (Shrestha, 2014). In order to understand how to remediate this situation best and encourage CAPS adoption in Nepal's hill region, it is very important to investigate how CAPS economics compares with the current farming system. To achieve this understanding, in this chapter we will: assess the socio-economic status and existing agriculture systems of Nepalese tribal farmers; estimate and compare the cost of production, labor requirements, and profitability of CAPS for these farmers; and determine, using a linear programming (LP) model, the best combination of CAPS and traditional systems to maximize profits for an average hill farmer.

7.2 Methods

The study populations were poor and marginalized Chepang farmers in Nepal's central hill region. Chepang farmers were deliberately selected for this analysis because they were one of Nepal's most marginalized and poor tribal communities. They live in the degraded lands throughout the central hill region, and have a total population of just above 100,000 people (CBS, 2012). These villages are geographically typical of villages in the hill region, though socio-economically more marginalized as compared to other regional villages.

7.2.1 Data and sources

We used two data sources to evaluate the economic returns from CAPS in this study: a socio-economic baseline survey and on-farm CAPS evaluation data. The first dataset includes information on household socio-economics, as well as existing farming and marketing practices. A socio-economic baseline survey was developed and 37 Chepang households were surveyed from three villages Thumka, Hyakrang and Kholagaum. Survey data were used to estimate the socio-economic conditions of the farmers, prevalence of cropping systems, baseline yields, and baseline labor requirements and labor availability, as well as the baseline crop yields, cost of production, and monthly labor requirements for the traditional system.

The on-farm CAPS performance data were used to determine the percentage change in crop yields, cost of production, and labor requirements from the traditional system to CAPS.

The on-farm CAPS evaluation trials were established in the three villages in the central part of Nepal's hill region from 2011 to 2014. Trials were held in the fields of eight to nine farmers from each of the three villages. In each study village, CAPS treatments were identified by focus group discussions attended by local researchers, development workers, and local farmers. The traditional treatment and three CAPS treatments identified by focus group discussion for the on-farm trials were: (i) maize followed by millet with full tillage (T1, the traditional practice); (ii) maize followed by legume with full tillage (T2); (iii) maize followed by millet and legume with full tillage (T3); and (iv) maize followed by millet and legume with strip tillage (T4) (Table 7.1). Cowpea was selected as the legume crop in 2011, but was replaced by black gram in 2012 because cowpea seemed to shade the millet (Table 7.1).

7.2.2 Research methods

To assess the socio-economic status, agriculture systems, and economic constraints of the Chepang farmers, average household size, education status, land availability, income and its sources, land allocations, and baseline crops yields were calculated via means and percentage values from the baseline survey. To evaluate the profitability of CAPS treatments, an enterprise budgeting technique was used. To determine the profitability of each CAPS treatment, the cost of production was calculated as the quantity of inputs multiplied by the price of inputs. Seed, fertilizer, insecticides, and pesticides were the main inputs. Labor requirement was calculated as the sum of labor hours required for practicing each CAPS treatment. Profit was calculated as crop yields multiplied by their price less total cost.

In this study, two types of profit were calculated: (i) "profit" (Π); and (ii) "profit before adjusting for labor cost" (Π_L). Historically, labor shortages did not exist in the study villages, as household labor was plentiful and traditional methods of workload sharing supplied sufficient labor for agricultural operations. As a result, hiring agricultural labor was uncommon, and farmers never counted labor in the

Table 7.1. Treatments for on-farm CAPS evaluation in Nepal.

	2011			2012		
	Crops and rotations			Crops and rotations		
Treatments	Spring–summer	Post-rainy	Tillage	Spring–summer	Post-rainy	Tillage
T1 (traditional)	Maize	Millet	Full	Maize	Millet	Full
T2	Maize	Cowpea	Full	Maize	Black gram	Full
T3	Maize	Millet/ cowpea	Full	Maize	Millet/black gram	Full
T4	Maize	Millet/ cowpea	Strip	Maize	Millet/black gram	Strip

cost of production. In recent years, however, higher out-migration rates and increased off-farm work opportunities, particularly for young males, have begun to cause labor shortages in villages. Therefore, profits were calculated for both scenarios. The formulas used for calculating Π and Π_L are presented below:

$$\Pi = \text{total revenue} - \text{total cost}$$

$$\Pi_L = \Pi + \text{labor cost}$$

$$\text{Labor cost} = \text{total labor (days)} \times \text{wage (US\$/day)}$$

Total revenue is the multiplication of individual crop yields by respective market prices. Crop yield per hectare was derived from on-farm trial plots, while market price was derived through a rapid market appraisal in 2011 in Muglin, which is the common local market for all the study villages. The labor opportunity cost was calculated by multiplying the total labor requirements by the market wage for farm labor in 2011. In order to find the optimal combination of CAPS treatments to maximize the average Chepang farmer's profits, an LP model was used. The mathematical formulation of the LP model for this study is presented below. The objective function was maximization of profit (Eqn 7.1), subject to a set of constraints. The first set of constraints (Eqn 7.2) regarded land constraints (two constraints in total, one per season). To account for multiple seasons in a year, land allocated to all crops under different production systems should not be higher than the available land for a typical farm in any single year. The second set of constraints (Eqn 7.3) dealt with labor (12 constraints in total, one per month). Total labor required for all crops under all systems in a month should not be higher than the available agricultural labor in a household on a monthly basis. The third set of constraints (Eqn 7.4) accounted for the cash constraints of typical farm households (two constraints in total, one per season), in that the total cash required to purchase agriculture inputs for a season should not be greater than the cash availability in the households. The fourth constraint (Eqn 7.5) allowed for the minimum consumption of millet per farm household (one constraint in total), which was required for household alcoholic production (millet is used in producing *raksi*, an integral part of Chepang culture). The fifth and final set of constraints (Eqn 7.6) controlled for crop rotation (three constraints, one for each cropping system). The crop rotation constraints ensured that the area allocated for one production system in the first season remained for the same system in the second season. Once all the constraints were accounted for, the LP model produced the optimal combination of CAPS treatments, which generated the highest profit in light of these constraints.

Mathematical formulation:

$$MAX\ \Pi = \sum_{i=1}^{4} \cdot \sum_{j=1}^{3} (A_{ij} \times Y_{ij} \times p_i - A_{ij} \times C_{ij}) \qquad (7.1)$$

where Π = profit; A_{ij} = area of crop *i* in *j* production system; Y_{ij} = yield of crop *i* in *j* production system; p_i = price of crop *i*; C_{ij} = cost of production of crop *i* in *j* production system; i = maize, millet, cowpea, or black gram; and j = sole cropping, intercropping with conventional tillage, or intercropping with minimum tillage.

Subject to

$$\sum_{i=1}^{4} \cdot \sum_{j=1}^{3} A_{ij} \leq A; \textit{for each cropping season} \tag{7.2}$$

$$\sum_{i=1}^{4} \cdot \sum_{j=1}^{3} A_{ij} \times L_{ijk} \leq L_k \tag{7.3}$$

$$\sum_{i=1}^{4} \cdot \sum_{j=1}^{3} A_{ij} \times C_{ijs} \leq M_s \tag{7.4}$$

$$\sum_{i=1}^{4} A_{ij} \times Y_{ij} \geq H_R \textit{ for millet} \tag{7.5}$$

$$\sum\nolimits_{s=1} A_{ij} - \sum\nolimits_{s=2} A_{ij} \geq 0 \tag{7.6}$$

where L_{ijk} = labor required for crop i in j production system in k month; L_k = monthly labor availability (hours); C_{ijk} = cash required for crop i under j production system in s season; M_s = available cash constraint; H_R = minimum household requirements (R) of millet; k = months (January to December); s = seasons (1) spring–summer, (2) post-rainy seasons.

7.3 Results and Discussion

7.3.1 Existing socio-economic status and agricultural systems of the Chepang farmers

Demography

Average family size in the study villages was 8.13 people per household (Table 7.2), which was higher than the national average of 4.88 (CBS, 2012). The average household size was highest in Thumka (9.6), while it was lowest in Kholagaun (5.75). The major reasons for larger family size in the villages were the existence of multigenerational families and lack of knowledge and facilities for family planning. Farmers do not use birth control, as an additional family member increases the household labor supply. The average number of agricultural laborers per household is between 2.6 and 3.6, with the lowest in Kholagaun.

About 80% of farmers in the villages were illiterate or nearly illiterate (Table 7.2), which was much higher than the national average of 34.1% (CBS, 2012). On average, about 10% of farmers had some secondary school education. Such extreme illiteracy among Chepang people is due to the lack of encouragement to attend school. There are primary schools in the villages, but parents expect their children to work in the house and on the farm from early childhood, and thus do not encourage children to go to school. However, there is an increasing trend of school attendance due to a change in the social mindset, and more schools being built in nearby villages. For that reason, the rate of literacy and the percentage of educated people would be expected to be higher among youths and children. There is a small difference in the education level of farmers in the three villages.

Despite having more mouths to feed, Chepang farmers have a smaller area of agricultural land to cultivate. In 2011, the average landholding of Chepang farmers was about 0.63 ha (Table 7.2), which was lower than the national average of 0.72 ha/household in 2008 (CBS, 2008). The average landholding in Kholagaun is lower (0.44 ha/household) compared to the other two villages (0.68 ha/household). Additionally, the majority of Chepang land was considered marginal for agriculture production because of the extreme slope and low soil quality. *Bari* and *khoria* land are very important to the Chepang people: 27% of land cultivated by Chepang is *khet*, 44% is *bari*, and about 28% is *khoria* (unpublished baseline data).

Though landholdings may be small and of marginal quality, agriculture still generates 60.6–69.8% of total cash income in Chepang villages, while the remainder comes from non-farm income sources (Table 7.3). The sale of cowpea and black

Table 7.2. Sociodemographic characteristics of the study villages.

Village	Thumka	Hyakrang	Kholagaun	Village average	National average
Average household size	9.6	7	5.75	8.13	4.88[a]
Level of education (% of total pop.)					
None	35.4	45.8	39.5	41.0	34.1[a]
Pre-school	45.0	32.2	34.9	39.0	
Primary	7.9	8.5	7.0	8.0	65.9[a]
Secondary+	11.1	8.5	9.3	10.0	
Non-formal	0.5	5.1	9.3	5.0	N/A
Average agricultural laborers/household (number of people)	3.6	2.8	2.6	3.1	2.42[b]
Average landholding (ha)	0.68	0.68	0.44	0.63	0.72[b]

[a]CBS, 2012.
[b]CBS, 2008.

Table 7.3. Average annual household cash income and sources in the Chepang tribal villages of Nepal.

Household income sources	Thumka	Hyakrang	Kholagaun
(a) Crop sales revenues			
(i) Cowpea (US$)	128	131	20
(ii) Black gram (US$)	31	7	176
(iii) Fresh vegetables (US$)	51		
(iv) Other crops (US$)	24		1
(b) Livestock sales (US$)	114	105	193
Total agriculture income (US$)	349	243	391
Percent of total income	69.8	60.6	66.9
(c) Total non-farm income (US$)	151	158	194
Total cash income per year (US$)	500	401	584

gram as well as livestock (especially goats and chickens) was the major source of cash income. Cowpea contributed more than one-third of households' agricultural income in Thumka and Hyakrang (US$128 and US$131, respectively), while black gram contributed more than half of the agricultural income in Kholagaun (US$176). Livestock sales contributed about 27% of the income in the study villages (US$349, US$243, and US$391, respectively, for Thumka, Hyakrang, and Kholagaun). Non-farm income sources included construction work in nearby town centers, agricultural labor in villages, and remittance from the foreign employment of household members. Whereas the income from agricultural sources was almost similar in the three villages, Kholagaun had the highest non-farm income generated from foreign employment remittances. The average cash income of the households (US$447/year) in Chepang villages was less than one-third of the average household income (US$1,445) in the hills of Nepal (CBS, 2011), which indicated that these villages were the poorest of the poor even by Nepalese standards.

Agricultural system

The allocation of spring-summer season crops in the Chepang villages was similar to Nepal as a whole, with maize covering 80.8% of the Chepang's total crop area, versus 60.3% maize coverage nationally (Table 7.4). However, the study site crop species in the post-rainy season differed from those seen throughout Nepal. Nationally, millet and black gram were the dominant post-rainy season crops, but cowpea was the dominant crop in the study villages. For example, in the study villages in 2010, about 56% of the upland maize production area was followed by cowpea in the post-rainy season, while about 30% was planted in black gram and only about 7% was planted in millet. These three villages were close to national highways and thus had easier access to markets compared to other villages in the hill region. Hence, farmers in these villages grew more cash crops such as cowpea and fresh vegetables compared to farmers in other hill villages.

In general, rice was the main crop in *khet*, while maize was the main crop for *bari* and *khoria* land. Maize, which was the main crop grown in the spring–summer

Table 7.4. Major crops and yields in the three study villages. ([a]From MoAC, 2011.)

	Percent of total area		Yield (t/ha)	
Crops	Study villages	National average[a]	Study villages	National average[a]
Spring–summer season (March–July)				
Maize	80.8	60.3%	2.17 (–10%)	2.41
Rice	12.0	NA	1.50 (–55%)	3.31
Post-rainy season (July–November)				
Finger millet	6.5	36.0	0.71 (–37%)	1.13
Cowpea	55.9	N/A	1.00 (+35%)	0.65
Black gram	29.9	11.0	0.76 (–6%)	0.81
Horse gram	2.6	0.7	0.49 (–34%)	0.74
Rice bean	4.3	N/A	0.65	N/A

Numbers in parentheses indicate the percent higher or lower than the national average.

season (March–July), was cultivated on more than 90% of *bari* and *khoria* land. The other major crops grown in the spring–summer season were upland rice, sesame, and groundnut. There was more crop diversity in the post-rainy season (July–November). Millet was the main post-rainy season crop for Nepal's hill region. Even though a large portion (>40%) of the maize area was left fallow in the post-rainy season, about 36% was grown in millet. Additionally, 11% and 14% of black gram and soybean, respectively, the most popular legume crops, were grown following the maize crop. Other legumes such as cowpea, horse gram, green gram, pea, ricebean, and groundnut were cultivated on about 10% of the total land (MoAC, 2011).

The yield of all the crops in the villages was generally lower than the national average (Table 7.4). Analysis of the baseline survey data indicated that in 2010 the maize yield was 2.17 t/ha (about 10% lower than the national average of 2.41 t/ha) and millet yield was 0.70 (37% lower than the national average of 1.13 t/ha). Black gram (0.76 t/ha) and horse gram (0.49 t/ha) yields were also 6% and 34% lower than the national averages of 0.81 and 0.74 t/ha, respectively. Although the average cowpea yield (1 t/ha) is higher than the estimated national average of cowpea (0.65 t/ha) reported by Shrestha *et al.* (2011), national production numbers in Nepal are misleading, as they do not record cowpea production and area separately from other legumes, but instead lump cowpea with "other legumes".

7.3.2 Yield, cost of production, and labor requirement for CAPS treatments

Crop yields under different CAPS treatments

Crop yield is the most important variable in terms of calculating profit. The on-farm CAPS evaluation plots established in the villages provided yield differences between crops grown under different CAPS treatments. Yields of different CAPS treatments were compared with the traditional system to gauge the percentage difference in the impact of CAPS treatments. The percent differences from CAPS evaluation plots were applied to baseline crop yields to derive the crop yields in farm scale. Table 7.5 shows the estimated crop yields, assuming constant return to scale from 2 years (2011 and 2012) on-farm trial data to farm scale.

Maize yields were higher for all CAPS treatments than for the traditional production system. The highest maize yield was possible from T2[a] (2.40 t/ha), followed by T2[b] (2.31 t/ha), with yield increased by 11% and 6% over the traditional system, respectively, due possibly to the nitrogen fixation capacity of the leguminous crops. The strip tillage treatments reduced maize yields by 8% over the traditional system, irrespective of the legume crops.

For millet yield in the post-rainy season, all CAPS treatments had lower yield than the traditional system, because millet was grown intercropped in the CAPS treatments, while it was a sole crop in the traditional system. Additionally, millet yield declined the most, by 51–59%, under millet–cowpea intercropping, while it declined less (22–26%) under millet–black gram intercropping. The higher decline of millet yield for millet–cowpea intercropping was because of the shading effect of cowpea on millet. Thus, it was evident that the adoption of any of these systems

Table 7.5. Maize, millet, cowpea, and black gram yields under different CAPS treatments.

Treatment code	Tillage Type	Season: Spring–summer	Season: Post-rainy[c]	Crop yield (t/ha): Maize		Millet		Cowpea[e]	Black gram[e]
T1 (traditional)	Full	Maize	Mi	2.17	(+0%)[d]	0.71	(+0%)[1]	–	–
T2[a]	Full	Maize	BG	2.40	(+11%)	0	(–100%)	–	0.76
T2[b]	Full	Maize	CP	2.31	(+6%)	0	(–100%)	1.00	–
T3[a]	Full	Maize	Mi/BG	2.21	(+2%)	0.55	(–22%)	–	0.37
T3[b]	Full	Maize	Mi/CP	2.26	(+4%)	0.35	(–51%)	0.75	–
T4[a]	Strip	Maize	Mi/BG	1.99	(–8%)	0.52	(–26%)	–	0.36
T4[b]	Strip	Maize	Mi/CP	1.99	(–8%)	0.29	(–59%)	0.68	–

[a]Black gram was used as the legume crop.
[b]Cowpea was used as the legume crop.
[c]Second season cropping pattern, where: Mi, millet; BG, black gram; CP, cowpea; /, intercropping.
[d]Percent change from the traditional system.
[e]All the yields of cowpea and black gram are additional over the traditional system because traditional systems typically do not have yields of these crops.

would reduce total millet production compared to the traditional system. However, since all the CAPS treatments also provided a novel, additional legume harvest, this more than compensated for the monetary loss resulting from decreased millet yields. Both cowpea and black gram yields were highest under sole cropping with full tillage (T2), and lower under intercropping. The black gram yield from T2[a] was 0.76 t/ha, which was reduced by 51% under T3[a] and by 55% under T4[a]. Similarly, cowpea yield from T2[b] was 1 t/ha, which was reduced by 25% and 32% under T3[b] and T4[b], respectively.

Labor requirement for different CAPS treatments

Labor requirements by CAPS treatments varied slightly for the Spring-Summer Season crop (for maize), but varied greatly for post-rainy season crops (millet, cowpea, black gram). For maize, the labor requirements for T2 were 225–229 person days versus 231–233 person days for T3 (Fig. 7.3). Neither was very different from the traditional system's labor requirement (226 person days), because all of these systems employed full tillage. However, the labor requirements of T4 (193 and 204 person days for T4[a] and T4[b], respectively) were lower than the traditional system, because they employed strip tillage. The strip tillage-based system (T4) required about 15% less labor than the traditional system. This reduction in labor required for T4 was due mainly to fewer labor hours required for land preparation and weeding of maize.

In general, during the post-rainy season, full tillage followed by legumes had lower and intercropping had higher labor requirements than the traditional system. Intercropping with legumes under full tillage had the highest labor requirement, with an increase of 33% and 63% person days for T3[a] and T3[b], respectively, over the traditional system. Sole cropping of legume crops required the least labor (20 and 38% less person days than the traditional system), because legumes did not require nursery management and transplanting as required by millet. Additionally, intercropping of millet with cowpea required 16–30% more

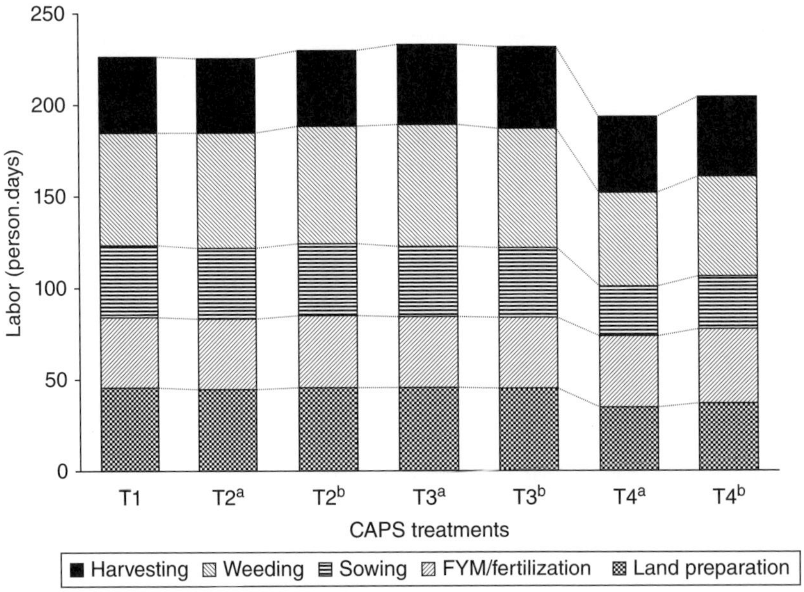

Fig. 7.3. Maize labor requirements under different CAPS treatments. Total labor requirements have been disaggregated by the agronomic operations of maize cultivation. [a]Denotes black gram was used as the legume crop; [b]denotes cowpea was used as the legume crop.

person days than intercropping with black gram, depending on the type of tillage. The highest labor requirement by function was transplanting, followed by harvesting, due to higher yields. Threshing reduced labor requirements substantially in the intercropping practice as legumes required less labor than millet for threshing (Table 7.6).

Labor requirements of CAPS treatments by months

Agricultural operations in the study villages were subject to monsoonal rain patterns, including a dry period during the post-rainy season. To maximize the advantage of the rainfall, Nepalese farmers try to grow as many crops as possible within the short monsoon season. Hence, the timing of the labor requirement is as important as the total labor requirement. Our results suggested that strip tillage systems (T4) required less labor at the beginning of the planting season (Fig. 7.4). However, treatments with intercropping (T3 and T4) had higher labor requirements than did the traditional system during the months of July, August, September, and November. Furthermore, the choice of specific legume crop affected the seasonal labor requirement greatly. All the cowpea-based systems (i.e. T2[b], T3[b], and T4[b]) demanded the most labor in August, while all black gram-based systems (i.e. T2[a], T3[a], and T4[a]) demanded the most labor in September.

These varying requirements have implications for the types of crops from which farmers will benefit most. CAPS treatments with black gram did not require as much labor in July, which was an advantage because during this time, farmers could earn additional income by being hired to work on *khet* land, with a relatively higher wage rate. In contrast, the higher August labor requirement

Table 7.6. Labor requirements for millet and legumes under CAPS treatments (person days/ha).

Crops	Farming operations	T1 (traditional)	T2[a]	T2[b]	T3[a]	T3[b]	T4[a]	T4[b]
Millet	Nursery	12.2			11.9	14.8	12.1	14.8
	Transplanting	56.7			46.0	60.6	52.1	59.5
	Weeding	15.6			15.5	16.4	17.0	15.8
	Harvesting	26.8			29.0	21.7	25.9	16.0
	Threshing	34.7			23.8	24.3	19.9	19.1
Legume	Sowing		28.7	22.1	8.5	25.4	8.6	17.7
	Weeding		15.6	17.2	15.6	11.5	15.6	13.8
	Harvesting		28.3	48.6	30.3	39.6	26.2	36.0
	Threshing		17.4	28.4	13.3	23.9	12.1	20.7
Total labor		146.0 (0%)	90.0 (–38%)	116.4 (–20%)	193.8 (+33%)	238.3 (+63%)	189.5 (+30%)	213.4 (+46%)

Numbers in parentheses show the percent change from traditional.
[a]Black gram was used as the legume crop.
[b]Cowpea was used as the legume crop.

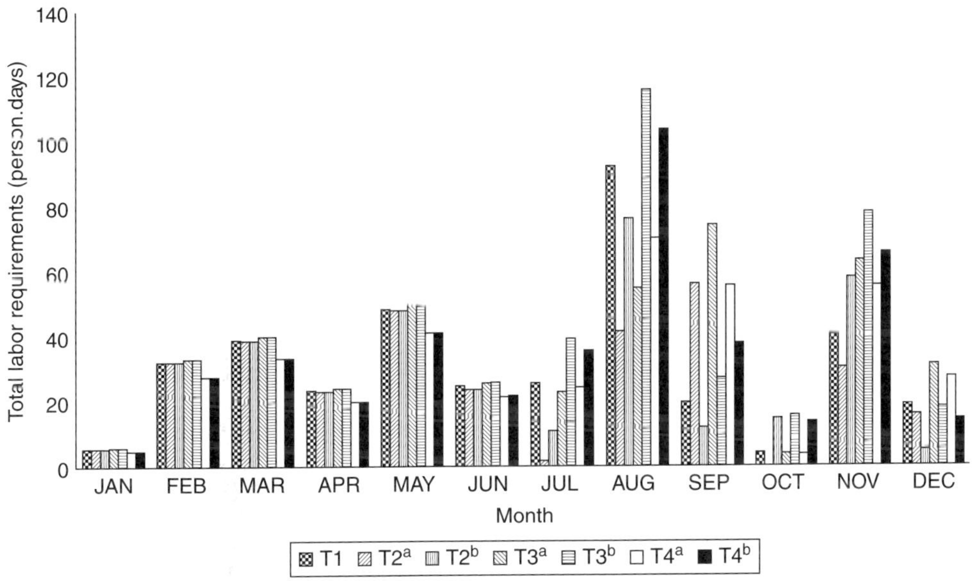

Fig. 7.4. Labor requirement of CAPS treatment by month. [a]Denotes black gram was used as the legume crop; [b]denotes cowpea was used as the legume crop.

of cowpea-based treatments (T3[b] and T4[b]) might present a problem, as labor constraints might preclude farmers from applying this treatment to all available lands during the limited period. Similarly, it might be difficult for farmers to cultivate black gram on all available lands, because of its higher September labor requirements. However, if both the black gram- and cowpea-based systems were adopted, the increased labor requirement due to intercropping would be spread over the 2 months (with cowpea requiring more labor in August and black gram requiring more in September), making it more possible to use family labor solely to cultivate the crops.

This factor was quite critical, as labor in general was an important variable cost affecting CAPS profitability. Most of the agricultural inputs (except labor) were fairly similar among all the different production systems. Although labor might not be very important to Chepang farmers, because they generally fulfilled demand by using the household labor available, the timing of when labor was needed was quite important for farm labor budgeting. It will be difficult to encourage farmers to adopt technology that requires more labor during July–September, which are the busiest months for Nepalese hill farmers because during this period spring–summer crops are being harvested and post-rainy season crops are being planted (Govinda Chepang, Hyakrang, 2013, personal communication). If T4 (strip tillage with intercropping) demanded labor earlier or later than the peak time, there would be more chance of adoption.

Production costs, total revenue and profit for CAPS treatments

There was very little variation in input costs (excluding labor) for all production systems. Costs ranged from US\$203/ha to US\$210/ha (Table 7.7). However, after adding the cost of labor, the total production cost varied greatly among the treatments, with the highest incurred for intercropping T3[b] (US\$1,442/ha), followed by T3[a] (US\$1,333/ha). The lowest production costs were seen in T2a (US\$1,038/ha), followed by T2b (US\$1,115/ha). Input costs included the cost of seeds, fertilizer, insecticides, and pesticides. In reality, most farmers do not purchase seed, unless farm-saved seed is damaged by pests or consumed for household use. However, for this analysis, the seed cost was included, based on its market price, to account for instances when farmers might need to purchase seed. The price of chemical fertilizers and insecticides were included as the prices the farmers paid in the nearby market. Farmers do not use farm machinery on sloping land, so

Table 7.7. Yearly cost of inputs, labor cost, revenue and profits of CAPS treatments in Nepal's central hill region.

Production system	Cost of inputs (US\$/ha)	Labor (person days/ha)	Labor cost (US\$/ha)	Total production cost (US\$/ha)	Revenue (US\$/ha)	Profit (Π_L)[c] (US\$/ha)	Profit (Π)[d] (US\$/ha)
T1 (Traditional)	203	373	982	1,185	1,314	1,111 (base)	129 (base)
T2[a]	209	315	829	1,038	2,248	2,039 (+84%)	1,210 (×9.3)
T2[b]	203	346	912	1,115	2,190	1,987 (+79%)	1,075 (×8.3)
T3[a]	210	427	1,123	1,333	1,821	1,611 (+45%)	488 (×3.7)
T3[b]	207	469	1,235	1,442	2,040	1,833 (+65%)	598 (×4.6)
T4[a]	210	383	1,007	1,217	1,822	1,612 (+45%)	605 (×4.6)
T4[b]	207	417	1,098	1,305	1,934	1,727 (+55%)	629 (×4.8)

[a]Black gram was used as the legume crop.
[b]Cowpea was used as the legume crop.
[c]Figures in parentheses show the percent change compared to the traditional system.
[d]Figures in parentheses show the multiple of the traditional system.

there is no petroleum or other machine-related costs. Labor requirements were also generally fulfilled through household labor, thereby incurring few actual costs.

In terms of total revenue, all CAPS treatments generated higher revenue than the traditional system. Maize followed by legume sole crop systems (T2[a] and T2[b]) generated the highest revenue (US$2,190–US$2,248/ha), due mainly to the higher market prices of black gram and cowpea compared to millet. The revenue from strip tillage systems is among the lowest revenues of all CAPS, but still higher than the traditional system.

Since there was not a lot of variation of input costs among the treatments, the profits before adjusting for labor cost were based mainly on the yield and price of the crops. Although there were variations between treatments in labor requirements, profits including and excluding labor costs did not have different trends by treatments. The lowest profits were from the traditional system. The highest profits came from treatments T2[a] and T2[b], which were about 80% higher not including labor costs, and over eight times higher after including labor costs, than the traditional farming practice. Profits using strip tillage systems were about 45–55% higher not including labor cost, and over five times higher including labor cost.

7.3.3 Profit maximization with CAPS

Using LP model results, the optimal allocation to maximize profit for a typical Chepang farmer with land and labor constraints and consumption needs was to place 28% of his land into strip tillage maize followed by intercropped millet and black gram (T4[b]), 63% of his land into full tillage maize followed by black gram (T2[a]), and 9% into the traditional system (T1) (Fig. 7.5). For a typical farm with land area

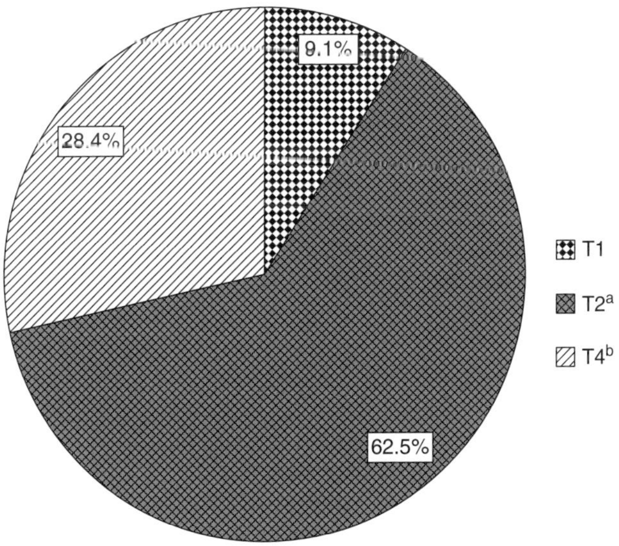

Fig. 7.5. Optimum combinations of CAPS treatments to maximize profit for a typical Chepang household. [a]Denotes black gram was used as the legume crop; [b]denotes cowpea was used as the legume crop.

of 4,670 m^2, this optimal treatment with associated constraints generated profit (after adjusting for labor cost) of US$489/farm/year (Table 7.8). The optimal land allocation provided about 88% higher profit than the traditional system. However, this profit was lower than those from the T2 systems. The reason that optimal allocation did not generate the highest profit was because of the millet consumption needs and peak season labor requirements. Practicing T2[a] generated about 16% higher profit, but it was not possible to practice all lands under T2[a] because of the August–September labor constraints and household consumption requirements for millet. Some labor would have to be hired to implement T2[b], which would reduce profit. Labor hiring is not a practice that currently exists for the farmers in the study sites.

Thus, an integration of strip tillage-based CAPS in combination with full tillage systems increased the annual profit of a typical Chepang household, compared to practicing the traditional maize–millet system. Therefore, it seems illogical to continue practicing the traditional maize–millet system (which still dominates the maize-based system covering about 36% of land) in the hill region of Nepal. However, currently the majority of the farmers of the study villages do not have sufficient economic incentive to adopt CAPS further, because they are already practicing a maize–legume system on a large percentage of their land. The baseline data of farms currently growing legumes and millet showed that the land allocation was about 56% for maize–cowpea, 30% for maize–black gram, and 7% for maize–millet, which generated an annual profit of US$454/farm/year from the land available. When compared to profits from the model-determined optimal mix of production systems (US$489), profit maximization output increased the profit by a mere US$39/household. Thus, if farmers were already cultivating legumes, adding strip tillage would not really increase profits in 1 year. However, as crop yields under strip tillage are expected to increase after a few years, while those under full tillage are expected to decrease, a longer-term planning analysis might show higher economic incentives for adopting CAPS treatments. Furthermore, as the traditional maize–millet system continues to dominate hill agriculture at the regional and national levels, it still seems that the average farmer can increase their income by changing to a CA-based system.

Table 7.8. Profits for a typical farm from optimum land allocation from an LP model with constraints compared with profits from other systems assuming no constraints.

Production system	Profit (Π) (US$/year)	Percent lower/higher compared to optimum allocation
T1 (Traditional)	60	(–87.7)
T2[a]	565	(+15.6)
T2[b]	502	(+2.7)
T3[a]	228	(–53.4)
T3[b]	279	(–42.9)
T4[a]	283	(–42.2)
T4[b]	294	(–39.9)
Optimum allocation (28% T4[b], 63% T2[a], 9% traditional)	489	0.0

[a]Black gram was used as the legume crop.
[b]Cowpea was used as the legume crop.

7.4 Conclusion

Sustainable intensification technologies such as CAPS can resolve the concurrent problems of low crop yields and high soil erosion in sloping areas of the hill region of Nepal. We conclude that CAPS with strip tillage is more profitable than the traditional farming system, which currently covers more than 36% of Nepal's hill region. Additionally, CAPS practices are compatible with the land and labor constraints and the consumption needs of a typical Chepang farmer. However, despite the economic feasibility of CAPS, its adoption faces many challenges, now and in the future.

One of the first challenges we are faced with is that the economic gains during the initial years after adoption are not sufficiently high to attract farmers. Other systems had a higher profitability than strip tillage CAPS, due, for instance, to the lower yield of high-value legume crops and inefficient use of labor. In addition, subsistence farmers prioritize fulfilling consumption needs, rather than practicing the most profitable and productive system. Reluctance among subsistence farmers to hire farm labor can also pose a challenge to CAPS adoption, because some CAPS require more labor due to intercropping or mixed cropping. This labor problem may increase in the future, due to increasing out-migration and to more children, who formerly provided the farm labor, attending schools. A final but perhaps most pressing challenge to CAPS adoption in Nepal's hill region is climate change. Climate uncertainties have become a pressing concern for regional farmers; In order to be attractive to hill farmers, CAPS must present higher climate resilience that allows for adaptation to future climate scenarios.

In order to ameliorate these challenges to CAPS adoption and find tenable, proactive solutions, we need to focus strategically on agricultural research, extension, and education promoting CAPS among Nepal's smallholder farmers. It is important to verify the longer-term economic gains of CAPS through evaluations covering a longer study period. Ongoing and future CAPS evaluation projects should be of sufficient tenure to overcome initial lower-yield periods. At the same time, farmers should be educated on the benefits of CAPS and the somewhat delayed realization of maximum benefits, which can be up to 4–5 years after adoption. To offset this initial lack of economic gain, the government could provide food and input subsidies to the farmers until maximum CAPS benefits are attained and sustained. Furthermore, the government should also improve market infrastructure and increase access to price information, so that farmers can sell their surplus at the highest price with low production costs. Finally, since resolving the labor problem is challenging, different options should be considered. One such option is to explore other legume crops with different timing of labor requirements. Therefore, identification of suitable crops (legumes and cereals) should be the priority for future research. Labor requirements for CAPS can also be reduced through the use of machinery (e.g. direct seed planters) or chemicals (herbicides). Though the option of expensive machinery or chemicals seems impractical given the economic condition of Nepal's smallholder farmers, the possibility of using locally made, cheap hand tools should be explored. At the same time, considering the rapid growth of the opportunity cost of labor, future research should explore the feasibility of promoting small machinery and chemical weed control. In short, even though the potential short-term gains for the adoption of CAPS by smallholder farmers seem small, governments should invest in research, education, and extension to promote CAPS for achieving greater societal welfare through the sustainable development of hill farming systems.

References

Acharya, G., McDonald, M., Tripathi, B., Gardner, R. and Mawdesley, K. (2007) Nutrient losses from rain-fed bench terraced cultivation systems in high rainfall areas of the mid-hills of Nepal. *Land Degradation and Development* 18, 486–499.

Amatya, S. and Newman, S. (1993) Agroforestry in Nepal: Research and practice. *Agroforestry Systems* 21(3), 215–222.

APP (Agriculture Perspective Plan) (1997) Agriculture Perspective Plan of Nepal 1995–2015. Ministry of Agriculture and Cooperatives, Government of Nepal, Kathmandu.

Atreya, K., Sharma, S., Bajracharya, R.M. and Rajbhandari, N.P. (2006) Applications of reduced tillage in hills of central Nepal. *Soil and Tillage Research* 88, 16–29.

Begum, F., Bajracharya, R.M., Sharma, S. and Staula, B.K. (2010) Influence of slope aspect on soil physic-chemical and biological properties in the mid hills of central Nepal. *International Journal of Sustainable Development and World Ecology* 17(5), 438–443.

CBS (Central Bureau of Statistics) (2008) Report of National Labor Force Survey, 2008. CBS, National Planning Commission, Government of Nepal, Kathmandu.

CBS (2011) Nepal Living Standards Survey 2010/11. Statistical Report Volume 2. CBS, Nepal Planning Commission, Government of Nepal, Kathmandu. Available at: http://cbs.gov.np/wp-content/uploads/2012/02/Statistical_Report_Vol2.pdf (accessed 3 November 2014).

CBS (2012) National Population Census. CBS, Nepal Planning Commission, Government of Nepal, Kathmandu.

Chaudhary, R.P. (2000) Forest conservation and environmental management in Nepal: A review. *Biodiversity and Conservation* 9, 1235–1260.

Craswell, E.T., Sajjapongse, A., Howlett, D.J.B. and Dowling, A.J. (1997) Agroforestry in the management of sloping lands in Asia and the Pacific. *Agroforestry Systems* 38(1–3), 121–137.

Das, R. and Bauer, S. (2012) Bio-economic analysis of soil conservation technologies in the mid-hill region of Nepal. *Soil and Tillage Research* 121, 38–48.

FAO (Food and Agriculture Organization) (2014) What is Conservation Agriculture? FAO. Available at http://www.fao.org/ag/ca/1a.html (accessed 20 March 2014).

FAOSTAT (FAO Statistics Division) (2014) Production. Available at: http://faostat.fao.org/site/339/default.aspx (accessed 20 March 2014).

Floyd, C., Harding, A.H., Paudel, K.C., Rasali, D.P., Subedi, K., *et al.* (2003) Household adoption and the associated impact of multiple agricultural technologies in the western hills of Nepal. *Agricultural Systems* 76(2), 715–738.

GADM (Global Administrative Areas) (2012) GADM (version 2). Database of Global Administrative Areas. Available at: http://www.diva-gis.org/gdata (accessed 3 November 2014).

Gardner, R. and Gerrard, A. (2003) Runoff and soil erosion on cultivated rainfed terraces in the Middle Hills of Nepal. *Applied Geography* 23, 23–45.

GLC (Global Land Cover) (2000) Global Land Cover, 2000. Available at: http://www.diva-gis.org/gdata (accessed 3 November 2014).

Hall, M., Dixon J., Gullivar, A. and Gibbon, D. (2001) *Farming Systems and Poverty. Improving Farmers' Livelihoods in a Changing World.* Food and Agriculture Organization of the United Nations, Rome.

IMF (International Monetary Fund) (2011) World Economic Outlook (WEO): Slowing Growth, Rising Risks. IMF. Available at: http://www.imf.org/external/pubs/ft/weo/2011/02/ (accessed 3 November 2014).

Jara-Rojas, R., Bravo-Ureta, B.E., Engler, A. and Díaz, J. (2013) An analysis of the joint adoption of water conservation and soil conservation in Central Chile. *Land Use Policy* 32, 292–301.

Kassam, A., Friedrich, T., Shaxson, F. and Pretty, J. (2009) The spread of conservation agriculture: justification, sustainability and uptake. *International Journal of Agricultural Sustainability* 7, 292–320.

Kerkhoff, E. and Sharma, E. (2006) *Debating Shifting Cultivation in the Eastern Himalayas: Farmers' Innovations as Lessons for Policy*: International Centre for Integrated Mountain Development (ICIMOD), Kathmandu.

Lee, D.R. (2005) Agricultural sustainability and technology adoption: Issues and policies for developing countries. *American Journal of Agricultural Economics* 87(5), 1325–1334.

McNamara, K.T., Wetzstein, M.E. and Douce, G.K. (1991) Factors affecting peanut producer adoption of integrated pest management. *Review of Agricultural Economics* 13(1), 129–139.

Makokha, M., Odera, H., Maritim, H., Okalebo, J. and Iruria, D. (1999) Farmers perceptions and adoption of soil management technologies in western Kenya. *African Crop Science Journal* 7(4), 549–558.

Malla, G. (2009) Climate change and its impact on Nepalese agriculture. *Journal of Agriculture and Environment* 9, 62–71.

Manandhar, G., Adhikary, S. and Sah, G. (2009) Sustainable agricultural practices and technologies in Nepal. United Nations Social and Economic Commission for Asia and the Pacific, Asian and Pacific Centre for Transfer of Technology Tech Monitor.

Marenya, P.P. and Barrett, C.B. (2007) Household-level determinants of adoption of improved natural resources management practices among smallholder farmers in western Kenya. *Food Policy* 32(4), 515–536.

Marra, M., Pannell, D.J. and Abadi Ghadim, A. (2003) The economics of risk, uncertainty and learning in the adoption of new agricultural technologies: Where are we on the learning curve? *Agricultural Systems* 75(2), 215–234.

Maskey, R., Sharma, B. and Joshi, M. (2003) Human Dimensions in Sustainable Land Use Management in Degraded Land Areas of Nepal. A paper prepared for presentation at the Open Meeting of the Global Environmental Change Research Community, Montreal, Canada, 16–18 October 2003.

MEST (Ministry of Environment, Science and Technology) (2006) Nepal: Third National Report on the Implementation of the UN Convention to Combat Desertification. MEST, HMG/N, Kathmandu.

MoAC (Ministry of Agriculture and Cooperatives) (2011) *Statistical Information in Nepalese Agriculture, 2010–11*. MoAC, Government of Nepal, Kathmandu.

MoAC/WFP/FAO (2011) *Crop Situation Update: A Joint Assessment of 2010 Summer Crops and Outlook for 2011 Winter Crops*. Ministry of Agriculture and Cooperatives, World Food Program and Food and Agriculture Organization, Kathmandu.

Neill, S.P. and Lee, D.R. (2001) Explaining the adoption and disadoption of sustainable agriculture: The case of cover crops in northern Honduras. *Economic Development and Cultural Change* 49(4), 793–820.

Neupane, R.P. and Thapa, G.B. (2001) Impact of agroforestry intervention on soil fertility and farm income under the subsistence farming system of the middle hills, Nepal. *Agriculture, Ecosystems and Environment* 84(2), 157–167.

Neupane, R.P., Sharma, K.R. and Thapa, G.B. (2002) Adoption of agroforestry in the hills of Nepal: A logistic regression analysis. *Agricultural Systems* 72(3), 177–196.

Pratap, T. (1999) Sustainable land management in marginal mountain areas of the Himalayan region. *Mountain Research and Development* 19(3), 251–260.

Quinton, J.N. and Catt, J.A. (2004) The effects of minimal tillage and contour cultivation on surface runoff, soil loss and crop yield in the long-term Woburn Erosion Reference Experiment on sandy soil at Woburn, England. *Soil Use and Management* 20(3), 343–349.

Rasul, G. and Thapa, G. (2003) Shifting cultivation in the mountains of South and Southeast Asia: Regional patterns and factors influencing the change. *Land Degradation and Development* 14(5), 495–508.

Raut, N., Sitaula, B.K., Aune, J.B. and Bajracharya, R.M. (2011) Evolution and future direction of intensified agriculture in the central mid-hills of Nepal. *International Journal of Agricultural Sustainability* 9(4), 537–550.

Sharma, R., Xu, J. and Sharma, G. (2007) Traditional agroforestry in the eastern Himalayan region: Land management system supporting ecosystem services. *Tropical Ecology* 48(2), 1–12.

Shrestha, R., Neupane, R.K. and Adhikari, N.P. (2011) Status and future prospects of pulses in Nepal. Paper presented at Regional Workshop on Pulse Production held at Nepal Agricultural Research Council (NARC), Kathmandu, Nepal, 24–25 October 2011. Available at: http://www.doanepal.gov.np/downloadfile/Current%20Status%20SAARC_paper-Nepal_1320838291.pdf (accessed 6 November 2014).

Shrestha, S.K. (2014) Decentralizing the farmer-to-farmer extension approach to the local level. *World Journal of Science, Technology and Sustainable Development* 11(1), 66–77.

Smale, M., Just, R.E. and Leathers, H.D. (1994) Land allocation in HYV adoption models: An investigation of alternative explanations. *American Journal of Agricultural Economics* 76(3), 535–546.

Soule, M.J., Tegene, A. and Wiebe, K.D. (2000) Land tenure and the adoption of conservation practices. *American Journal of Agricultural Economics* 82(4), 993–1005.

Templeton, S.R. and Scherr, S.J. (1999) Efects of demographic and related microeconomic change on land quality in hills and mountains of developing countries. *World Development* 27(6), 903–918.

Thapa, S. (2009) *Three Essays on Nepalese Development: Technology Adoption, Agricultural Risk and the Effects of Remittances on Recipient Households*. University of Trento, Italy.

Thierfelder, C., Cheesman, S. and Rusinamhodzi, L. (2013) Benefits and challenges of crop rotations in maize-based conservation agriculture (CA) cropping systems of southern Africa. *International Journal of Agricultural Sustainability* 11(2), 108–124.

Tiwari, T., Brook, R. and Sinclair, F. (2004) Implications of hill farmers' agronomic practices in Nepal for crop improvement in maize. *Experimental Agriculture* 40, 397–417.

Uaiene, R.N., Arndt, C. and Masters, W. (2009) Determinants of agricultural technology adoption in Mozambique. Discussion Papers (67E). National Directorate of Studies and Policy Analysis, Ministry of Planning and Development, Republic of Mozambique.

UNDP (United Nations Development Program) (2009) *Nepal Human Development Reports 2009: State Transformation and Human Development*. UNDP, Kathmandu.

WORLDSTAT (2014) Nepal: Agriculture Land. Available at: http://en.worldstat.info/Asia/Nepal/Land (accessed 20 March 2014).

8 Evaluation of Tillage and Farmyard Manure on Soil Properties and Maize Yield in the Mid-hills of Nepal

ROSHAN PUDASAINI[1]* AND KESHAV RAJ PANDE[2]

[1]*Local Initiatives for Biodiversity Research and Development, Pokhara, Nepal;* [2]*Agriculture and Forestry University, Chitwan, Nepal*

8.1 Introduction

In Nepal, maize (*Zea mays* L.) is the major staple crop after rice, both in terms of area and production. Grain is used as a staple food by people, as well as used as animal feed. Maize stover is also used as bedding material for livestock and as fuel for cooking. Maize is currently grown on 875,660 ha of land, with a total production of 1,855,184 Megagrams (Mg) and an average yield of 2.119 Mg/ha (MoAC, 2010), and therefore plays an important role in national food security. About 70% of Nepal's total maize production area is within the country's east to west oriented mid-hills region, where the crop is grown in rainfed conditions during the summer months, i.e. April–August (MoAC, 2010).

Maize yield in Nepal is lower than world levels. There are several reasons associated with low productivity of maize, including low nutrient supply, poor irrigation facilities, poor yield varieties, poor weed management practices, and most seriously, rapidly degrading soil quality, particularly in Nepal's mid-hill region (Paudyal *et al.*, 2001). Maize growing on *bari* lands (rainfed uplands) are characterized by sloping terraces, excessive drainage, shallow soil depth, moisture deficit, and acidic reactions. Maize grown during the summer season in the mid-hill region is prone to soil erosion, which is a serious problem through time. Due to excessive soil loss, plant nutrients are lost, soil structure deteriorates, and the production capacity of the soil is reduced (Troeh *et al.*, 1980). The annual loss of soil from agricultural land ranges from a mere 0.1 Mg/ha to a very high 105 Mg/ha in Nepal (Chalise and Khanal, 1997). Such continuous soil loss is adversely

*Corresponding author; e-mail: rospud@gmail.com

affecting maize productivity as well as the environment, specifically the quality of downstream water resources in the mid-hills of Nepal (Atreya *et al.*, 2005). Furthermore, soil fertility management plays a vital role in increasing crop productivity and production (Ponsica *et al.*, 1983; Barsukov, 1991; Das *et al.*, 1991; Gajri *et al.*, 1994; Minhas and Sood, 1994; Belay *et al.*, 2001), which is, in turn, an important tool for addressing the problem of food security and income generation of farming households.

Nepalese rural farmers have little awareness of appropriate tillage and nutrient management systems to match the soil type for long-term sustainability. They follow the traditional tillage method of ox-driven plows to till the land, which heavily disturbs the structure of soil. With traditional plowing, rainfall easily washes out the soil particles from the slopping landscape of the mid-hill region of Nepal. Applying additional fertilizer to these lands to enhance soil nutrients is rare. In most situations, farmers cannot afford commercial fertilizer. In fact, only 38% of maize growers in Nepal use chemical fertilizer on summer maize (NLSS, 2011). Those who are applying chemical fertilizer use mainly urea fertilizer and apply it haphazardly at the surface. The most commonly applied fertilizer in Nepal is farmyard manure (FYM). It is applied by almost all farmers, as 64% raise cattle and produce FYM on their own farms (NLSS, 2011). However, the current rate of application of fertilizers is not adequate to make up for the nutrient loss due to crop removal and erosion. Most farmers apply only a small amount of FYM on the surface of their entire field. Such non-optimal rates of application reduce fertilizer efficiency, as well as degrade soil quality. In order to conserve soil and water resources in the mid-hills of Nepal, it is necessary to identify appropriate tillage methods for the region, as well as the level and method of FYM application that could be most beneficial, both in terms of soil health and maize production.

8.1.1 Conservation agriculture

One of the suggested methods to address the problem of soil degradation is to apply conservation agriculture (CA) technologies. CA technology offers optimal growth conditions to crops for increased yield, along with a balance between long-term agricultural, economic, and environmental benefits (Lal, 1982). The concept suggests that the combined environmental and economic benefits gained from reduced input, soil erosion, and optimal cropping pattern are more sustainable than current production practices. There are three major components of CA that have the greatest potential to protect and enhance soil and yield productivity in the mid-hills of Nepal. The first of these is minimum tillage, which has now been applied to more than 95 million (Dumanski *et al.*, 2006) of the 1,365 million ha of arable land in the world. Conservation or minimum tillage enhances crop production by decreasing soil bulk density, increasing infiltration, decreasing surface runoff, and conserving soil moisture (Chichester and Richardson, 1992; Fortin, 1993; Kettler *et al.*, 2000; Bhatt *et al.*, 2004; Licht and Al-Kaisi, 2005; Lipic *et al.*, 2005). Conservation tillage methods like zero tillage, minimum tillage, and reduced tillage that support conservation agricultural production provide the best opportunity for halting degradation and for restoring and improving soil productivity (Lal, 1983; Parr *et al.*, 1990). The second and third components are intercropping and optimal crop rotation. Compared to conventional agriculture, CA offers the potential to increase crop productivity (Sayre and Hobbs, 2010), reduce production costs, increase soil organic carbon (Lal *et al.*, 2010), and decrease soil salinity in the long run. At the same time, in spite of the lower yield performance, due initially to reduced tillage methods, conservation tillage practices are essential to reduce soil erosion on highly erodible, sloping silt-loam soils like the mid-hills of Nepal (Howard and Essington, 1998). In addition to conservation tillage and crop rotation, CA also deals with promoting a healthy and living soil through high organic matter content and the use of integrated nutrient and pest management technologies. Such practice reduces requirements for chemically sourced fertilizers, pesticides, and herbicides, which consequently helps to control off-site pollution and enhances biodiversity (Ahmad *et al.*, 2008).

Tillage

Tillage aims to create a favorable soil environment for plant growth by influencing soil physical conditions and nutrient availability, and consequently growth and yield of crops (Ojeniyi and Agboola, 1995). Tillage operations generally loosen the soil, as well as decrease soil bulk density and penetration resistance by increasing soil porosity. Appropriate soil tillage or land preparation methods, in addition to manure use, are basic practices required to maximize crop yield, while avoiding soil degradation and maintaining ecosystem stability. In recent years, mechanized and heavy tillage in developed countries has resulted in unfavorable soil disturbance, which has been linked to the destruction of soil aggregates (Hulugalle *et al.*, 1985; Ohu *et al.*, 1994). Soil compaction, a common effect of tilled soils, decreases the macroporosity of soil and contributes to higher bulk density and more dense soil than under no-tillage methods. Conservation tillage methods can help in controlling soil erosion, especially by reducing surface runoff (Lee *et al.*, 2010) and altering physical, chemical, and biological soil properties. Conservation tillage increases water storage in the soil profile, thereby increasing crop productivity (Pelegrin *et al.*, 1990; Brandt, 1992; Moreno *et al.*, 1997). In places like the mid-hills of Nepal, with sloping agricultural lands, nutrient retention is a major challenge. Conservation tillage could be a solution, since it has been found to promote soil quality. Reduced tillage systems cause minimal soil disturbance and increase the build-up of surface residue, which may slow down nitrogen (N) mineralization and prevent nitrogen losses from leaching and denitrification (Gilliam and Hoyt, 1987).

Despite the fact that conservation tillage enhances both crop yield and soil properties, farmers are not easily convinced to adopt it. This is often because using the conservation tillage method without improving proper management of organic manure reduces crop yield. On the other hand, deep conventional tillage generally increases soil aeration, residue decomposition, organic N mineralization, and the availability of N for plant use, resulting in higher maize grain yield (Rice *et al.*, 1987; Sainju and Singh, 2001; Dinnes *et al.*, 2002;

Halvorson *et al.*, 2005; Ahmad *et al.*, 2010). To achieve the same level of grain yield from reduced tillage, N availability and other nutrients must be increased, and the soil physical properties improved, through adding organic manure. Unless these conditions are met, the yield obtained from conservation tillage might not be able to exceed that of conventional tillage in the initial years of adopting minimum tillage (Opoku and Vyn, 1997; Adeyemo and Agele, 2010; Ahmad *et al.*, 2010). However, researchers throughout the world are providing increasing evidence that in most soil types, maize produced under minimum or no tillage can be similarly or even more profitable than maize produced under conventional tillage (Ahmad *et al.*, 2010).

Farmyard manure (FYM)

It is well known that farmyard manure (FYM) directly supplies N, phosphorus (P), sulphur (S), and many other elements in plant-available forms through biological decomposition (Brady and Weil, 1996). Indirectly, it improves physical properties of soil such as aggregation, aeration, permeability, and water-holding capacity. Overall, it promotes crop yields as well as soil health (Negassa *et al.*, 2005; Khan *et al.*, 2010). Farmyard manure acts as an alternative to inorganic fertilizers for soil fertility enhancement, as manure releases nutrients slowly and steadily over longer periods of time, and also improves soil fertility status by activating the soil microbial biomass (Ayuso *et al.*, 1996; Belay *et al.*, 2001). In most experiments, FYM supported maize grain yield (Parmar and Sharma, 2001; Purushottam and Puri, 2001; Vadivel *et al.*, 2001; Kumar *et al.*, 2005). For example, Ponsica *et al.* (1983) recorded a grain yield of 2.67 Mg/ha of maize with 12.0 Mg/ha of cow manure, which was significantly higher than the control yield of 1.2 Mg/ha.

Similar to conservation tillage, the addition of FYM is supportive for nutrient retention in the soil. The addition of FYM reduced the loss of nitrates through leaching from the soil under maize by providing a significant amount of plant nutrients, which created a balancing effect on the supply of N, P, and potassium (K) (Singh *et al.*, 1979). The uptake of K, calcium (Ca), magnesium (Mg), and iron (Fe) by maize shoots was increased over the control, due to the addition of FYM. Similarly, K, Ca, Mg, and Fe contents in grains and maize stover were also increased significantly over organic manures alone, due to P enrichment (Das *et al.*, 1992). High rates of manure application increased soil pH, and content of K, Mg, and P were observed in acidic soil (Lungu *et al.*, 1993). Increased soil organic carbon as a result of the application of FYM and compost has also been observed (Grewal *et al.*, 1981; Badanur *et al.*, 1990). All of this evidence supports the idea that FYM can contribute to soil nutrient retention. Similarly, higher nutrient uptake by maize with higher levels of FYM use has been noted, due to FYM producing a higher availability of mineral N and other plant nutrients (Minhas and Sood, 1994). FYM application also increased nutrient supply and supported a well-developed root system, resulting in better absorption of water and nutrients (Brar *et al.*, 2001; Parmar and Sharma, 2001; Datt *et al.*, 2003; Kumar and Thakur, 2004; Singh *et al.*, 2011; Islam and Munda, 2012).

8.2 Methods

8.2.1 Effect of tillage and FYM on maize productivity and soil properties

To provide evidence for discussion of the scope of CA in the mid-hill region of Nepal, a field experiment was conducted at Kabilash, a tribal village in the Chitwan District of Nepal. The experiment had specific focus on the most efficacious combination of tillage and FYM as applied to crops grown under analogous conditions. The research was part of the activity of the Sustainable Management of Agroecological Resources in Tribal Societies (SMARTS) project, which intended to identify sustainable technology for resource management in such communities.

The Kabilash village of Chitwan was selected as the experiment site, as it represented a typical maize-based farming community in the mid-hills of Nepal. Kalibash is composed primarily of the *Chepang*, an economically resource-poor ethnic group in Nepal that practices sloping land agriculture. They farm maize utilizing traditional methods, and cannot afford chemical fertilizers and other farming inputs. Full tillage is practiced using ox-driven plow or manual digging, and FYM is the major source of soil nutrients. According to area farmers, the average rate of FYM application is <5 Mg/ha. As a result, the existing conventional tillage practice and low organic fertilizer input might be affecting not only maize production but also the productive capacity of the soil. Furthermore, farmers have been unable to apply CA measures to enhance the overall system of productivity as they are lacking any site-specific recommendation of appropriate tillage and fertility management techniques. This is due mainly to the limited availability of pertinent and topical research studies, coupled with poor extension services to disseminate what little information is available.

To evaluate the effect of tillage and different doses of FYM on soil properties and yield of maize crop, a field experiment was conducted from May–July 2012. The experiment was laid out in a two-factor, randomized complete block design (RCBD), with the factors tillage (dibbling, strip tillage, and conventional or full tillage) and FYM levels (2 Mg/ha, 5 Mg/ha, 10 Mg/ha, and 20 Mg/ha) allocated randomly to 4.8×2 m^2 plots in three replications. The soil in the experimental site was acidic (pH 5.56), clay-based soil, with 10% gravel, 0.813% organic matter, and 1.75 mg/cm^3 bulk density. It contained 0.095% available N, 17.73 mg/kg available P, and 82.83 mg/kg available K. The FYM used for the experiment contained 1.23%, 0.31%, and 0.71% N, P, and K, respectively. One control plot was established having chemical source NPK at the rate of 120:60:40 and FYM at 15 Mg/ha (the recommended dose for the particular variety of maize, viz. Arun 2), to compare the economic benefit with regards to each treatment. Dibbling as a method of minimum or reduced tillage comprised a hole dug 10 cm deep with a 20 cm^2 area by a traditional spade. FYM, as per treatment, was placed into the resulting hole, which was then refilled with the soil. In strip tillage, furrows of 10 cm breadth and 10 cm depth were dug by traditional spade, and FYM was placed into the furrow and then refilled with the soil while seeding. Conventional tillage comprised plowing two times with an ox-driven plow, once 2 weeks before seeding and again while seeding. Maize seeds were sown on 5 May 2012, with 60 cm between-row and 25 cm between-plant spacing. Arun-2, a short-duration maize variety (80–90 days) recommended for the mid-hills and inner *terai* (east–west southern plain) of Nepal, was selected for this research as it was considered suitable for marginal lands.

A total of 1,212 mm rain was received during the monsoon season, and the average temperature during the period of research was recorded as 26.73°C. These values are on a par with average seasonal values (rainfall 1,000 mm and temperature 24°C) for the region during the particular months. Dry matter yield and plant NPK content were recorded by taking two plants randomly from the destruction row (an extra row of crop planted to be uprooted at different stages to analyze plant chemicals) of each plot and air dried, followed by oven drying at 65 ± 10°C, to a constant weight. The harvested crop was separated into grain and stover, and yield was recorded. The grain yield was adjusted at 12% grain moisture content. The data collected were subjected to statistical analysis using "MSTAT-C", whereas the means were compared through Duncan's Multiple Range Test (DMRT) at $P < 0.05$ (Gomez and Gomez, 1984).

8.3 Results

8.3.1 Grain and stover yield

Adoption of dibbling and strip tillage was not associated with a significant decrease in maize yield ($P > 0.05$). On the other hand, FYM level was associated with a highly significant ($P < 0.05$) increase in both grain and stover yield. The highest grain yield (3.81 Mg/ha) and stover yield (9.87 Mg/ha) were recorded by supplying 20 Mg/ha FYM. Similarly, thousand-grain weight (index of size and plumpness of dry grain that is used especially to evaluate the quality of seed) was also highest (270.78 g) at 20 Mg/ha FYM. The higher rate of FYM was associated with a higher value of thousand-grain weight, which means the quality of maize grain improved with increased rate of FYM application (Table 8.1).

We found no significant change in maize yield associated with altering tillage regimes. Studies suggest that the reduced tillage may cause delay in the early crop growth and development but does not have a detrimental effect on the final crop yield (Mehdi *et al.*, 1999; Beyaert *et al.*, 2002). Regardless of differing yields during early growing seasons, Kihara *et al.* (2012) found improved performance of reduced tillage after several years. Thus, current research suggests that the adoption of reduced tillage does not reduce maize yield in the short term, and in fact the continuous application of such practice can increase the yield in the long run.

Our findings suggest that an increase in both maize grain and stover yield can be achieved by applying higher levels of FYM to any of the three types of tillage used in this study. Others have found similar results (Mahmood *et al.*, 1997; Adeyemo and Agele, 2010; Verma, 2011; Islam and Munda, 2012). This improvement in yield is due mainly to improved soil conditions and a favorable environment for seedling development and subsequent growth, due to the increased FYM (Aggarwal *et al.*, 1995).

Table 8.1. Effect of tillage and farmyard manure (FYM) on grain and stover yield of maize at Kabilash, Chitwan, Nepal, 2012.

Treatments	Grain yield (Mg/ha)	Stover yield (Mg/ha)	Thousand-grain weight (g)
Tillage			
Dibbling	3.17	8.05	299.37
Strip tillage	2.98	8.27	212.99
Conventional tillage	3.29	8.45	237.72
SEm	0.13	0.22	5.455
LSD	NS	NS	NS
FYM			
2 Mg/ha	2.41[c]	6.79[c]	182.36[d]
5 Mg/ha	2.93[bc]	7.62[c]	209.49[c]
10 Mg/ha	3.45[ab]	8.74[b]	244.16[b]
20 Mg/ha	3.81[a]	9.87[a]	270.78[a]
SEm (±)	0.14	0.26	6.30
LSD (= 0.05)	0.58**	1.02**	25.11**
CV %	13.96	9.31	8.34

Mean values within a column followed by the same letter(s) in superscript are not significantly different at the given level of significance.
**, highly significant ($P < 0.01$); SEm, standard error of mean; LSD, least significant difference; CV, coefficient of variation; NS, not significant.

8.3.2 Nutrient uptake

In addition to the method of tillage, soil nutrient availability also affects plants' capacity to take up nutrients. Nutrient uptake by the maize plant is directly associated with yield quantity as well as quality, as it supports plant growth and development. Nitrogen (N) is considered to be a crucial element among plant nutrients, because it is one of the primary nutrients required by plants in a larger amount. At the same time, it is removed easily from soil by rainwater, so soils can quickly become N deficient, thereby not allowing plants enough time to absorb the necessary nutrient.

In this study, we investigated if maize N uptake efficiency could be increased by managing tillage and FYM use. Tillage method was found to affect significantly the level of N uptake by maize ($P < 0.05$). A higher level of N uptake (134.67 kg/ha) was recorded where dibbling was applied, as compared to strip (119.43 kg/ha) and conventional tillage (110.99 kg/ha). Levels of P and K uptake were not affected significantly by different tillage methods. However, different levels of FYM application were associated with different levels of N, P, and K uptake. The highest N uptake (174.71 Mg/ha) was observed in the 20 Mg/ha FYM treatment and the lowest N uptake (78.31 Mg/ha) was found in the 2 Mg/ha FYM treatment. P and K uptake were also highest in the 20 Mg/ha FYM treatment (Table 8.2).

Higher N uptake in the case of reduced tillage might be due to the extended availability of soil N in the less disturbed soil than in conventionally tilled soils (Thiagalingam *et al.*, 1996; Ozpinar, 2009; Lee *et al.*, 2010). Similarly, higher

Table 8.2. Effect of tillage and FYM on nutrient uptake of maize in Kabilash, Chitwan, Nepal, 2012.

Treatments	Nutrient uptake (kg/ha)		
	Nitrogen (N)	Phosphorus (P)	Potassium (K)
Tillage			
Dibbling	134.67[a]	31.73	137.69
Strip tillage	119.43[b]	32.92	117.06
Conventional tillage	110.99[b]	34.17	129.04
SEm (±)	4.50	1.46	6.86
LSD (= 0.05)	13.21*	NS	NS
FYM			
2 Mg/ha	78.31[d]	17.65[d]	73.27[c]
5 Mg/ha	101.78[c]	25.30[c]	117.70[b]
10 Mg/ha	131.97[b]	33.79[b]	123.79[b]
20 Mg/ha	174.71[a]	55.02[a]	196.97[a]
SEm (±)	5.20	1.69	7.92
LSD (= 0.05)	20.73**	6.74**	31.56**
CV %	12.8	15.41	18.56

Mean values within a column followed by the same letter(s) in superscript are not significantly different at the given level of significance.

*, significant (P < 0.05); **, highly significant (P < 0.01); SEm, standard error of mean; LSD, least significant difference; CV, coefficient of variation; NS, not significant.

availability of mineral N and other plant nutrients was caused by an increased level of FYM (Minhas and Sood, 1994), which supported the supply of nutrients and well-developed root systems, resulting in better absorption of water and nutrients (Datt *et al.*, 2003; Singh *et al.*, 2011; Islam and Munda, 2012). Thus, better plant nutrient availability can be ensured by reduced tillage and a higher rate of FYM application, rather than conventional tillage and lower rates of FYM application.

8.3.3 Soil properties

Soil property generally describes the overall condition of the soil with respect to its intended use. It integrates soil physical and chemical properties and reflects the effects of management. Among various factors that govern the overall quality of a soil, soil pH, bulk density, soil moisture, organic matter, and soil NPK, which are directly related with the nutrient availability for plants, were considered to examine the effect of tillage and FYM on that particular soil. Soil pH was not affected significantly by tillage or FYM levels (Table 8.3). Soil moisture, however, was affected significantly ($P < 0.01$) by tillage method, and was observed to be highest in dibbled plots (14%); soil moisture was not affected by the FYM levels. Soil bulk density was affected by higher levels of FYM application and significantly lower ($P < 0.05$) bulk density was observed in both 20 Mg/ha FYM (1.49 g/cm^3 bulk density) and 10 Mg/ha FYM (1.51 g/cm^3) treated plots. Higher residual N was recorded in dibbled (0.15%) and strip tillage (0.13%) plots as compared to

Table 8.3. Effect of tillage and FYM on soil properties at Kabilash, Chitwan, Nepal, 2012.

Treatments	pH	Bulk density (g/cm³)	Soil moisture (%)	Organic matter (%)	Nitrogen (N) (%)	Phosphorus (P) (mg/kg)	Potassium (K) (mg/kg)
Tillage							
Dibbling	5.67	1.60	17.28[a]	1.55	0.15[a]	22.30	80.57
Strip tillage	5.60	1.58	14.85[b]	1.61	0.13[a]	21.72	80.30
Conventional tillage	5.35	1.55	13.89[b]	1.63	0.10[b]	21.12	87.70
SEm (±)	0.07	0.03	0.35	0.07	0.01	2.12	5.07
LSD (0.05)	NS	NS	1.41**	NS	0.03*	NS	NS
FYM							
2 Mg/ha	5.44	1.66[a]	15.28	1.42[b]	0.12[b]	16.40[b]	63.20[b]
5 Mg/ha	5.53	1.64[a]	14.82	1.45[b]	0.12[b]	20.81[ab]	67.52[b]
10 Mg/ha	5.45	1.51[b]	15.24	1.56[b]	0.11[b]	21.36[ab]	80.30[b]
20 Mg/ha	5.75	1.49[b]	16.02	1.95[a]	0.16[a]	28.28[a]	120.41[a]
SEm (±)	0.08	0.03	0.41	0.09	0.01	2.45	5.85
LSD (= 0.05)	NS	0.10*	NS	0.35**	0.03*	7.19*	23.33**
CV %	4.26	6.75	8.03%	16.30	25.06	33.86	21.19

Mean values within a column followed by the same letter(s) in superscript are not significantly different at the given level of significance.
*, significant (P < 0.05); **, highly significant (P < 0.01); SEm, standard error of mean; LSD, least significant difference; CV, coefficient of variation; NS, not significant.

conventionally tilled (0.1%) plots. Residual soil organic matter, P, and K were not affected by tillage. The highest levels of NPK and organic matter were observed in the treatments having 20 Mg/ha FYM. All the nutrients decreased gradually in accordance to the decreased level of FYM application.

Other studies have also found that adding organic matter decreases soil bulk density. For example, organic matter promotes soil aggregation, resulting in decreased bulk density (Arriaga and Lowery, 2003; Khan *et al.*, 2010). Similarly in the case of N, higher residual soil N may be due to a larger and longer-lasting pool of labile N in the minimally tilled soil. Minimum disturbances to the soil might have caused less availability of oxygen, which slowed down the process of oxidative decomposition of organic matter N, thereby providing a more continuous supply of nitrogen. Similarly, higher organic matter and NPK concentration in soil with a higher level of applied FYM have been reported (Halvin *et al.*, 1990; Negassa *et al.*, 2005; Khan *et al.*, 2010; Ahmed *et al.*, 2012), suggesting more labile organic matter results in short-term as well as long-term influences on soil physical and chemical properties, including aggregate stability and nutrient availability.

The combined effect of tillage and FYM was also associated with changes in soil properties. The interaction effect of various tillage methods with FYM levels was significant ($P < 0.05$) for N retention in soil. The best environment for soil N conservation was provided by dibbling with 5 Mg FYM/ha, followed by strip tillage with 20 Mg FYM/ha, and dibbling with 20 Mg FYM/ha. Higher soil moisture in dibbling might be due to less surface exposure for evaporation loss, better infiltration, and better water-holding capacity of the soil. Hence, the combination of reduced tillage with a higher level of FYM was found to be quite favorable for retention of soil N.

8.3.4 Net returns

To calculate the cost of cultivation, labor costs at the rate of US\$0.99/h, seed costs of US\$1.38/kg and FYM costs of US\$17.24/Mt were used, based on prices at the Chitwan market in May 2012. The cost of labor for maize cultivation was similar for conventional and strip tillage, but was slightly higher in the case of dibbling. Since the dibbled plots required manual digging, it took slightly more labor (1.3 times more than strip tillage and 1.2 times more than conventional tillage) than the other two methods, which were tilled using a cattle plow (Table 8.4). Since the results from the experiment suggested grain yields of 3.29 Mg/ha and straw yields of 8.45 Mg/ha from conventional tillage (Table 8.1), the net return was found to be US\$477.17 by multiplying the yields with the unit selling rates (US\$0.29/kg for grain and US\$0.01/kg for stover) and subtracting the cost of cultivation. The net return was found to be highest in conventional tillage, followed by strip tillage (US\$379.27) and dibbling (US\$318.29). However, the average value of the revenue:cost ratio (R:C) (gross return divided by total cost of cultivation) was similar for dibbling (1.50), strip tillage (1.67), and conventional tillage (1.74) practices. While analyzing the combinations, the highest ratio was found in conventional tillage, with FYM 5 Mg/ha (1.91), followed by conventional tillage with FYM 10 Mg/ha and strip tillage with FYM 10 Mg/ha (1.84). Similarly, dibbling

Table 8.4. Net return analysis of maize cultivation for different tillage methods and levels of FYM at Kabilash, Chitwan, Nepal, 2012.

Treatment combinations	Cost (US$/ha)	Gross return (US$/ha)	Net return (US$/ha)	R:C
Dibbling + 2 Mg FYM/ha	517.24	701.32	184.08	1.36
Dibbling + 5 Mg FYM/ha	568.97	922.26	353.30	1.62
Dibbling + 10 Mg FYM/ha	655.17	1,071.55	416.38	1.64
Dibbling + 20 Mg FYM/ha	827.59	1,146.99	319.40	1.39
Strip tillage + 2 Mg FYM/ha	448.28	712.53	264.25	1.59
Strip tillage + 5 Mg FYM/ha	500.00	862.59	362.59	1.73
Strip tillage + 10 Mg FYM/ha	586.21	1,076.87	490.67	1.84
Strip tillage + 20 Mg FYM/ha	758.62	1,157.41	398.79	1.53
Conventional tillage + 2 Mg FYM/ha	471.26	726.44	255.17	1.54
Conventional tillage + 5 Mg FYM/ha	522.99	999.71	476.72	1.91
Conventional tillage + 10 Mg FYM/ha	609.20	1,123.31	514.11	1.84
Conventional tillage + 20 Mg FYM/ha	781.61	1,324.31	542.70	1.69
Control (conventional tillage + 120:60:40 chemical source NPK)	938.45	1,465.40	526.95	1.56

Exchange rate considered: US\$1 = 87 NRs (May 2013). R:C, revenue : cost ratio.

with FYM 5 and 10 Mg/ha had an R:C ratio of 1.62 and 1.64, respectively. The R:C for the control (15 Mg FYM/ha and chemical source NPK at the rate of 120:60:40) was found to be 1.56, which was lower than all the 5 and 10 Mg/ha of FYM treatments of all methods of tillage. Hence, from the revenue:cost perspective, replacing chemical fertilizer with increased FYM was quite effective. Although the cost of cultivation was higher for dibbling, as compared to other types of tillage, there was little difference in R:C, because of its higher gross return. Treatments with FYM 5 Mg/ha and 10 Mg/ha have shown higher R:C than the lowest and highest levels of FYM application, including the control, in all types of tillage practices.

8.4 Conclusion and Implications

Overall findings of the experiment suggested that, among the three different tillage methods, dibbling was the most effective tillage method in terms of increased uptake of N and K, improved holding of soil N, and improved moisture retention, followed by strip tillage. The untilled surface in reduced tillage keeps crop residue on the surface and helps to trap water in the soil by providing shade, which reduces water evaporation. It also helps to slow runoff and increases the opportunity for water to soak into the soil. Depending on the amount of residues present in the surface, soil erosion can be reduced by up to 90% compared to an unprotected and intensively tilled field. Similarly, once adopted, reduced methods of tillage such as dibbling and strip tillage will increase soil particle aggregation, making it easier for plants to establish roots, which is directly related to plant growth and development. All of these ultimately help to increase crop production sustainably.

Regarding the revenue to cost ratio, FYM application at 5–10 Mg/ha had the highest returns. Applying FYM at 5–10 Mg/ha was found to be most cost-effective compared to the lowest (at 2 Mg/ha) and the highest (at 20 Mg/ha) levels of FYM. This middle level of FYM application reduced the cost of cultivation by 65%, to increase the R:C. On the other hand, the R:C ratios were similar for dibbling, strip tillage, and conventional tillage. However, dibbling required a little more labor as compared to the other two methods of tillage. Therefore, in the particular areas where labor availability is a serious problem, farmers can go for strip tillage, which comes second in terms of soil conservation and offers even better revenue than from dibbling. In the case of farmers adopting either dibbling or strip tillage, tillage could be done manually, thus avoiding the requirement of an ox-driven plow. This means the requirement for male labor might be avoided. Hence, reduced tillage by either dibbling or strip with FYM applied at 5–10 Mg/ha FYM is preferable to conventional farming. Without reducing economic yield, this approach enhances soil properties, which can lead Nepali farmers to sustainable improvement in maize production, and thereby help to ensure national food security.

Resource-poor farming households from the hill area of Nepal who farm sloping lands and have restricted access to chemical fertilizer can adopt reduced tillage with the use of FYM at 5–10 Mg/ha to increase maize returns and make the system sustainable in the long term. In the sloping hills of Nepal, where maize is cultivated as the main crop, this technology could help to sustain production by restricting soil degradation, which is perhaps the most serious threat to the agricultural system as a whole. FYM in place of chemical fertilizer is more sustainable, because it can be produced by resource-poor farmers on individual farms at the household level, and in fact most of the households are already doing so. However, the manner in which farmers are applying FYM is not appropriate, and there is a great need to reinforce the production of quality manure and maximize its use. Farmers need to be aware of efficient methods of FYM preparation and application/placement methods, as well as timing and concentrations of FYM for specific crops. Education and building capacity of resource-poor farmers in terms of appropriate and efficient uses of FYM should be prioritized by both governmental and non-governmental extension programs. Additional research and on-farm demonstrations should be pursued in order to persuade regional farmers of the utility of these new ideas in FYM application.

Nepal's current agricultural policy (2004) focuses primarily on increased production and commercialization. It also addresses resource conservation by minimizing the negative impacts of using agrochemicals in soil and by promoting the use of organic fertilizer. However, the effort seems not to be enough to overcome the problem of soil degradation in the mid-hills region. There is a lack of planning and implementation of specific programs that could address CA at the grassroots level. There are, however, steps that can be taken to address these concerns. First, there should be agriculture programs in place that focus on the needs and challenges of resource-poor farmers. One aspect of this would be to concentrate on local resource-based management techniques like simple modifications in tillage and FYM application, rather than on costly machines or chemicals that could preclude poor households from adopting novel techniques and methods easily. Furthermore, the Nepal Department of Agriculture, Nepal Agriculture Research Council (NARC), and other non-governmental institutions working for agricultural development in Nepal should provide mid-hills maize producers unaware of alternative tillage methods with information about dibbling and strip tillage. Similarly, information on the correct application rates of FYM (5–10 Mg/ha) should be disseminated via the same channels. If the farmers who follow the traditional tillage method replace these techniques with reduced tillage and applied the proper amount of FYM to grow maize, a great saving of soil and nutrients would be achieved in the sloping landscape of Nepal's mid-hills region, without reduction in maize production. Though the process could be led by the government, it should be complemented by NGOs working in Nepal with similar interests in food security and livelihood.

Acknowledgements

This research was supported by the Sustainable Management of Agroecological Resources in Tribal Societies (SMARTS) project of the Sustainable Agriculture and Natural Resource Management Collaborative Research Support Program (SANREM CRSP), which is funded by USAID and works in Nepal through the University of Hawai'i in collaboration with the Institute of Agriculture and

Animal Science (IAAS) at Tribhuvan University, Nepal, and Local Initiatives for Biodiversity Research and Development (LI-BIRD) in Pokhara, Nepal. We owe our sincere gratitude to all the agencies involved in this project. In addition, we would like to acknowledge the support from LI-BIRD for laboratory analysis of the study plant and soil samples.

References

Adeyemo, A.J. and Agele, S.O. (2010) Effects of tillage and manure application on soil physicochemical properties and yield of maize grown on a degraded intensively tilled alfisol in southwestern Nigeria. *Journal of Soil Science and Environmental Management* 1(8), 205–216.

Aggarwal, G.C., Sekhon, N.K., Sidhu, A.S. and Mahant (1995) Plant water status of maize and succeeding wheat grown on organically amended soils. *Journal of Indian Society of Soil Science* 43(2), 152–155.

Ahmad, I., Mohammad, T.J. and Muhammad, A. (2010) Tillage and nitrogen management impact on maize. *Sarhad Journal of Agriculture* 26(2), 157–167.

Ahmad, R., Arsad, M., Khalid, A. and Zahir, A. (2008) Effectiveness of organic-/biofertilizer supplemented with chemical fertilizers for improving soil and water retention, aggregate stability, growth and nutrient uptake of maize (*Zea mays* L.). *Journal of Sustainable Agriculture* 31(4), 57–77.

Ahmed, A.R., Abbass, M.M. and El-Fayoumy, M.E. (2012) Enhance soil quality and increasing yield of wheat crop followed by maize crop grown on calcareous soils. *Nature and Science* 10(11), 212–220.

Arriaga, F.J. and Lowery, B. (2003) Soil physical properties and crop productivity of an eroded soil amended with cattle manure. *Soil Science* 168(12), 888–899.

Atreya, K., Sharma, S. and Bajracharya, R.M. (2005) Minimization of soil and nutrient losses in maize-based cropping systems in the mid-hills of central Nepal. *Kathmandu University Journal of Science, Engineering and Technology* 1(1), 1–9.

Ayuso, M.A., Pascal, J.A., Garcia, C. and Hernandez, T. (1996) Evaluation of urban wastes for urban agricultural use. *Soil Science and Plant Nutrition* 142, 105–111.

Badanur, V.P., Poleshi, C.M. and Naik, B.K. (1990) Effect of organic matter on crop yield and physical properties of vertisols. *Journal of the Indian Society of Soil Science* 38(3), 426–429.

Barsukov, S.S. (1991) Effect of organic and mineral fertilizer on yield of maize grown according to cereal technology. *Agrokhimiya* 12, 43–48.

Belay, A.A., Classens, S., Wehner, F.C. and De Beer, J.M. (2001) Influence of residual manure on selected nutrient elements and microbial composition of soil under long term crop rotation. *South African Journal of Plant and Soil* 18, 1–6.

Beyaert, R.P., Schott, J.W. and White, P.H. (2002) Tillage effects on corn production in a coarse-textured soil in southern Ontario. *Agronomy Journal* 94, 767–774.

Bhatt, R., Khera, K.L. and Arora, S. (2004) Effect of tillage and mulching on yield of corn in the submontaneous rainfed region of Punjab, India. *International Journal of Agriculture and Biology* 6(1), 126–128.

Brady, N.C. and Weil, R.R. (1996) *The Nature and Properties of Soils*, 11th edn. Prentice Hall, Upper Saddler River, New Jersey, pp. 445–628.

Brandt, S.A. (1992) Zero vs. conventional tillage and their effects on volatilization of N, which is an important mechanism of crop yield and soil moisture. *Canadian Journal of Plant Science* 72, 679–688.

Brar, B.S., Dhillon, N.S. and China, H.S. (2001) Integrated use of farmyard manure and inorganic fertilizers in maize (*Zea mays*). *Indian Journal of Agricultural Sciences* 71(9), 605–607.

Chalise, S.R. and Khanal, N.R. (1997) Erosion processes and their implications in sustainable management of watersheds in Nepal Himalayas. In: Gustard, A., Blazkova, S., Brilly, M., Demuth, S., Dixon, J., van Lanen, H., Llasat, C., Mkhandi, S. and Servat, E. (eds) *Regional Hydrology: Concepts and Models for Sustainable Water Resource Management*. Proceedings of the Postojna, Slovenia, Conference, September–October 1997. International Association of Hydrological Science, Wallingford, UK, pp. 325–334.

Chichester, F.W. and Richardson, C.W. (1992) Sediment and nutrient loss from clay soils as affected by tillage. *Journal of Environmental Quality* 21, 587–590.

Das, M., Singh, B.P, Ram, M. and Prasad, R.N. (1991) Response of maize due to phosphorus enriched manures grown in P-deficit Alfisols on terraced land in Meghalaya. *Indian Journal of Agricultural Sciences* 61(6), 383–388.

Das, M., Singh, B.P., Ram, M. and Prasad, R.N. (1992) Mineral nutrition of maize and groundnut as influenced by P-enriched manures on acid Alfisols. *Journal of Indian Society of Soil Sciences* 40, 580–583.

Datt, N., Sharma, R.P. and Sharma, G.D. (2003) Effect of supplementary use of farmyard manure along with chemical fertilizers on productivity and nutrient uptake by vegetable pea (*Pisum sativum* var arvense) and build up of soil fertility in Lahaul valley of Himachal Pradesh. *Indian Journal of Agricultural Science* 73(5), 266–268.

Dinnes, D.L., Karlen, D.L., Jaynes, D.B., Kaspar, T.C., Hatifield, J.L., *et al.* (2002) Nitrogen management strategies to reduce nitrate leaching in tile-drained Midwestern soils. *Agronomy Journal* 94, 153–171.

Dumanski, J., Peiretti, R., Benites, J.R., McGarry, D. and Pieri, C. (2006) The paradigm of conservation agriculture. *Soil and Water Conservation* 1, 58–64.

Fortin, M.C. (1993) Soil temperature, soil water and no-till corn development following in row residue removal. *Agronomy Journal* 85, 571–576.

Gajri, P.R., Arora, V.K. and Chaudhary, M.R. (1994) Maize growth responses to deep tillage, straw mulching and FYM in coarse textured soils of North Western India. *Soil Use and Management* 10(1), 15–20.

Gilliam, J.W. and Hoyt, G.D. (1987) Effect of conservation tillage on fate and transport of nitrogen. In: Logan, T.J., Davidson, J.M., Baker, J.L. and Overcash, M.R. (eds) *Effects of Conservation Tillage on Groundwater Quality*. Lewis Publishers, Inc, Chelsea, Michigan, pp. 217–240.

Gomez, K.A. and Gomez, A.A. (1984) *Statistical Procedures for Agricultural Research*. Wiley, New York.

Grewal, J.S., Sharma, R.G. and Sud, K.G. (1981) Effect of continuous application of P, K fertilizers and FYM on potato yield and soil properties. *Journal of Indian Society of Soil Science* 29, 129–131.

Halvin, J.L., Kissel, D.E., Maddux, L.D., Classen, M.M. and Long, J.L. (1990) Crop rotation and tillage effects on soil organic carbon and nitrogen. *Soil Science Society of America Journal* 54, 161–166.

Halvorson, A.D., Schweissing, F.C., Bartula, M.E. and Reule, C.A. (2005) Corn response to nitrogen fertilization in a soil with high residual nitrogen. *Agronomy Journal* 97, 1222–1229.

Howard, D.D. and Essington, M.E. (1998) Effects of surface-applied limestone on the efficiency of urea-containing nitrogen sources for no-till corn. *Agronomy Journal* 90, 523–528.

Hulugalle, N.R., Lal, R. and Opara-Nadi, O.A. (1985) Effect of tillage system and mulch on soil properties and growth of yam (*Dioscorea rotundata*) and cocoyam (*Xanthosoma sagillifolium*) on an ultisol. *Journal of Root Crops* 11, 9–22.

Islam, M. and Munda, G.C. (2012) Effect of organic and inorganic fertilizer on growth, productivity, nutrient uptake and economics of maize (*Zea mays* L.) and toria (*Brassica campestris* L.). *Agricultural Science Research Journal* 2(8), 470–479.

Kettler, T.A., Drew J.L., Doran, J.W., Powers, W.L. and Stroup, W.W. (2000) Soil quality assessment after weed control tillage in a no-till wheat-fallow cropping system. *Soil Science Society of America Journal* 64, 339–346.

Khan, N.I., Malik, A.U., Umer, F. and Bodla, M.I. (2010) Effect of tillage and farmyard manure on physical properties of soil. *International Research Journal of Plant Sciences* 1(4), 75–82.

Kihara, J., Martius, C., Bationo, A., Thuita, M., Lesueur, D., *et al.* (2012) Soil aggregation and total diversity of bacteria and fungi in various tillage systems of sub-humid and semi-arid Kenya. *Applied Soil Ecology* 58, 12–20.

Kumar, A. and Thakur, K.S. (2004) Effect of integrated nutrient management on promising composite maize (*Zea mays* L) varieties under rainfed mid hills conditions of Himachal Pradesh. *Indian Journal of Agricultural Sciences* 74(1), 40–42.

Kumar, A., Gautam, R.C., Singh, R. and Rana, K.S. (2005) Growth, yield and economics of maize (*Zea mays* L) wheat (*Triticum aestivum*) cropping sequence as influenced by integrated nutrient management. *Indian Journal of Agricultural Sciences* 75(11), 706–711.

Lal, R. (1982) Tillage research in the tropics. *Soil and Tillage Research* 2, 305–309.

Lal, R. (1983) No-till farming: Soil and water conservation and management in the humid and sub-humid tropics. IITA Monograph No 2, Ibadan, Nigeria.

Lal, R., Reicosky, D.C. and Hanson, J.D. (2010) Evolution of the plow over 10,000 years and the rationale for no-till farming. *Soil and Tillage Research* 93(1), 1–12.

Lee, J.T., Lee, G.J., Ryu, J.S., Jin, Y.I., Park, S.H., *et al.* (2010) Application of reduce tillage with a strip tiller and its effect on soil erosion reduction in highland agriculture. *19th World Congress of Soil Science, Soil Solutions for a Changing World,* 1–6 August 2010, Brisbane, Australia. Published on DVD.

Licht, M.A. and Al-Kaisi, M. (2005) Corn response, nitrogen uptake and water use in strip-tillage compared with no tillage and chisel plow. *Agronomy Journal* 97, 705–710.

Lipic, J., Kus, J., Slowinska-Jurkiewicz, A. and Nosalewicz, A. (2005) Soil porosity and water infiltration as influenced by tillage methods. *Soil and Tillage Research* 89, 210–220.

Lungu, O.I., Temba, J. Cherwa, B. and Lungu, B. (1993) Effects of lime and farmyard manure on soil acidity and maize growth on an acidic alfisol from Zambia. *Tropical Agriculture* 70(4), 309–314.

Mahmood, T., Azam, F., Hussain, F. and Malik, K.A. (1997) Carbon availability and microbial biomass in soil under an irrigated wheat–maize cropping system receiving different fertilizer treatments. *Biology and Fertility of Soils* 25(1), 63–68.

Mehdi, B.B., Madramootoo, C.A. and Mehuys, G.R. (1999) Yield and nitrogen content of corn under different tillage practices. *Agronomy Journal* 91, 631–636.

Minhas, R.S. and Sood, A. (1994) Effect of inorganic and organics on yield and nutrients uptake by three crops in rotation on in acid alfisol. *Journal of Indian Society of Soil Science* 42, 257–260.

MoAC (Ministry of Agriculture and Cooperatives) (2010) *Statistical Information on Nepalese Agriculture, 2009–10.* Government of Nepal, Ministry of Agriculture and Cooperatives Kathmandu, pp. 1–4.

Moreno, F., Pelegrin, F., Fernandez, J. and Murillo, J.M. (1997) Soil physical properties, water depletion and crop development under traditional and conservation tillage in southern Spain. *Soil Tillage Research* 41, 25–42.

Negassa, W., Gebrekidan, H. and Friesen, D.K. (2005) Integrated use of farmyard manure and NP fertilizers for maize on farmers' fields. *Journal of Agriculture and Rural Development in the Tropics and Subtropics* 106(2), 131–141.

NLSS (2011) *Nepal Living Standards Survey, 20010/11*, Statistical Report. Volume One and Two. Central Bureau of Statistics, Kathmandu.

Ohu, J.O., Ekwue, E.I. and Folorunso, O.A. (1994) The effect of addition of organic matter on the compaction of a vertisol from northern Nigeria. *Soil Technology* 7(2), 155–162.

Ojeniyi, S.O. and Agboola, A.A. (1995) A review of tillage requirement of crops in Africa. *Africa Soils* 28, 259–266.

Opoku, G. and Vyn, T.J. (1997) Wheat residue management option for no-till maize. *Canadian Journal of Plant Science* 78, 207–213.

Ozpinar, S. (2009) Tillage and cover crop effects on maize yield and soil nitrogen. *Bulgarian Journal of Agriculture Science* 15, 533–543.

Parmar, D.K. and Sharma, V. (2001) Nitrogen requirement of single hybrid maize (*Zea mays* L) wheat (*Triticum aestivum*) system under rainfed conditions. *Indian Journal of Agricultural Sciences* 71(4), 252–254.

Parr, J.F., Papendick, R.I., Hornickand, S.B. and Meyer, R.E. (1990) The use of cover crops, mulches and tillage for soil water conservation and weed control. In: Pushparajah, E. and Elliot, C.R.

(eds) *Organic Matter Management and Tillage in Humid and Sub-humid Africa*. IBSRAM Proceeding No 10. IBSRAM, Bangkok, pp. 246–261.

Paudyal, K.R., Ransom, J.K., Rajbhandari, N.P., Adhikari, K., Gerpacio, R.V., *et al.* (2001) *Maize in Nepal: Production Systems, Constraints and Priorities for Research*. NARC and CIMMYT Nepal, Kathmandu, p. 48.

Pelegrin F., Moreno, F., Martin-Aranda, J. and Camps, M. (1990) The influence of tillage methods on soil physical properties and water balance for a typical crop rotation in SW Spain. *Soil Tillage Research* 16, 345–358.

Ponsica, E.P., Escalada, R.G. and Quirol, B.F. (1983) Effect of animal manure application on the growth and yield of corn. *Annals of Tropical Research* 5, 110–116.

Purushottam, K. and Puri, U.K. (2001) Effect of nitrogen and farm yard manure application on maize (*Zea mays* L.) varieties. *Indian Journal of Agronomy* 46(2), 255–259.

Rice, C.W., Grove, J.H. and Smith, M.S. (1987) Estimating soil net nitrogen mineralization as affected by tillage and soil drainage due to topographic position. *Canadian Journal of Soil Science* 67, 513–520.

Sainju, U.M. and Singh, B.P. (2001) Tillage, cover crop, and kill-planting date effects, on corn yield and soil nitrogen. *Agronomy Journal* 93, 878–868.

Sayre, K.K. and Hobbs, P. (2010) The raised-bed system of cultivation for irrigated production conditions. In: Devkota, M.K., Martius, C., Sayre, K.D., Egomberdiev, O., Devkota, K.P., *et al.* (eds) *Performance of Maize under CA on Salt-affected Irrigated Croplands in Uzbekistan, Central Asia*. Conference on International Research on Food Security, National Resource Management and Rural Development, pp. 337–355. Available at: http://www.tropentag.de/2010/abstracts/full/1022.pdf (accessed 13 November 2014).

Singh, B.K., Sharma, N., Rana, D.S. and Kapur, M.L. (1979) Effect of applying farmyard manure to maize-wheat rotation on the distribution of nitrate nitrogen in soil profiles. *Journal of Research, Punjab Agricultural University* 16(1), 150–154.

Singh, C.M., Sharma, P.K., Kishor, P., Mishra, P.K., Singh, A.P., *et al.* (2011) Impact of integrated nutrient management on growth, yield and nutrient uptake by wheat (*Triticum aestivum* L.). *Asian Journal of Agricultural Research* 5, 76–82.

Thiagalingam, K., Dalgliesh, N.P., Gould, N.S., McCown, R.L. and Cogle, A.L. (1996) Comparison of no-tillage and conventional tillage in the development of sustainable farming systems in the semi-arid tropics. *Australian Journal of Experimental Agriculture* 36, 995–1002.

Troeh, F.R., Hobbs, J.A. and Donahue, R.L. (1980) *Soil and Water Conservation: For Productivity and Environmental Protection*. Prentice Hall, Inc, Englewood Cliffs, New Jersey.

Vadivel, N., Subbian, P. and Velayutham, A. (2001) Effect of integrated nitrogen management practices on the growth and yield of rainfed winter maize (*Zea mays* L.). *Indian Journal of Agronomy* 46(2), 250–254.

Verma, N.K. (2011) Integrated nutrient management in winter maize (*Zea mays* L.) sown at different dates. *Journal of Plant Breeding and Crop Science* 3(8), 161–167.

9 Soil Quality in Conservation Agriculture Production Systems (CAPS) of Rainfed, Sloping Land Farming in the Central Mid-hills Region of Nepal

SUSAN CROW,[1*] OLIVIA SCHUBERT,[1] BIR BAHADUR TAMANG,[2] THEODORE RADOVICH,[1] BIKASH PAUDEL,[1] JACQUELINE HALBRENDT[1] AND KESHAB THAPA[2]

[1]*University of Hawai'i at Mānoa, Honolulu, Hawaii, USA;* [2]*Local Initiatives for Biodiversity, Research, and Development (LI-BIRD), Pokhara, Nepal*

9.1 Introduction

9.1.1 Nepal

Nepal is a populous country (30.4 million people) of small size (147,181 km^2), with a growth rate of 1.35% per annum (NPHC, 2012; CIA, 2013). Agriculture is the main sector contributing to people's livelihood and the national economy, and nearly three-quarters of the total population depends primarily on agriculture (CSB, 1999). Nepal is landlocked on the southern slopes of the central Himalayas in South Asia. The topography, elevation, and climatic conditions of Nepal are all wide-ranging. The country is commonly divided into three ecological zones: mountains, hills, and lowland areas called *terai* (Khatri-Chhetri and Maharjan, 2006). Approximately 86% of the total area of Nepal is high mountains and rolling hills ranging from 70 to 8,850 m in elevation, while the other 14% of land is classified as *terai* (Chaudhary, 2000). This grand elevation gradient provides for a multitude of ecological zones including tropical lowlands, temperate valleys, and alpine meadows (Basnet, 1992). The country's climate is divided into five primary zones that span the extremes: tropical, subtropical, temperate, subarctic, and arctic. Almost equal portions of the population reside in the *terai*, 46.7%,

*Corresponding author; e-mail: crows@hawaii.edu

and the hills, 45.5%, with the remainder distributed throughout the highlands (Chaudhary, 2000). The average landholding in Nepal, including both productive and unproductive land, is approximately 0.8 ha (CSB, 1999).

The cultural diversity of Nepal is as extensive as its ecology, with over 100 recognized languages and more than 100 ethnic groups. Nepal has seen vast political shifts throughout time, which have influenced not only its social and economic atmospheres but also the connection between the Nepalese and their natural environment. From the 11th century to the 19th century, Nepalese agriculture was primarily pastoralism dominated by shifting cultivation. Farming practices transitioned to a more permanent type of agriculture during the 1800s (Rasul and Thapa, 2003; Raut *et al.*, 2010). The country's population was fairly stable from the late 1700s to the 1900s and the growth rate was at or below 1% (Schroeder, 1985), which was sustainable with minor shifts in agricultural production systems. For example, crops such as rice, barley, legumes, and several species of millet are native to this region, but diversification increased, particularly in the upslope *bari* lands (slightly sloping or plain upland without bound, which is often non-irrigated, is called *bari* land), in the 18th century with the introduction of maize and potatoes (Joshi *et al.*, 2012). During the political rule of the Rana dynasty (1846–1951), the feudal elite blocked modern influences from entering Nepal, simultaneously seizing control of a quarter of the country's farmland to pursue personal endeavors such as the construction of extravagant palaces, thus limiting access to land and thereby stunting agricultural growth. Some claim that the Rana rulers caused development in Nepal to fail by monopolizing power and wealth, impoverishing Nepal's majority (Metz, 1995).

With a prosperous natural environment and increased government backing for immigration (Blaikie *et al.*, 1980), Nepal's population began increasing after WWI, growing from 5 million in 1930 to 11.6 million in 1971 (Schroeder, 1985). In response to the burgeoning population, an economic development agency named Udyog Parishad was created in the 1930s to promote the growth and extension of agricultural, industrial, and commercial activities in Nepal (Raut *et al.*, 2010). The goals of the agency, including improving farming techniques and irrigation, were carried further after the termination of the Rana rule and the arrival of foreign development agencies such as the Food and Agriculture Organization (FAO) and the US Agency for International Development (Raut *et al.*, 2010). Nepal initiated formal development planning in 1954 with the first 5-year plan, with agricultural development as one of the primary objectives, receiving 20% of budget disbursements to develop villages and improve irrigation. The borders of Nepal were closed to outsiders until 1949; hence, the people and economy have historically been largely reliant on internal resources (Basnet, 1992).

Currently, roughly 83% of the people in Nepal remain dependent on subsistence farming and forests to provide food, energy, shelter, medicine, and wood; therefore, the economy counts on efficient use of natural resources (Acharya, 2003). Approximately 32,000 km^2 of land is cultivated, and an additional 10,000 km^2 could be converted to agriculture (LRMP, 1985). To increase supply and create more employment opportunities, hilled areas have been intensely developed. However, this movement has been challenging, due to rugged topography, poor infrastructure facilities, and limited cultivated areas (Sharma, 1997). Deforestation and soil degradation are both pressing environmental concerns in Nepal, where population pressure has driven many to convert forested areas to farm and pasture land (Sitaula *et al.*, 2005), and even the steepest of terrain has been cultivated to meet the growing population's food demand (Maskey *et al.*, 2003). Nepal experienced greater flow of resources across its borders during the latter half of the 20th century, yet resource production and utilization within the country remained imperative.

9.1.2 Middle-hills region

Nearly three-quarters of the land area of Nepal is hilly or mountainous, and approximately half the human population inhabits these areas. Therefore, it is essential to understand methods of food production and basic survival in this challenging terrain. The middle-hills, or mid-hills, region of Nepal rests between 1,000 and 1,600 m in elevation (Pilbeam *et al.*, 2000), and stretches across the country from east to west between the Himalayan Mountains to the north and the *terai* plains to the south (Adhikari *et al.*, 2004). The middle-hills physiographic region makes up 4,350 km^2 (or 29.5% of the land area) of Nepal. Water availability in the hills region is often a limiting factor on agricultural output, due to the strong seasonality of rainfall in areas affected by monsoon rainfall, which typically occurs between June and October (Schroeder, 1985). In the past, the majority of the Nepalese population inhabited this high elevation region because of the fear of malaria in low-lying areas. After eradication of malaria in the late 1950s, the rate of hill to *terai* migration increased (Gartaula and Niehof, 2013), which resulted in the clearing and settlement of *terai* and inner-*terai*. Moreover, the soils in the hills were suitable for terracing, the climate generally allowed for two cropping cycles, and adjacent forested areas provided most customary household needs (Carson, 1992). With favourable climatic and environmental conditions, small-scale agriculture in the hills was prosperous. Traditional subsistence farming has been in place for hundreds of years; however, cropping intensification to accommodate growing populations in recent decades has stressed the land's natural resources (Begum *et al.*, 2010). Today, traditional agricultural production systems are still practiced by marginalized, impoverished farmers belonging to smallholding tribal communities. The average household in the mid-hills region owns 0.5–1 ha of land (Thapa, 1989), 3.3–4.4 units of livestock (e.g. one bullock, one buffalo, two cows, two goats, and poultry) (Sharma, 1996), and a stand of 30–150 trees (Wyatt-Smith, 1982; Thapa, 1994). Since the mid-hills region is often in food deficit, food is commonly imported from the plains to areas accessible by road.

There is a complex bond between humans and the environment in the mid-hills, due to the remoteness of the area, traditions, and the high level of poverty. A historical relationship exists amid forestry, livestock, and crop production that is crucial for subsistence farming. The forest supplies fodder and bedding for livestock, which provide manure to fertilize the soil and power to work the fields that grow food (Desbiez *et al.*, 2004). A recent imbalance in resource flow, largely due to declining soil quality, between the forest, livestock, and cropping schemes has affected the agricultural system negatively (Pilbeam *et al.*, 2005). Consequently, biodiversity and soil fertility are in decline (Shrestha *et al.*, 2000). Furthermore, intensification of agriculture is resulting in a decline of soil health, climate change is altering the growing season and environment, and lack of land tenure or right to harvest forested lands is decreasing incentives for responsible use of forest resources. After the 1980s, the community forestry program handed over the management of forests to local communities. This program was successful, but due to new regulations enforced by community forestry units, free access to forests by people has been controlled. This shift in forestry management has led to a reduced number of livestock because it created scarcity of pastureland and reduced fodder availability. The decreased availability of manure reflects reduced livestock numbers. Together, these factors, among others, present challenges to the development of meaningful soil fertility and sustainable management programs.

9.1.3 Soil fertility

Land productivity is influenced by soil fertility, which includes both the physical and chemical properties of the soil. The average farm in the hills yields 1.2–2.8 t/ha of crops such as rice, wheat, maize, and millet, which is slightly lower than adjacent terrain regions (MoAC, 2011). Diminishing soil fertility (Tuladahar, 1994), due to poor soil management and crop husbandry as a result of lack of knowledge, contributes to this low crop yield (Tiwari *et al.*, 2004). Addressing the current issues of declining crop yields and soil fertility status in the mid-hills region of Nepal requires a fundamental understanding of the complex nature of the mountain farming systems. A dynamic Nepalese landscape, characterized by continental plate tectonics and resulting in frequent landslides, debris flows, floods, and high natural erosional rates, has led to the development of high maintenance and risk-prone agricultural systems (Carson, 1992). This ever-changing landscape is formative to soil fertility, in both loss and regeneration, and soil management must be considered within its context.

A central component of the mountain farming system in the middle-hill region is the use of farmyard manure (FYM) and compost (in few instances) as a soil amendment. The use of tree fodder and forage as animal feed facilitates the net movement of nitrogen (N) from non-agricultural land to agricultural land through the production of manure (Pilbeam *et al.*, 2000). Availability of FYM currently is in decline due to a reduction in the number of livestock, partly caused by fodder shortages and labor constraints as villagers increasingly leave to seek employment off site (Pilbeam *et al.*, 2005). Goats, which are common as household livestock in the mid-hills, have high concentrations of potassium (K) and phosphorus (P) in their manure, which aid in the growth of a healthy crop (Peacock, 1996). Some farmers utilize chemical fertilizers in an effort to maintain crop production and compensate for tree fodder and manure losses; however, total natural and synthetic nutrient inputs remain low (Brown, 1997).

The application of FYM and compost in these farming systems serves many integrated purposes for soil management and sustainability. These purposes include improvement in the soil water regime (e.g. improved infiltration and water-holding capacity, resulting in reduced runoff and erosion), nutrient content (particularly N, P, and to a lesser degree sulfur (S) and other plant micronutrients), and amelioration of negative impacts due to soil acidity and coarsely textured soil (Carson, 1992). Natural fertilizers such as manure and compost can provide four times the amount of nitrogen to crops than chemical fertilizers (Pilbeam *et al.*, 2000). Chemical fertilizers often offer rapid improvements in yield, which may be desirable to farmers in the short term, but are harmful to soil quality and fertility in the long term (Tiwari *et al.*, 2008). As the organic matter levels drop because manure compost is no longer added for nutrients, soil structure (the way individual particles are arranged within a soil) breaks down. Loss of soil structure results in decreased water-holding capacity, drainage, and increased erosion. In the mid-hills region, increased decomposition of organic matter following plowing, declining FYM and compost availability, and diversion of FYM and compost to the irrigated lands (Carson, 1992) all cause declines in organic matter, health, and loss of soil quality. But soil fertility maintenance is vital to meet the minimum food and resource needs of the Nepalese people (Brown *et al.*, 1999). For this reason, implementing sustainable agricultural practices throughout the mid-hills region of Nepal is a well-advised approach to conserve soil and improve crop yields.

9.1.4 Sustainable intensification of agriculture

The challenge of sustainable intensification of production systems despite significant global change pressures that face many developing countries is exacerbated in the mid-hills region of Nepal by soils that generally are characterized by poor nutrient content and great susceptibility to erosion (Schwab, 2012). Incorrect farming techniques, along with poor management of natural resources in watersheds, has accelerated soil erosion, in turn harming agriculture productivity and increasing downstream sedimentation of dams, reservoirs, and irrigation systems (Chaudhary, 2000). When yield is subsequently low, many individuals and whole families choose to outmigrate to urban areas in search of work and more readily available food (CIMMYT, 2001). Many people leaving the hills region have moved to the flat *terai* lands to farm, sought wage employment in India, or entered the mercenary service in the British and Indian armies (Schroeder, 1985); and more recently, migrated as low-paid construction and agriculture workers in Arab and Southeast Asia. That is why current agricultural practices and patterns must be modified to improve food security for a growing population, ensure employment opportunities, and conserve the land for future generations.

The objective of conservation agriculture is to use agricultural resources more efficiently than conventional agriculture through the integrated management of available soil, water, and biological resources, while minimizing external inputs (FAO, 2000; Garcıa-Torres *et al.*, 2003; Knowler and Bradshaw, 2007). As such, conservation practices that minimize the rate of soil degradation and improve soil fertility while increasing yields must be employed (Neupane and Thapa, 2001). Conservation agriculture practices, such as reduced (i.e. minimal or zero) tillage, cover crops, the use of high-yield crop varieties, organic and inorganic fertilizers, and intercropping, are employed to enhance agriculture in a sustainable way (Raut *et al.*, 2010) and are integral to sustainable intensification of production.

The addition and maintenance of organic matter in soil is a fundamental component of successful sustainable agriculture (Shrestha *et al.*, 2009). Methods by which agriculture can be intensified sustainably include organic matter amendments and reducing tillage, both of which decrease erosion and increase nutrient retention. According to Troeh and Thompson (1993), the annual average rate of soil formation is estimated to be 1 t/ha under favorable climatic and topographic conditions; therefore, the rate of soil erosion should not surpass 1 t/ha in order to maintain sustainable productivity (Thapa, 1996).

Changes in soil organic matter (SOM) content commonly correlate with changes in water-stable aggregation (Puget *et al.*, 2000). The formation of water-stable aggregates (WSA) initiates with the physical aggregation of clay, followed by the amassing of macroaggregates clumped together by fungal hyphae, fine roots, and bacterial secretions (Paudel *et al.*, 2011). The fertility of soil depends on a stable structure, particularly the presence of macroaggregates (larger than 250 μm) (Puget *et al.*, 2000). Aggregates resist dispersion and reduce erosion (Chakrabarti, 1969), but type, vegetation, soil and crop management, and the duration of management, may influence the stability of soil aggregates (Paudel *et al.*, 2011). In particular, macroaggregation is sensitive to land-use change and cultivation practices (Chaney and Swift, 1984). Soil cultivation increases the loss of soil organic carbon (SOC) via the break-up of macroaggregates (Franzluebbers and Arshad, 1997). Better maintenance of SOC has been observed under zero tillage as opposed to conventional tillage at depths of 0–5 cm (Franzluebbers and Arshad, 1997).

Many farmers in Nepal still use labor-intensive customary tilling methods such as locally fabricated bullock-drawn plows and hand implements (Bajracharya, 2001), which accomplishes tillage to depths of 15 cm (dry land) or 20 cm (paddy soils) (Laryea *et al.*, 1994). So *et al.* (2001) suggest that although tillage leads to increased crop yield in the short term, it also contributes to the destruction of soil structure and associated decline in crop yield over time. Primary tillage such as plowing increases surface roughness, breaks crusts, and increases infiltration (Doolette and Smyle, 1990), but also escalates biodegradation rates in native soils (Balesdent *et al.*, 2000). Thus, in frequently tilled areas, soil erodes and nutrients are depleted at unsustainable rates (Thapa, 1996). Discarding tillage reduces losses of organic matter and physical structure and allows for the collection of crop residues on the soil, which form mulch and protect the soil surface from physical erosion from wind and rain. The presence of residues further improves soil aggregation and decreases losses of SOM (Beare *et al.*, 1994).

A goal of sustainable agriculture is to maximize the efficient utilization of inputs, while integrating long-term environmental and social uncertainties related to the outputs (Maglinao, 2000). Some Nepalese farmers are taking steps to increase the sustainability of their agricultural practices, such as using sediment-laden runoff from rivers and streams, *in situ* manuring, terrace riser slicing, and incorporating N-fixing plants (Maskey *et al.*, 2003). Other conservation techniques involve technological innovations. Many models of the sloping agriculture land technology (SALT) farming system have been tried in Nepal. The goals of SALT are to use N-fixing hedgerows to facilitate soil conservation, and to improve soil fertility via bionitrogen fixation and biomass application after hedgerow pruning (Maskey *et al.*, 2003). In light of the deteriorating agricultural situation in the mid-hills, the government of Nepal started offering a package of improved practices through its agriculture extension system, which focused on the preparation and utilization of organic matter such as FYM (Shrestha *et al.*, 2009). However, the scale and accessibility of government extension is too limited, while the scale of projects directed by non-governmental organizations (NGOs) is too small to have any significant effect.

9.2 Case Study Background

9.2.1 Site description

The Chepang, also called Praja, are one of several ethnic minorities of the western hills of Nepal, who inhabit the sloping landscapes of the Tanahun, Chitwan, Makawanpur, Dhading, and Gorkha districts. This study focused on the Chepang communities of Hyakrang village of the Dhading district; Thumka village of the Gorkha district; and Kholagaun village of the Tanahun district (Fig. 9.1). These villages are located in the Trishuli River watershed area and near the Prithivi highway, which links Kathmandu to Pokhara (Fig. 9.2). All three villages are located in the Lesser Himalayan geologic zone, which have unfossiliferous, sedimentary, and metasedimentary rocks such as slate, phyllite, schist, quartzite, limestone, dolomite, etc., ranging in age from Precambrian to Eocene (Upreti, 1999; Dahal, 2006). There is no road connecting the villages with the highway;

Fig. 9.1. Location of Nepal and the three districts where the study was conducted.

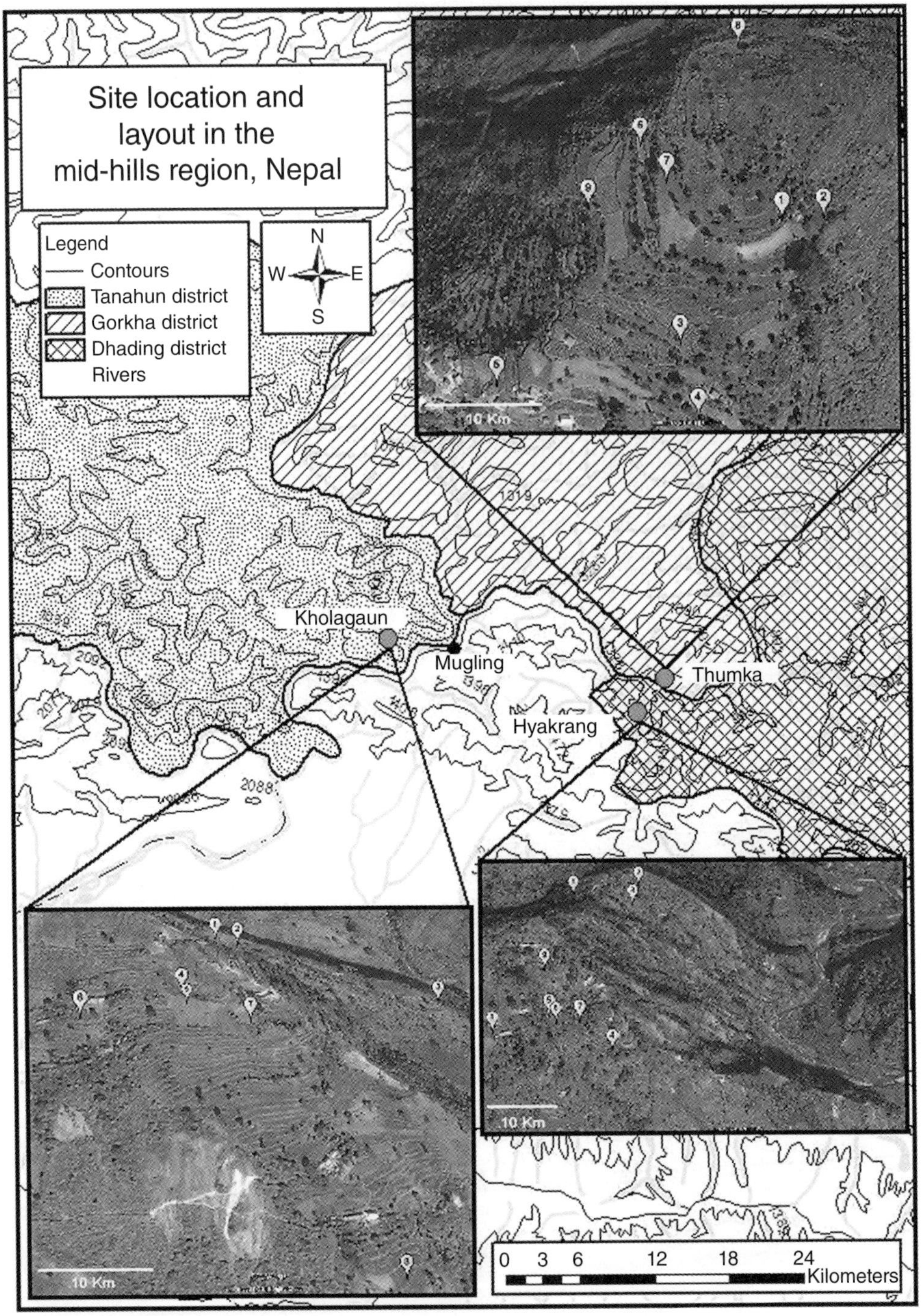

Fig. 9.2. Location of study villages, Kholagaun, Hyakrang, and Thumka, and the farmers' fields selected within each village.

consequently, the research villages are only accessible by foot. This isolation limits the inflow of goods and services such as new technologies, health resources, and food. The Chepang generally live in poor socio-economic conditions and have few livelihood alternatives and limited access to public services such as health, sanitation, and education. Women and children are especially marginalized when it comes to health, education, and workload (Joshi *et al.*, 2012).

Two common categories of agricultural lands in the mid-hills region are *bari*, or dry crop lands, and *khet*, or paddy lands (Thapa, 1996). *Bari* are unbunded (without ridges or embankments) rainfed lands, typically located on the upper slopes, while *khet* land is generally bunded (ridged) land on irrigated lower slopes (Tiwari *et al.*, 2004). Otherwise, available land is most often used as shifting cultivation plots, as they are small-scale, less productive, and sloping. Shifting cultivation fields are prepared by clearing natural forested vegetation, allowing the land to dry, and then burning it (Sharma, 2011). The ash infuses nutrients into the soil, which promotes crop growth until the fertility of the soil is depleted. The space is then left to regenerate while cultivation is moved to another area. Thus, a third category of agricultural land, termed *khoria*, involves the cultivation of crops followed by a fallow period between 3 and 5 years in length and lastly the practice of slashing and burning the land. Approximately 71% of shifting cultivation takes place on land with a slope greater than 30% (Balla *et al.*, 1999). *Khoria* land is commonly located above or below the villages adjacent to the forest line (Adhikari and Bohle, 1999). Rainfed *bari* land is usually more fertile than *khoria* land, and many farmers transition *khoria* areas into *bari* areas (Gautam, 2006). Nevertheless, all land classifications are influenced by the presence and health of the nearby wooded areas.

Farmers utilize *bari*, *khet*, and *khoria* land to grow a wide variety of crops to meet not only their household consumption requirements but also cultural and social needs, including medicinal ingredients. Local varieties of maize and millet cereal crops are staple foods in the villages. Most farmers plow the land using a traditional plow driven by a pair of oxen at least twice annually. The first plowing is done immediately after the first winter showers to till the fallow land left after harvesting the last crop (millet and other leguminous crops such as black gram, soy bean, rice bean, and horse gram). The second plowing is done in March to sow maize seeds in the furrows behind them; however, some farmers with sufficient labor and oxen plow the land twice in this season. The incorporation of manure into the soil often takes place during the plowing process. In some cases, manure is continually piled then allowed to ripen for 3–4 months before it is carried to the fields and applied during cropping. Alternatively, manure and bedding that have accumulated over the previous year are dug by women and transported to the field several weeks prior to the spring cultivation. Weeding of maize fields is usually done twice (at 30 and 60 days after sowing) using a hand spade. Before 30 days of maize harvesting, millet seedlings are transplanted in the maize field. In several areas, farmers may apply chemical fertilizer (urea and diammonium phosphate (DAP)) at the base of the plants while weeding. The second weeding is done especially to transplant millet seedlings prepared separately in a nursery bed. The maize cobs are harvested in August and stover is collected immediately to use as cattle fodder. The remaining millet crop then stands in the *bari* land until harvested in October. The millet stover is also removed for cattle feed. Finally, the land is left fallow as free-grazing land until tilled for the next year's maize crop in March. In recent years, some village farmers have cultivated cowpea instead of millet after maize. In this case, farmers do minimum tillage and sow cowpea seeds before harvesting the maize (75% maturity) with a hand spade. During cowpea sowing, farmers slash the leaves of the maize plants below the cob. Generally, farmers remove the upper part of the biomass from its cob for green fodder. Cowpea is harvested with its pods at the end of October or in the first week of November. The cowpea biomass is often removed from the field and used as cattle fodder.

9.2.2 On-farm evaluation of CAPS

A participatory research approach was employed with farmers of the central mid-hills region of Nepal to identify different conservation agriculture practices suitable for maize-based farming in the three selected villages. At each of the three villages, the identified conservation agriculture production systems (CAPS) were implemented in on-farm evaluation trials. Conservation agriculture practices that were selected for the first research phase included reduced tillage and intercropping of a legume species with millet in the second cropping season of the year (see below for research method details).

According to the FAO (2000), reduced till is "tilling the whole soil surface but eliminating one or more of the operations that would otherwise be done with a conventional tillage system", or any farming practice that involves less soil disturbance than used in conventional practices in a given region (Atreya *et al.*, 2006). Benefits of tillage include weed control to reduce competition for water and nutrients (So *et al.*, 2001), and also N mineralization and subsequent breakdown of organic matter where inadequate fertilizer is applied (Laryea *et al.*, 1994). On the other hand, avoiding tillage may improve soil aggregation and retention of SOM (Beare *et al.*, 1994). Obtaining more technologically advanced machinery to modernize farming practices is often alluring to farmers, researchers, and policy makers, although guaranteed benefits to crop yield do not outweigh the potential disadvantages (So *et al.*, 2001). By implementing reduced tillage, soil and nutrient losses are minimized without compromising economic yields (Atreya *et al.*, 2006). In some cases, reduced tillage allows for harvest residue build-up and hardening of the soil surface, both of which prohibit plant establishment and consequently lower yield (Koch *et al.*, 2009). Ideally, implementing reduced or zero tillage would produce only desirable results. However, as with many new techniques, it often takes many years to see optimum outcomes. Atreya *et al.* (2006) found a decrease in plant height, cob height, and dry stover when reduced tillage was implemented in maize fields as opposed to conventional tillage. They attributed this shortcoming to lack of root development and hence nutrient access. Conversely, grain yield in the same study was not affected, and overall soil nutrient losses from conventional tillage were much higher than from reduced tillage.

Cropping patterns are another method that affects soil fertility and crop yield. Intercropping is the practice of growing at least two crops simultaneously in the same field, which is often a more sustainable method than monocropping. The benefits of intercropping include reduced risk of total crop failure and reduction of weeds, as well as physical aids such as one plant type providing shade to another (Horwith, 1985). Implementing intercropping may also decrease the need for synthetic fertilizers by alleviating soil erosion (Siddoway and Barnett, 1976) and by incorporating N-fixing species as the intercrop. Upadhayay *et al.* (1990) suggest that intercropping can provide production advantages over monocropping in instances without added external inputs, due to more efficient utilization of resources.

Intercropping is common in the hills of Nepal, especially among farmers in areas where land is limited (Prasad and Brook, 2005). Maize is often intercropped or relayed with millet or legumes, and is an important source of protein and calories. An estimated 80% of cultivated land in the mid-hills is maize (CIMMYT, 2001). Nearly two-thirds of the maize produced in Nepal is for human consumption (Tiwari *et al.*, 2004). The growth period of maize is normally 125–135 days at elevations between 1,200 and 2,000 m, making it difficult to raise successive crops. Therefore, farmers typically employ intercropping and relaying crops to make effective use of the short rainy season (Prasad and Brook, 2005).

9.2.3 Research methods

Environmental data

A HOBO Data Logger (Onset Computer Corporation, Massachusetts, USA) Microstation system was installed at each research site to observe environmental conditions during the 2-year research period. The measurements recorded included air temperature, soil temperature (Tidbit Data Logger v2, Onset Computer Corporation, Massachusetts, USA), precipitation (tipping bucket), photosynthetically active radiation (PAR), relative humidity, and soil moisture. The automated information was recorded and regularly downloaded and averaged or totaled, as appropriate for the individual parameter, weekly. Station malfunctions and download failures occurred often, due to the remote location and logistics of replacing parts. When data are missing due to technical failure, weekly data are not reported and a break occurs between data points. For PAR and precipitation, a horizontal line indicates missing data in the weekly totals.

Soil collection

Within the three study villages (Hyakrang, Kholagaun, and Thumka), nine *bari* fields were chosen for study participation. Baseline soil samples were collected beginning in March 2011 from each of the selected fields prior to the establishment of experimental units (see below). Fifty-four composited samples (two depths for each of the nine fields in each of the three villages) were taken from the *bari* land after the preparation tillage following the first rain but prior to cultivation of the first-season maize. Within each of the 27 experimental plots, approximately 200 g of soil was collected randomly from five locations in an X pattern using an auger with a diameter of 2.5 cm and a length of 10 cm. Edge effects on the soil sampling were avoided by establishing a 2 m buffer around the plot perimeter and samples were collected within the interior of the plot. Two samples per coring location were taken in increments of 0–5 cm and 5–10 cm below the soil surface. Soil samples from each depth at each field were spread on a large clean plastic sheet and mixed together. Clods were thoroughly broken and vegetative material was removed to create a uniform mixture. The mixture was then divided into quarters and two diagonal parts were retained. The process was repeated until the composite sample was approximately 200 g. Moist field samples were placed in numbered plastic bags, transported to Pokhara, Nepal, and air-dried for 3–6 weeks. Two additional volumetric cores (one from each depth) were collected to determine the bulk density (BD) of soil from each plot.

Soil analysis

Air-dried subsamples were ground and analysed using standard methods in the Local Initiatives for Biodiversity and Research (LIBIRD) laboratory in Pokhara (Table 9.1). Subsamples of each soil were ground and transported to the Virginia Tech Soil Testing Laboratory at Virginia Technological University, Blacksburg, Virginia, USA, for additional analyses. The pH was determined using a wet pH method (Kalra, 1995), and buffer pH was determined using the Mehlich Buffer pH method (Mehlich, 1976). Concentrations (mg/kg) of phosphorus (P), potassium (K), calcium (Ca), magnesium (Mg), zinc (Zn), manganese (Mn), copper (Cu), iron (Fe), and boron (B) were determined using the Mehlich-1 extractant by means of inductively coupled plasma (ICP). Effective cation exchange capacity (ECEC) was determined and base saturation (BS), Ca^{2+} saturation, K^{+} saturation, and Mg^{2+} saturation were calculated from the ICP results.

CAP treatments

To test the effect of CAPS treatments including intercropping and reduced tillage on crop yield and soil quality parameters, a randomized complete block design with three villages as blocks and each farmer's field in a village as a replication was employed. Hence, there were eight replications of the treatments in each village, and each replicate (field) contained all four CAPS treatments, including the control. Thus, this study examined four treatments in total (Table 9.2). Initially, the treatments were assigned randomly to avoid bias, but once assigned, the treatment plots were constant throughout the entire experimental period (2011–2013). Treatment 1 was a maize crop in the first season (March–June), followed by millet transplantation in the second season (July–October), and then a fallow period.

Table 9.1. Soil analyses and methodology completed at LIBIRD.

Analysis	Method of measurement
Bulk density	Replacement method
Organic matter (%)	Walkley–Black
Nitrogen (%)	Kjeldahl titration
Available potassium (K) and phosphorus (P)	Ammonium acetate extraction
pH	Electronic pH meter in water
Texture	Hydrometer

Table 9.2. Experimental treatments for the on-farm evaluation of the conservation agriculture production system (CAPS). In year 1 of the trial, the legume was cowpea, in year 2 the legume was switched to black gram.

	Cropping pattern			
Treatment	1st season (March–June)	2nd season (July–October)	Practice	Tillage
1	Maize	Millet	Sole crop	Conventional
2	Maize	Legume	Sole crop	Conventional
3	Maize	Legume + millet	Intercropping	Conventional
4	Maize	Legume + millet	Intercropping	Strip tillage

Conventional tillage, in which farmers plow the land twice a year, was implemented in this treatment. After thorough consultation with local farmers, it was determined that this was the traditional agricultural production system in the village practiced by the farmers and represented the experimental control. Treatment 2 was maize followed by cowpea or black gram, which essentially was done with similar cultural practices as maize–millet, except that farmers dug and sowed cowpea seeds as opposed to transplanting the millet, and broadcasting black gram, with conventional tillage. This system has grown in popularity among the farmers in recent years, due to market influence. Treatment 3 was maize followed by intercropping of millet and legume with conventional tillage. The intercropping was done in alternate rows of millet and legume. Maize followed by millet and a legume intercropping and strip tillage (i.e. a reduced tillage treatment that decreased the area tilled by two-thirds) was treatment 4. Each treatment plot consisted of 20 m^2 (4 m × 5m).

Soil samples from each plot were collected at 6 months (following two crop cycles) and at 2 years after the onset of the experimental trials. Samples were collected and processed in the same manner as the baseline samples. Organic matter concentration, N concentration, and extractable P and K were determined at the LIBIRD facility in Pokhara.

Statistical analysis

The baseline soil physical and chemical properties were examined using a multivariate statistical approach to identify suites of characteristics that distinguished villages from one another. This approach allows the simultaneous analysis of multiple, covarying responses that synergistically combine to define the soil environment (Peck, 2010). A principal components analysis (PCA), which has an underlying linear model form appropriate for relatively homogeneous data (Peck, 2010), was performed using PC-ORD 6 (McCune and Mefford, 2011). Multiple "stopping rules", for example observed eigenvalue as compared to randomizations and broken-stick eigenvalue, concurred that the first three axes should be interpreted (Peck, 2010). Pearson and Kendall correlations (r values) with the ordination axes were used to aid in the interpretation. For each direction of each axis, the variables with absolute r values greater than 0.75 were discussed. If no variables had r values above 0.75 for a given axis, then the three variables with the greatest correlations for each direction were discussed.

A randomized block design (with each village as a block), one-way ANOVA (PROC MIXED in SAS v. 9.2) was used to compare means for differences in soil parameters (i.e. N, P, K, and OM) among villages after 6 months and 2 years. For this analysis, the mean of eight farmer plots in each village for each treatment was used, thus the sample size was three. The 6-month and 2-year soil samples were taken during different points in the cropping season, so no direct comparisons were made between sampling dates. Because no significant differences were detected among CAPS treatments, the mean of treatment 1 to treatment 4 was taken for each farmer plot and used as a replicate to test for significant differences among villages for the same soil parameters using completely randomized design, one-way ANOVA (PROC MIXED, SAS v. 9.2). For this analysis, the sample size was eight. Significant differences among means were determined by a least significant difference (LSD) *post-hoc* test. Qualitative comparisons were also made and discussed for observed patterns in results among villages, treatments, and comparing baseline to subsequent sampling dates.

9.3 Results and Discussion

9.3.1 Environmental variables

Climatic variables during the study period were dynamic across the three villages. The average total weekly photosynthetically active radiation (PAR), which is positively correlated to yield (Brown and Rosenberg, 1997), for the study period ranged from 185,748 μE in Kholagaun, 191,078 in Thumka to 322,426 μE in Hyakrang (Fig. 9.3a–c). Temperatures in Hyakrang and Kholagaun were lower from October 2011 to February 2012 than in the spring and summer months, but seasonal changes were generally low, consistent with subtropical and tropical climate. It is difficult to discern patterns in the Thumka data due to missing data points. Relative humidity was fairly consistent throughout the trial period, ranging from 44.16% to 99.54% in Hyakrang, from 44.56% to 99.75% in Kholagaun, and from 19.06% to 77.60% in Thumka (Fig. 9.3). Temperature and relative humidity exhibited a typical inverse relationship, as evident in Kholagaun, when relative humidity tended to decrease between January and June while temperature increased. Precipitation and water content were greatest during the monsoon months in both Hyakrang and Kholagaun. Mean water content parallels rainfall, increasing with almost no lag time as precipitation increased. There was a decrease in PAR between June and September 2012 in Kholagaun, which coincided with increased precipitation during the monsoon season (Fig. 9.3).

Timing of the seasons, occurrence of early rains, onset, intensity, and duration of monsoon rains dictate the planting cycles and crop success in rainfed farming. Even with the adoption of sustainable practices, yield is dependent on favorable weather conditions (Pandey *et al.*, 2009).

Climate change threatens this rhythm and adds to the uncertainty faced by subsistence farming communities. Intensifying crop production will inherently be limited by environmental factors such as available crop-growing days, landforms, and soil moisture (Chhetri, 2011). Furthermore, reports predict that developing countries in low latitudes may be the most vulnerable to climatic variation in temperature, precipitation, and frequency and magnitude of extreme weather events (Poudel and Kotani, 2013). An increase in temperature could reduce crop yield and encourage pest populations (Nelson *et al.*, 2009), and enhanced variation in rainfall increases crop susceptibility to disease (Ludi, 2009), while events of prolonged

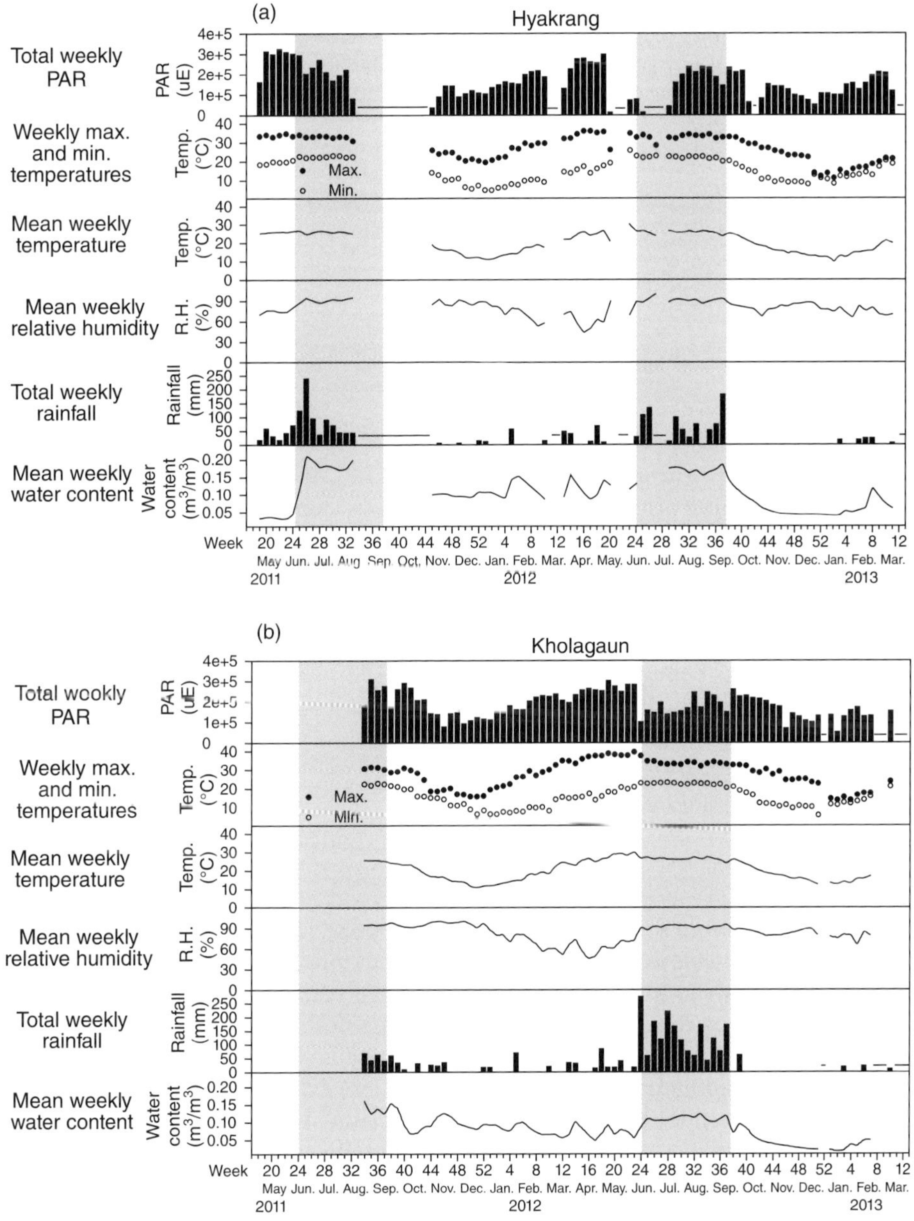

Fig. 9.3. Climate data for Hyakrang (a), Kholagaun (b), and Thumka (c). Shaded bars represent the typical period of the monsoon season.

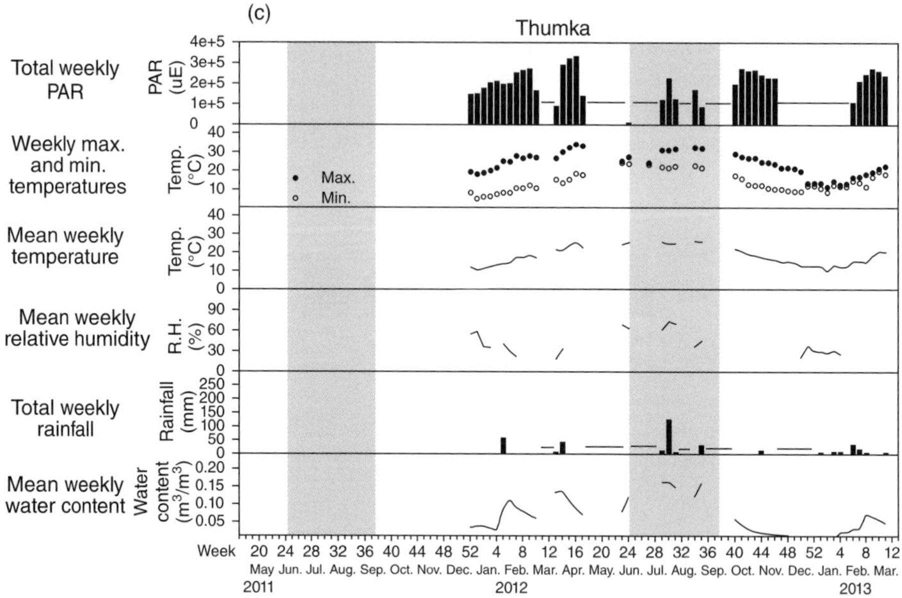

Fig. 9.3. Continued

drought reduce crop yield due to soil moisture shortages (Bates *et al.*, 2008). On the other hand, as approximately 67% of agricultural land in Nepal is rainfed, increased precipitation may benefit productivity in some cases (Poudel and Kotani, 2013). In summary, the findings of these studies suggest that the relationship between climate and agriculture is not consistent throughout Nepal and differs among our villages, and further study of the mid-hills of Nepal may elucidate how climate impacts productivity.

9.3.2 Baseline soil quality conditions

Physical, chemical, and biological measures

Coarse-textured soils with very little clay content characterized all of the villages (Table 9.3). This is consistent with the distribution of loam and sandy loam soils throughout the hill regions in Nepal, where the soil is predominantly derived from micaceous parent material (e.g. phyllites and schists; Carson, 1992). Loamy soils can be plowed easily whether moist or dry, and maintain good water-holding capacity and aeration for roots (Carson, 1992). Stoniness was quite variable throughout the region, consistent with the study villages, and could cause problems with tillage if severe but generally did not impede plant growth (Carson, 1992). Thumka was the only village with a substantial percentage of silt-sized particles with red-colored soil, which should increase the water-holding capacity of the soil compared to more coarse-textured soils. Kholagaun and Hyakrang were particularly rocky and coarse in texture.

Table 9.3. Measured soil physical, chemical, and biological parameters for the baseline samples collected from farmers' plots at each village in the study. Values are means ± one standard error.

	Hyakrang		Kholagaun		Thumka	
	0–5 cm	5–10 cm	0–5 cm	5–10 cm	0–5 cm	5–10 cm
Organic matter (%)	3.21 ± 0.29	2.68 ± 0.18	4.78 ± 0.25	4.24 ± 0.27	2.05 ± 0.14	1.87 ± 0.14
Nitrogen (%)	0.28 ± 0.02	0.26 ± 0.01	0.30 ± 0.01	0.30 ± 0.01	0.23 ± 0.01	0.23 ± 0.01
Phosphorus (ppm)	75.28 ± 11.91	65.77 ± 12.13	203.61 ± 33.24	203.08 ± 33.00	13.38 ± 5.96	7.27 ± 3.95
Potassium (ppm)	187.33 ± 44.02	111.93 ± 26.93	330.48 ± 45.66	311.38 ± 39.53	147.29 ± 26.40	72.10 ± 14.90
Bulk density (g/cm^3)	1.53 ± 0.11	1.55 ± 0.05	1.41 ± 0.13	1.52 ± 0.09	1.12 ± 0.05	1.24 ± 0.06
CEC (cmol/kg)	8.46 ± 1.56	8.56 ± 1.61	20.33 ± 1.36	22.00 ± 1.36	5.60 ± 1.13	6.14 ± 0.84
Cations (K^+)	7.17 ± 1.87	4.14 ± 0.78	11.68 ± 2.05	10.74 ± 1.72	5.59 ± 0.73	3.44 ± 0.21
Exch. acidity (cmol/kg)	0.13 ± 0.01	0.11 ± 0.01	0.17 ± 1.01	0.18 ± 0.01	0.11 ± 0.01	0.12 ± 0.01
pH	6.69 ± 0.18	6.71 ± 0.19	6.34 ± 0.09	6.44 ± 0.10	6.72 ± 0.14	6.82 ± 0.13
Sand (%)	68.72 ± 3.84	73.56 ± 2.68	75.67 ± 2.05	76.06 ± 1.87	54.89 ± 1.81	54.89 ± 1.35
Silt (%)	21.56 ± 4.18	17.11 ± 2.82	13.39 ± 1.53	13.00 ± 1.22	32.67 ± 1.89	32.11 ± 1.00
Clay (%)	9.72 ± 0.64	9.33 ± 0.17	10.94 ± 1.09	10.94 ± 1.09	12.44 ± 0.56	13.00 ± 1.18
Soil type (range)	SL, L, LS	SL, LS	SL, LS	SL, LS	SL, L	SL, L
Ca (ppm)	1,582.65 ± 142.68	1,437.14 ± 146.39	1,820.74 ± 102.46	1,863.44 ± 104.90	1,149.12 ± 123.17	1,114.69 ± 124.11
Mg (ppm)	203.45 ± 18.21	155.75 ± 16.30	199 70 ± 16.30	198.98 ± 14.69	292.03 ± 38.07	272.05 ± 37.02
Zn (ppm)	3.14 ± 0.35	2.91 ± 0 37	2.34 ± 0.43	2.19 ± 0.39	3.00 ± 0.54	2.82 ± 0.52
Mn (ppm)	46.40 ± 3.73	32.51 ± 3.43	23.41 ± 2.13	21.73 ± 1.84	56.28 ± 7.92	48.34 ± 6.16
Cu (ppm)	1.04 ± 0.17	1.26 ± 0.22	0.45 ± 0.07	0.44 ± 0.07	1.20 ± 0.14	1.41 ± 0.20
Fe (ppm)	6.18 ± 0.81	7.73 ± 1.23	4.62 ± 0.32	4.56 ± 0.31	5.81 ± 0.53	5.74 ± 0.59
B (ppm)	0.54 ± 0.10	0.41 ± 0.09	0.44 ± 0.06	0.43 ± 0.06	0.42 ± 0.06	0.34 ± 0.06
Acidity (%)	6.87 ± 2.08	6.46 ± 2.30	9.98 ± 1.47	9.58 ± 1.35	3.21 ± 0.73	4.93 ± 1.38
Base saturation (%)	96.19 ± 1.90	94.97 ± 2.21	90.02 ± 1.47	91.48 ± 1.65	97.50 ± 0.79	96.71 ± 1.37
Ca saturation (%)	75.56 ± 2.04	78.43 ± 2.19	71.48 ± 0.99	73.40 ± 1.37	66.59 ± 0.72	68.00 ± 1.11
Mg saturation (%)	16.09 ± 0.67	14.01 ± 0.55	12.86 ± 0.78	12.93 ± 0.87	27.32 ± 0.87	26.76 ± 1.13
K saturation (%)	4.54 ± 0.76	2.59 ± 0.42	5.68 ± 0.71	5.18 ± 0.66	3.55 ± 0.42	2.02 ± 0.24

CEC, cation exchange capacity; L, loam; LS, loamy sand; SL, sandy loam

When all measured values for soil physical and chemical properties (Table 9.3) were considered together in a multivariate approach, the first three axes of the PCA explained a total of 71.7% of the variation within the data (Fig. 9.4). Axis 1 described 38.9% of variability within the data set and resulted in a wide separation among the centroids of each village (Fig. 9.4, top). Many response variables were strongly negatively correlated with Axis 1, including P concentration, percent organic matter (OM), percent sand, ECEC, and percent N; whereas percent silt, Mg^{2+} saturation, and Cu concentration were strongly positively correlated with Axis 1 (Table 9.4). Axis 2 further described 20.4% of the variability within the data set, but largely did not capture differences among villages (Fig. 9.4, top). Rather, the variation in these factors, including Zn concentration, B concentration, BS, pH, and to a lesser degree, percent sand and Fe concentration (Table 9.4), were likely due to inherent landscape heterogenity or differences among individual farmers' practices. Axis 3 explained another 12.4% of the variability within the data set, and the 2D representation of Axis 1 together with Axis 3 provided the most useful distinction among the villages (Fig. 9.4, bottom). Bulk density, Ca^{2+} saturation, and Cu concentration were most negatively correlated with Axis 3, whereas porosity, percent clay, and K^{+} saturation were most positively correlated with Axis 3 (Table 9.4).

The value of the multivariate, PCA approach was in the identification of a summary set of variables that best defined differences in the soil enviroment among the villages. The response vectors on the 2D comparison of Axes 1 and 3 provided the most useful multidimensional view of differences among villages (Fig. 9.4, bottom), but consideration of all axes was necessary for identifying suites of response variables that characterized each village. Thumka was characterized by greater percent silt, Mg^{2+} saturation, and Mn concentration than the other villages (Fig. 9.4 top and bottom). In addition, Cu concentration was high, but was similar to many fields in Hyakrang. Relatively high Ca^{2+} saturation of exchange sites, BD, and low porosity and percent clay separates Hyakrang from the other villages. In other aspects, however, the data suggest Hyakrang had many overlapping qualities with both Kholagaun and Thumka. Generally, Kholagaun and Thumka had very different soil characteristics than one another. Kholagaun was characterized by a high percent of sand and K^{+} saturation of exchange sites, as well as a high percent OM, percent N, and ECEC.

Plant mineral nutrients and organic matter

Nitrogen and P are the most productivity-limiting plant nutrients in the mid-hill region of Nepal. Both are supplied by compost amendments, and additional N may be added through biological N fixation in tree and crop species that host symbiosis. Potassium, Mg^{2+}, and Ca^{2+} are rarely deficient in Nepalese soils, whereas plant micronutrients such as boron (B), chlorine (Cl), copper (Cu), iron (Fe), manganese (Mn), molybdenum (Mo), and zinc (Zn) deficiencies are often recognized when N and P are plentiful (Carson, 1992). Thumka was characterized by greater base saturation, followed by Hyakrang and Kholagaun, potentially because of lower precipitation and less leaching in Thumka. Concentrations of plant available P and K^{+} were substantially different among the villages. In Kholagaun, and to a lesser degree in Hyakrang, P and K^{+} were present in excess of plant requirements,

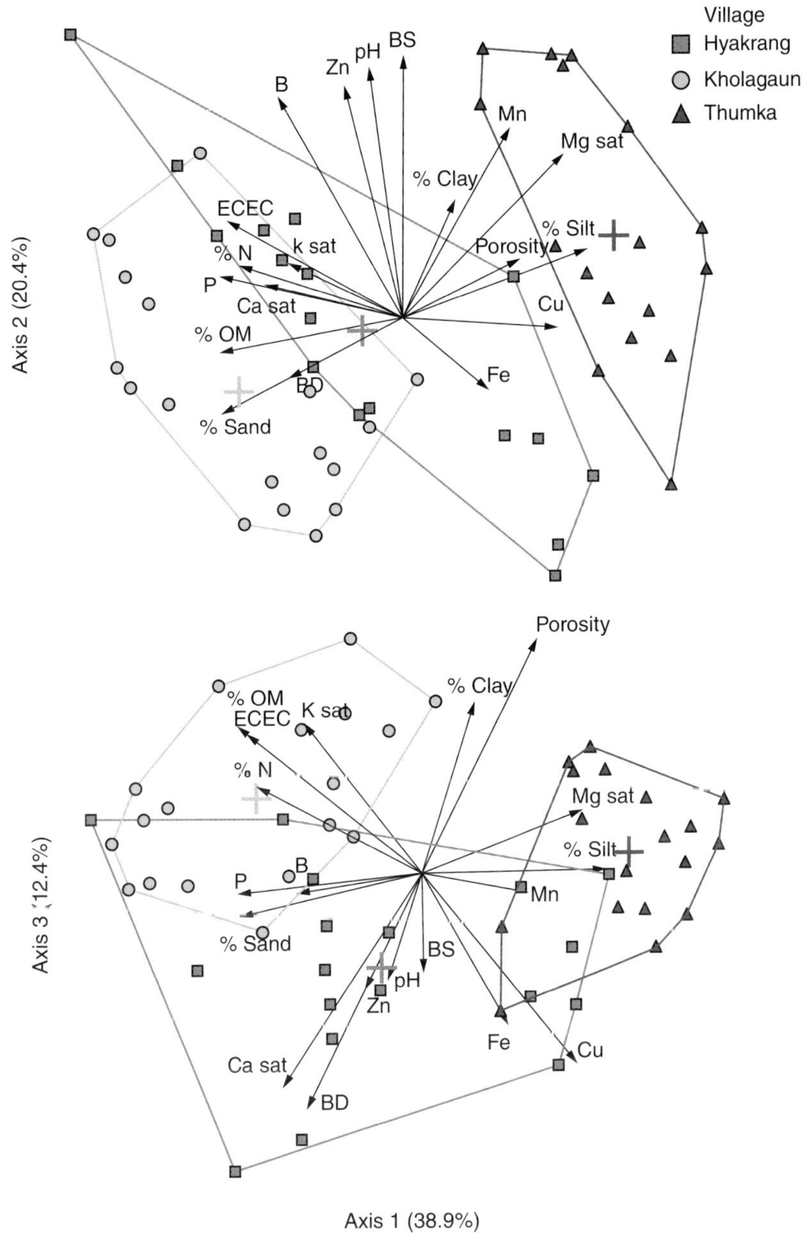

Fig. 9.4. Principal components analysis of baseline soil biological, chemical, and physical properties collected from both 0–5 cm and 5–10 cm depths at each field within the three villages prior to the implementation of the experimental plots.

while other factors, for example N, plant micronutrients and/or available water, likely limited yield. Desbiez *et al.* (2004) reported an average N percentage between 0.145 and 0.216; average P and K^+ content between 15.6–37.5 ppm and 90–204 ppm, respectively; and average pH range of 4.216–4.905. The higher potassium

Table 9.4. Correlations between quantified soil physical and chemical parameters and the first three axes of the principal components analysis (PCA).

Variable	Axis 1	Axis 2	Axis 3
Bulk density (kg/m³)	–0.535	***–0.189***	***–0.607***
Sand (%)	***–0.849***	***–0.308***	–0.113
Silt (%)	***0.851***	0.229	0.006
Clay (%)	0.245	0.389	***0.437***
Porosity (%)	0.537	0.191	***0.601***
pH	–0.154	***0.832***	–0.275
ECEC (cmol/kg)	***–0.817***	0.323	0.36
Base saturation (%)	0.004	***0.861***	–0.258
Mg^{2+} saturation (%)	***0.746***	0.537	0.159
Ca^{2+} saturation (%)	–0.648	0.113	***–0.552***
K^+ saturation (%)	–0.541	0.185	***0.386***
Organic matter (%)	***–0.852***	–0.108	0.378
Nitrogen (%)	***–0.766***	0.176	0.222
P (ppm)	***–0.853***	0.141	–0.051
Cu (ppm)	***0.715***	–0.035	***–0.492***
B (ppm)	–0.577	0.731	–0.053
Fe (ppm)	0.393	***–0.233***	–0.39
Mn (ppm)	0.49	0.627	–0.049
Zn (ppm)	–0.263	***0.766***	–0.293

Bold and italicized coefficients are the three most positive or negative values for each axis, or those r values >0.75 in either direction.

levels in Kholagaun may be a result of more potassium-rich bedrock constituents such as biotite and muscovite. However, K^+ deficiency is most common in light, sandy soils, because there is not enough clay in sandy soil to hold the potassium and leaching commonly occurs.

Soil parent material inherently governs soil nutrient and mineral composition, though environmental factors such as vegetation and animals also contribute to soil nutrient levels. Goats are the most populous animal providing farmyard manure to the village plots. Goat manure has high contents of N and phosphoric acid, while goat urine is rich in N and K. On average, goat manure is composed of 40–46% moisture, 1.0–3.0% N, 0.2–0.8% P, and 0.4–0.8% K (Peacock, 1996). Ojeniyi and Akanni (2007) reported an increase in soil OM, total N, available P, and moisture content as goat manure application increased. The average flock of goats per family in the mid-hills consists of seven goats, but the number of goats per family varies from one to twenty (Abington, 1992). Kholagaun may have had, on average, a greater application rate of manure compost. In fact, the topsoil tended to have more K than the subsoil, suggesting the incorporated manure as the source (Table 9.3).

Together with soil nutrients, sufficient plant micronutrients are also necessary to ensure a healthy crop. Essential plant micronutrients constitute less than 1% of total dry weight in plant biomass, but deficiencies contribute to low yields. The solubility of many plant micronutrients is highly pH dependent;

in soil pH greater than 6.0–6.5, micronutrients become less available for plant uptake. In particular, Cu, Fe, Mn, and Zn often are deficient in high pH soils. Soil pH ranged from 5.6 to 7.2 among all fields and villages, and mean soil pH was 6.8, 6.7, and 6.4 at Thumka, Hyakrang, and Kholagaun, respectively, suggesting that these micronutrients might be deficient, especially if they were present in low concentrations. Greater concentrations of Cu and Mn were among the variables that differentiated Thumka from the other villages, particularly Kholagaun (Fig. 9.4). These micronutrients may be less limiting in Thumka than the other villages. However, a plant tissue analysis for these micronutrients is a better indicator of deficiency than soil data.

Soil organic matter (OM) concentration was greatest in Kholagaun, likely contributing to other properties such as soil water-holding capacity, nutrient availability, ECEC, lower soil pH, and aggregate structure, which are associated with the presence of OM. A decrease in pH associated with greater use of manure amendment will have the added benefit of increasing the solubility, and therefore availability, of plant micronutrients. OM can play a very important role in productivity, especially when clay content is low. Kholagaun soil, even though it was coarse-textured, was characterized by the highest ECEC, percent OM, and percent N, followed by Hyakrang and then Thumka (Fig. 9.4). Desbiez *et al.* (2004) studied soils in the mid-hills of the Parbat district of Nepal, just west of the three districts of this study, and found an average of 2.04–3.03% OM, which was consistent with our findings.

Productivity and soil quality

People of the mid-hills have adjusted their agricultural practices as the population has continued to grow, although in recent decades it has become increasingly difficult to cope with the rapid population increase, land degradation, and exhaustion of mountain ecosystems (Maskey *et al.*, 2003). Moreover, limited access to technologies has stunted agricultural growth in the mid-hills, subsequently causing a decrease in farm productivity. Soil erosion and decreased fertility are detrimental to crop yield, and are unfortunately two of the greatest problems in the mid-hills of Nepal. Productivity of maize, millet, and cowpea crops produced by the farmers in the studied villages via traditional practices of soil tillage and cropping system were compared to the national averages prior to this study (Fig. 9.5). The satisfactory yield of cowpea is presumably due to its N fixation and utilization characteristics.

Without ready access to technology and information, farmers in the mid-hills rely on observation and traditional knowledge to assess their agricultural practices. Existing conditions help form farmers' perceptions of soil quality and productivity. Astute observation and experience working the land give the farmers an invaluable perspective on the likelihood of conservation agriculture practices to improve productivity. In Kholagaun, the soil quality and properties make tillage important. Therefore, it is logical that they be skeptical of the success of CAPS such as reduced tillage (Halbrendt *et al.*, 2014). In Hyakrang, the wide variety of soil properties makes it difficult to suggest practices that will work across the gradient.

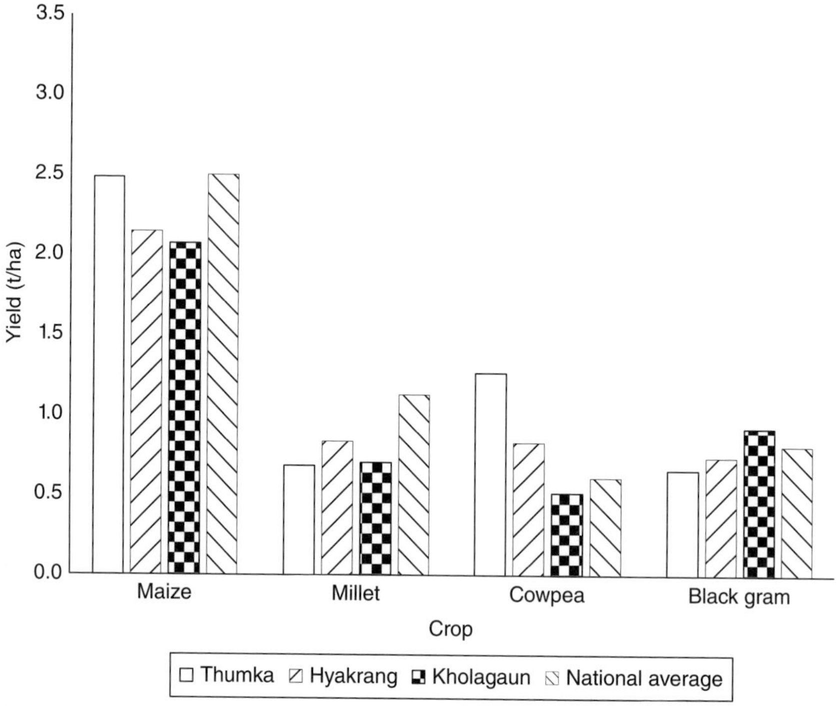

Fig. 9.5. Farm yields using traditional practices prior to the study. Village estimates are from household surveys and are shown compared to national averages. National averages reported for maize and millet are from the Government of Nepal Ministry of Agricultural Development (2012) and those for cowpea are from Shrestha *et al.* (2000).

Coarseness of soils is a property, common in Kholagaun, which is often difficult to improve. Organic matter amendment is the primary source of fertility and CEC, physical structure, and water-holding capacity. A positive link between compost availability and household wealth would likely be strongest in the communities involved in this study. Gurung *et al.* (2005) identified a relationship between ethnicity, wealth, and livestock keeping, documenting that higher income groups such as Brahmins more commonly raised large ruminants. This relates to soil fertility in that a greater amount of large animals produce more FYM. Lower income and disadvantaged ethnic groups such as the Tamang and Chepang tend to raise small livestock such as pigs, goats, and chickens, because the initial investment cost is small, the turnover of these enterprises is quick, and the required time input is minimal. Households with greater income often focus their time and capital in fewer farm activities (either crops or livestock), while poorer households tend to diversify their activities to achieve a balance between risks such as climate and market trends (Pica-Ciamarra *et al.*, 2011). This translates into a positively correlated feedback, raising less livestock produces less manure, which decreases the fertility of the soil, and ultimately causes low yield. Better soil management would promote nutrient retention, thereby decreasing the need for more livestock input.

9.3.3 CAPS treatments

No significant differences among means for the CAPS treatments were detectable for either the 6-month or 2-year sampling dates for N, available P, available K, or OM (Table 9.5). Percent OM was greater in the top 5 cm than in the 5–10 cm depth for all treatments in both the 6-month and 2-year data. Organic matter values ranged from 2.65 to 2.92% at 0–5 cm depth and from 2.38 to 2.67% at 5–10 cm depth. Surface soil N increased by 0.02–0.06% across all treatments and soil depth. Phosphorus levels in the topsoil increased from the 6-month to 2-year measurements in the order of 0.16–13.76 ppm. Surface levels of P were greater than those at 5–10 cm for all treatments, regardless of time. Potassium levels varied between 134.50–173.95 ppm in 0–5 cm and 92.92–141.04 ppm in 5–10 cm. Potassium consistently decreased at the 5–10 cm depth in the 6-month to the 2-year measurements. As a result of, or perhaps coinciding with, the minimal differences in nutrients between treatments, no significant differences in yield by treatment resulted. In year 2, however, there was an observed decline in maize production for reduced tillage.

9.3.4 Village differences

Organic matter ranged from 1.66 to 3.65% across villages and depths (Table 9.6). Kholagaun was the only village that had an increase in OM at both depths from the 6-month and 2-year measurements. Organic matter in Thumka decreased over time at both 0–5 cm and 5–10 cm, while Hyakrang OM increased in the topsoil but decreased at 5–10 cm. Nitrogen did not increase in Thumka topsoil, but increase by 0.07 in Kholagaun and Hyakrang. There was a great range in phosphorus values, 9.47–175.27 ppm for 0–5 cm and 4.90–162.67 ppm for 5–10 cm. Phosphorus levels for 0–5 cm and 5–10 cm were appoximately 15 and 25 times greater, respectively, than Thumka. Potassium decreased across all villages and depths from 6 months to 2 years. The greatest declines were in Kholagaun, where the topsoil decreased 16.96 ppm and at 5–10 cm fell by 80.28 ppm.

9.3.5 Qualitative comparisons among villages, treatments, and sampling dates

Intersampling variability of organic matter and nutrients

Fertility in agricultural systems is linked closely to OM amendment. Thus, fluctuations in OM and plant available nutrients depending on the crop cycle, season (wet/dry), and interannual availability of manure compost were expected. The baseline N values were higher than the 6-month and 2-year percentages at both depths across all villages (Fig. 9.6). Nitrogen increased from the 6-month measurement to the 2-year measurement in all villages at all depths, except in Thumka at 5–10 cm (Fig. 9.6). Although not statistically significant, treatment 2 in Hyakrang showed an increase in N from the 6-month reading to the

Table 9.5. Soil parameters measured 6 months and 2 years following initiation of the on-farm evaluations by treatment. Values are means ± one standard error.

	Organic matter (OM) (%)		Nitrogen (N) (%)		Phosphorus (P) (ppm)		Potassium (K) (ppm)	
	0–5 cm	5–10 cm	0–5 cm	5–10 cm	0–5 cm	5–10 cm	0–5 cm	5–10 cm
Treatment 1								
6 month	2.74 ± 0.19	2.52 ± 0.15	0.10 ± 0.01	0.09 ± 0.01	83.63 ± 14.87	73.78 ± 13.09	162.97 ± 28.54	136.10 ± 29.32
2 year	2.79 ± 0.85	2.51 ± 2.51	0.14 ± 0.05	0.13 ± 0.05	85.70 ± 88.23	77.07 ± 97.81	144.15 ± 98.13	94.16 ± 64.70
Treatment 2								
6 month	2.66 ± 0.17	2.38 ± 0.15	0.10 ± 0.01	0.10 ± 0.00	81.60 ± 14.12	75.55 ± 14.29	161.18 ± 26.42	126.81 ± 27.75
2 year	2.86 ± 0.98	2.46 ± 1.00	0.16 ± 0.06	0.14 ± 0.06	82.43 ± 88.40	71.20 ± 86.81	164.48 ± 145.45	106.64 ± 96.21
Treatment 3								
6 month	2.72 ± 0.19	2.42 ± 0.13	0.11 ± 0.01	0.10 ± 0.01	82.74 ± 14.42	77.11 ± 14.67	167.45 ± 27.04	133.30 ± 30.65
2 year	2.74 ± 0.93	2.49 ± 1.02	0.15 ± 0.05	0.14 ± 0.06	82.90 ± 87.19	71.39 ± 89.05	166.75 ± 185.66	118.47 ± 132.55
Treatment 4								
6 month	2.65 ± 0.15	2.67 ± 0.15	0.11 ± 0.01	0.11 ± 0.01	81.44 ± 14.32	77.27 ± 14.57	173.95 ± 28.67	141.04 ± 30.77
2 year	2.92 ± 0.99	2.65 ± 1.13	0.16 ± 0.05	0.13 ± 0.05	95.20 ± 100.86	77.42 ± 94.71	134.50 ± 101.87	92.92 ± 68.38

Table 9.6. Soil parameters measured 6 months and 2 years following initiation of the on-farm evaluations by village. Values are means ± one standard error.

	Organic matter (OM) (%)		Nitrogen (N) (%)		Phosphorus (P) (ppm)		Potassium (K) (ppm)	
	0–5 cm	5–10 cm	0–5 cm	5–10 cm	0–5 cm	5–10 cm	0–5 cm	5–10 cm
Hyakrang								
6 month	2.77 ± 0.13	2.49 ± 0.12	0.10 ± 0.01	0.09 ± 0.00	75.06 ± 4.89	66.94 ± 5.93	80.99 ± 9.62	47.63 ± 3.60
2 year	3.17 ± 0.10	2.28 ± 0.09	0.17 ± 0.01	0.14 ± 0.00	74.94 ± 4.88	55.24 ± 5.17	69.23 ± 3.59	39.47 ± 3.64
[a]Tukey test								
6 month	AB	AB	AB	A	B	B	B	B
2 year	A	B	A	B	B	B	B	B
Kholagaun								
6 month	3.16 ± 0.16	2.94 ± 0.12	0.13 ± 0.01	0.12 ± 0.00	160.43 ± 8.40	153.55 ± 7.67	295.47 ± 26.34	279.61 ± 29.69
2 year	3.49 ± 0.13	3.65 ± 0.14	0.20 ± 0.01	0.18 ± 0.01	175.27 ± 17.23	162.67 ± 18.50	278.51 ± 30.33	199.33 ± 18.81
[a]Tukey test								
6 month	A	A	A	A	A	A	A	A
2 year	A	A	A	A	A	A	A	A
Thumka								
6 month	2.15 ± 0.10	2.06 ± 0.09	0.09 ± 0.01	0.10 ± 0.01	11.56 ± 3.05	7.30 ± 2.03	122.71 ± 9.37	75.71 ± 5.53
2 year	1.83 ± 0.08	1.66 ± 0.06	0.09 ± 0.00	0.01 ± 0.04	9.47 ± 2.33	4.90 ± 1.07	109.67 ± 6.59	70.34 ± 2.64
[a]Tukey test								
6 month	B	B	B	A	C	C	B	B
2 year	B	B	B	C	B	B	B	B
P value								
6 month	0.018	0.019	0.018	0.056	0.000	0.000	0.001	0.000
2 year	0.000	0.000	0.000	0.000	0.000	0.000	0.000	0.000

[a]The Tukey test used was a one-way multiple comparisons, family error rate. The different letters indicate means that were significantly different.

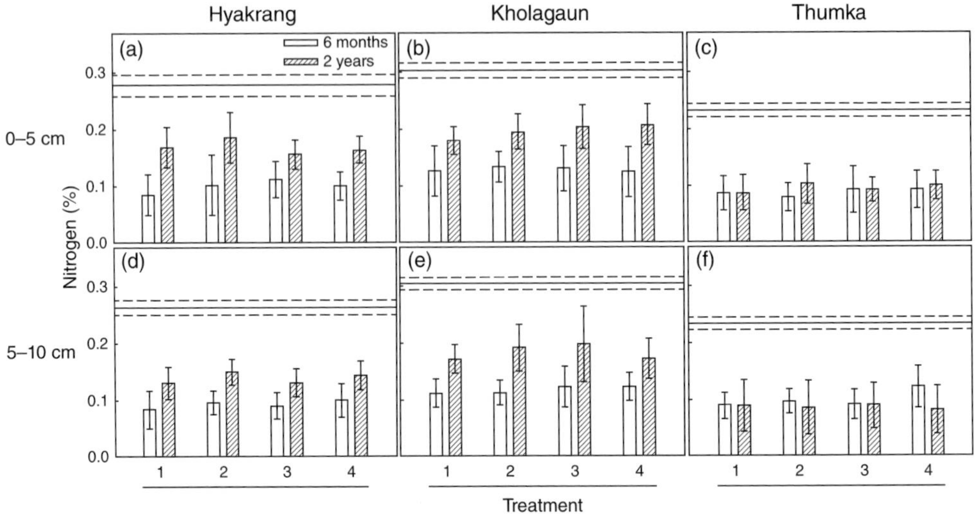

Fig. 9.6. Nitrogen concentration for each village by treatment. Bars are means ± one standard error. For reference, baseline measurements and standard errors are denoted by horizontal solid and dashed lines, respectively. (a), (b), and (c) are 0–5 cm depth, while (d), (e), and (f) are 5–10 cm depth.

2-year reading, presumably due to the N-fixing capabilities of the legumes in the trial. Kholagaun had the highest overall percentage OM, and therefore N, as OM was a major source of N for plants.

Kholagaun had the greatest amounts of P and K at the 6-month reading (Figs 9.7 and 9.8, respectively). Overall, P levels remained relatively constant across villages and sampling times. It appears that the intercrop with reduced tillage treatment in Hyakrang and Kholagaun produced the highest amount of P. Phosphorus levels were low in Thumka compared to the other two villages, and generally greater in the topsoil than the subsoil (Fig. 9.7). Treatment 4, intercrop with reduced tillage, produced the greatest P content, potentially as a result of more residues being left on the soil. Thumka had the lowest OM percentage, and naturally also had the lowest P content. Perhaps the pH in the Thumka soils was low, limiting the available P due to fixation by aluminum, iron, and calcium (Arai and Sparks, 2007). Runoff and erosion may also explain the low amount of P in the Thumka and Hyakrang soils. Phosphorus can be a major limiting factor for plant growth, because of its low availability due to slow diffusion and high fixation in soils (Shen *et al.*, 2011).

Kholagaun consistently had the greatest OM content between villages from the baseline to year 2 measurements, thus exhibiting the highest N, P, and K levels as well. Organic matter is relatively consistent across treatments within each location (Fig. 9.9). Organic matter increased from the 6-month reading to the 2-year reading in the topsoil in Hykrang, and for the most part in Kholagaun. The amount of OM in the soils of Thumka increased slightly from the baseline to 6-month measurements, then decreased to below baseline levels at year 2.

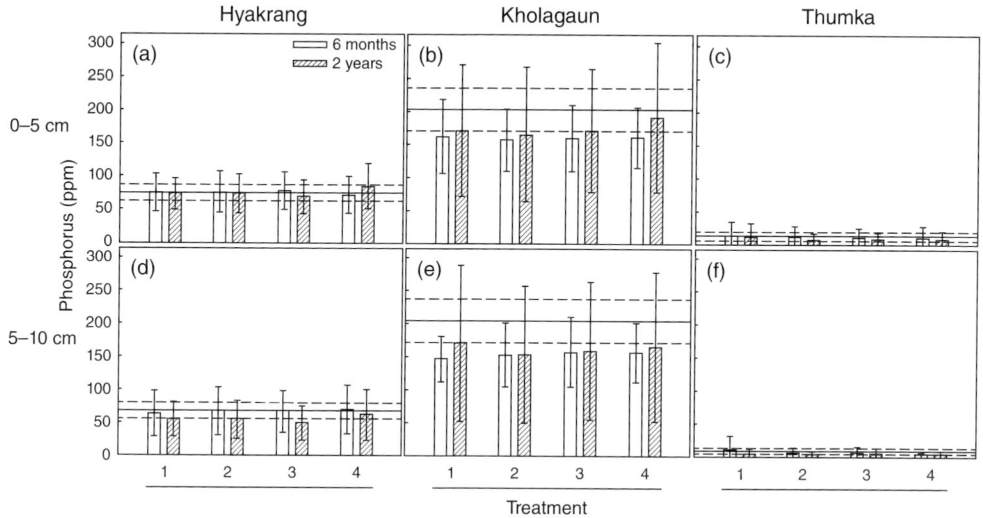

Fig. 9.7. Extractable phosphorus concentration for each village by treatment. Bars are means ± one standard error. For reference, baseline measurements and standard errors are denoted by horizontal solid and dashed lines, respectively. (a), (b), and (c) are 0–5 cm depth, while (d), (e), and (f) are 5–10 cm depth.

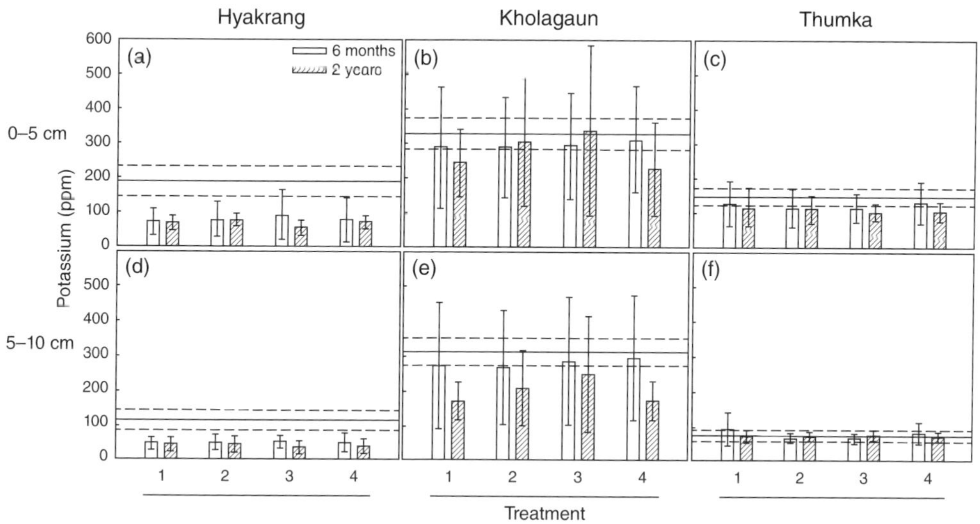

Fig. 9.8. Extractable potassium concentration for each village by treatment. Bars are means ± one standard error. For reference, baseline measurements and standard errors are denoted by horizontal solid and dashed lines, respectively. (a), (b), and (c) are 0–5 cm depth, while (d), (e), and (f) are 5–10 cm depth.

Water-stable aggregates (WSA)

Water-stable aggregates do not show any significant differences among treatments (Fig. 9.10). Between villages, the percentage WSA ranged from 25.92 to 37.25% for the >250 μm range and from 0.93 to 5.7% for the 53–250 μm range

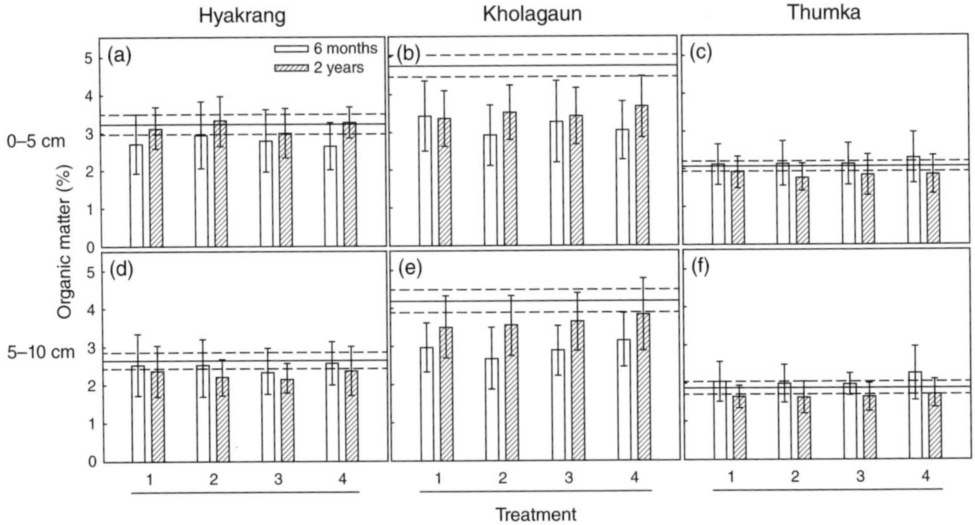

Fig. 9.9. Organic matter content for each village by treatment. Bars are means ± one standard error. For reference, baseline measurements and standard errors are denoted by horizontal solid and dashed lines, respectively. (a), (b), and (c) are 0–5 cm depth, while (d), (e), and (f) are 5–10 cm depth.

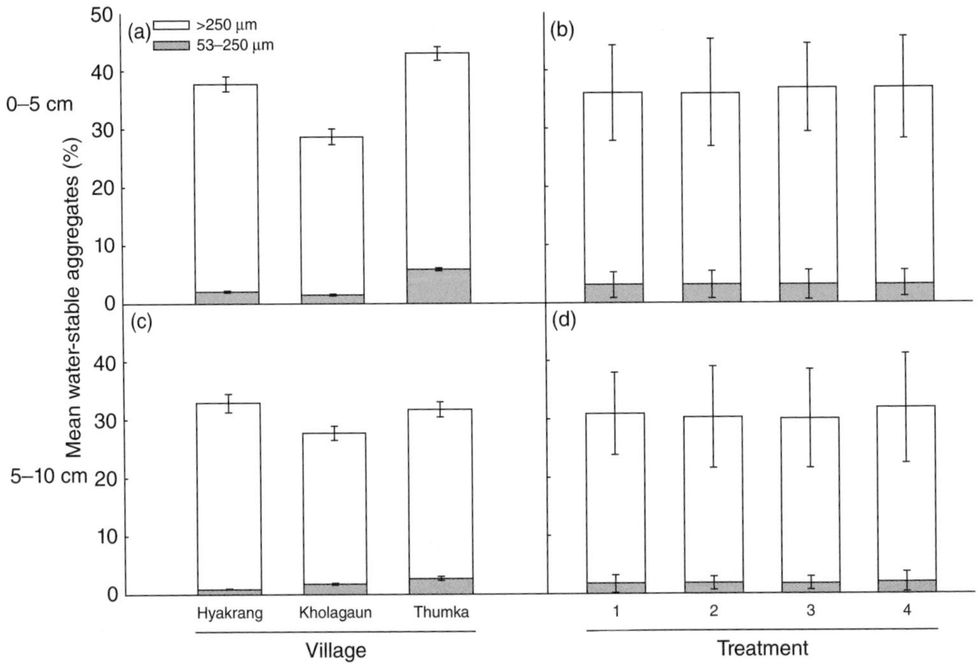

Fig. 9.10. Water-stable aggregates by village and treatment. (a) and (b) are 0–5 cm depth, while (c) and (d) are 5–10 cm depth. Values are means ± one standard error.

(Fig. 9.10a,c). The WSA variation is less between treatments, as the >250 μm category ranged from 28.36 to 33.90%, while the 53–250 μm category ranged from 1.72 to 3.23% (Fig. 9.10b,d). Mean WSA percentages were slightly greater for treatment 4 than for treatments 1, 2, and 3. The 53–250 μm range had a greater WSA percentage in the 0–5 cm depth than the deeper soil. Thumka had a slightly greater percentage in both the >250 and 53–250 μm ranges for the 0–5 cm depth than the other two villages.

The average WSA percentage for both >250 μm and 53–250 μm was greater for the 0–5 cm depth than the averages of the 5–10 cm depth across both villages and treatments (Fig. 9.10). These results agree with the hypothesis of Paudel *et al.* (2011) that surface soil has greater WSA percentage than subsurface soil. The average WSA for 0–10 cm, >250 μm was 31.21% between treatments and villages. Chakrabarti (1969) found an average WSA content of six arable soils in eastern Nepal to be 29.10%. Kholagaun's soils were the least resistant to breakdown, perhaps because rainfall, and thus destabilization, was greatest at this location. Perfect *et al.* (1990) found soil aggregate stability decreased when soil water content increased. Root activity enhances aggregate stability (Jirku *et al.*, 2013), which also could have resulted in a lower WSA percentage because Kholagaun had a low yield of millet and cowpea compared to the other locations. Although Kholagaun had the greatest OM percentage, it also received the most rainfall, which may have contributed to the breakdown of the soil aggregates. Jirku *et al.* (2013) found aggregate stability increased during periods of low rainfall. The soils of treatment 4 were the most resistant to dispersion, possibly due to the implementation of strip tillage rather than conventional tillage (Fig. 9.10b,c).

Kholagaun had the lowest WSA, highest intersampling variability of OM, and lowest yield, which suggested that coarse soils had the most to gain, and to lose, from the availability of compost manure. Organic matter is not well protected within soil aggregates in these soils and therefore is open to fluctuations due to oxidation during wet–dry cycles, decomposition, and leaching. Organic amendments decay quickly; therefore, any decrease in their availability, incorporation, or longevity in the soil will impact yields negatively. Through positive feedbacks between household income, livestock, compost availability, soil quality, and yield, farmers in Kholagaun potentially are most at risk to erosion, outmigration of labor, and climate change.

9.4 Implications and Future Study

Soil takes time to respond to change, and we have found that the CAPS treatments are not yet manifest in soil parameters or yields. Similarly, other studies suggest a period of 5–7 years for the benefits of conservation techniques to materialize and for yields to increase (Quinton and Catt, 2004; Thierfelder *et al.*, 2012). However, even without direct, measurable benefits to soil quality or yield, CAPS still benefit crop diversification and increase household nutrition and/or income. If those improvements increase labor availability or investment in livestock to increase manure compost, they may feedback positively into the production system. Paudel *et al.* (Chapters 3 and 7, this volume) have evaluated the efficacy of CAPS in this same study and have determined that by implementing CAPS, it is possible to improve overall nutrition in the villages through increasing protein-rich foods. It was also found that food availability in the villages might improve using CAPS as opposed to traditional agricultural practices.

Despite human-influenced alterations to land and agricultural methodology, climatic changes such as increase in annual temperature and erratic rainfall may further soil erosion and degradation (Bastakotia *et al.*, 2011). Malla (2008) reported an increase of temperature in Nepal between 1975 and 2006 of 0.042°C/year, while precipitation patterns have become increasingly unpredictable. It is evident that in order to provide food and employment in the mid-hills region into the future, alternative agricultural practices need to be adopted. Sustainable practices are any methods that contribute to a stable yield of a crop over a long time with minimal soil degradation (Kang *et al.*, 1990). Eco-farming for instance, is a sustainable methodological alternative that promotes soil and water conservation and biodiversity by excluding the use of agrochemicals and genetically engineered seeds (Sharma, 1997). Regmi *et al.* (2004) point out that people of the mid-hills region have developed and utilized effective traditional conservation agricultural practices in the past. However, many have lost sight of these practices in the wake of rapid population growth and subsequent need to produce more food. Furthermore, some communities possess rich local knowledge of optimal farming practices. Certainly, modern and traditional conservation techniques, such as the community forest management movement in the last 30 years, may be reconciled in an effort to preserve soil nutrients and rehabilitate the land.

The increased population in the mid-hills region of Nepal and degraded productivity has increased most households' dependency on wage laboring for additional income, particularly during the food-deficient months of March–July. During this period, food is often bought with wages earned from laboring and selling agricultural products, including livestock. A few households also earn their living from services within the village, and a small number of them perform services outside the country as well. The outmigration of the younger generations in search of income and education is resulting in labor deficits that endanger the subsistence of families remaining in the villages. The preparation and selling of homestead liquor, as well as occupational enterprises like carpentry, making ironware, and masonry works are additional cash income sources exploited to support the livelihoods of many people. In addition, the majority of the Chepang in Tanahun and Gorkha depend on wild, uncultivated food crops collected from the forest, riverbanks, or from their own land to see them through food shortages (Regmi *et al.*, 2004). Moreover, agriculture still provides the bulk income for the majority of the Nepalese population and must therefore be managed properly to yield maximum returns. Thus, to better understand optimal farming practices in the mid-hills of Nepal, further research is needed to improve crop yields, and ultimately the population's health and prosperity.

Agriculture in the mid-hills has long been dependent on the health and presence of forests. Forests are a major source of food, fodder, and firewood. Severe deforestation took place in Nepal from the 1950s to the 1980s. Sitaula *et al.* (2005) reported that forest cover in Nepal decreased from 39% in 1980 to 28% in 1993. Hill (1999) reports that more than 50% of the original forests in Nepal have been destroyed. The Forest Act of 1993 addressed the worsening situation of Nepal's forests by outlining the benefits of proper management by community forestry groups (Purvis and Grainger, 2013). The Forest Rules of 1995 further articulated to forest users their rights and duties within forests. Many successes have come from the community forestry programs, though; wealth disparities between community forest member households, as well as inefficient implementation due to political inconsistencies, have complicated this program for many people across Nepal (Thoms, 2008). Community forest management has been widely successful in some regions of Nepal. However, continued degradation of forests and increased stress on already marginal and sloping lands have attributed to increased population pressure and has reduced the period of food sufficiency to less than 5 months a year. The majority of households are located near forest and practice subsistence-based, traditional agriculture for their means of livelihood. The increased pressure on land can be observed in the reduction of fallow period in shifting cultivation lands by 2.5 years only in the past three decades. Ultimately, this change has led to the conversion of shifting cultivation land to annual cropping. The local consequences of global climate change have also been observed in the form of prolonged drought, the drying up of water sources, a seasonal shift in precipitation, and enhanced landslide and loss of topsoil (Kafle *et al.*, 2009). In brief, traditional practices have optimized agriculture, even on the most difficult terrain. However, changing climatic conditions and population dynamics demand further adaptations in the mid-hills.

Over the past few decades, investments were made into the research and development of agricultural sciences to meet food security needs. The Green Revolution was achieved in many parts of the developing world, but the challenge in the agricultural sector to increase food production and feed the rapidly growing population, especially in developing countries, still persists. A popular trend in agriculture was to increase the use of high external inputs and agrochemicals, which often caused adverse effects to environmental quality and, by extension, to ecological systems. Consequently, there is an immense need to increase the sustainability of agricultural production in order to feed the ever-growing population while minimizing negative environmental impacts.

Conservation agriculture is a holistic system based on interactions among households, crops, and livestock, bringing about a sustainable agricultural system that meets the needs of farmers (Hobbs, 2007). Conservation agriculture promotes a healthy environment while enhancing economically sustainable production conditions (FAO, 2000). It is a resource-efficient strategy that aims to conserve, improve, and make the best use of available natural resources such as soil and water through integrated management. The facilitation of ecosystem services such as clean water, carbon sequestration, and avoidance of landscape degradation on agricultural landscapes and surrounding areas have reportedly resulted from conservation agriculture (FAO, 2000; Swinton *et al.*, 2007). A key goal of conservation agriculture is maintaining a permanent or semi-permanent organic soil cover, which can be a growing crop or dead mulch of crop residues. Studies have shown that mulch protects the soil against compacting and the erosive effects of heavy rain, wind, and sun exposure, which improves the physical and biological qualities of the soil. Frequent mechanical tillage disturbs the soil microorganisms and soil fauna that play important roles in balancing soil nutrients. Conservation agriculture is based on minimal soil disturbance (no till) and permanent soil cover (mulch) (Hobbs, 2007) among crop rotations, as required by subsistence farmers. Implementation and long-term adoption of CAPS could provide a cultivation system integral to future sustainable intensification of crop production.

Acknowledgements

Funding for this project came from the US-AID Sustainable Agriculture and Natural Resource Management Program (Dr Catherine Chan, lead Principal Investigator). The project has been implemented in Chepang tribal communities in the western hills of Nepal by LI-BIRD, a local NGO, with Tribhuwan University as its research partner. We thank Puspa Poudyal and Mariko Panzella for their efforts in support of this research.

References

Abington, J. (1992) The role of small ruminants. In: Abington, J.B. (ed.) *Sustainable Livestock Production in the Mountain Agro-ecosystem of Nepal.* Food and Agriculture Organization of the United Nations, Rome, pp. 82–84.

Acharya, K. (2003) Conserving biodiversity and improving livelihoods: The case of community forestry in Nepal. Paper presented at The International Conference on Rural Livelihoods, Forests and Biodiversity, 19–23 May 2003, Bonn.

Adhikari, B., Di Falco, S. and Lovett, J.C. (2004) Household characteristics and forest dependency: Evidence from common property forest management in Nepal. *Ecological Economics* 48, 245–257.

Adhikari, J. and Bohle, H. (1999) *Food Crisis in Nepal: How Mountain Farmers Cope.* Adroit Publishers, Delhi.

Arai, Y. and Sparks, D. (2007) Phosphate reaction dynamics in soils and soil minerals: A multiscale approach. *Advances in Agronomy* 94, 135–179.

Atreya, K., Sharma, S., Man Bajracharya, R. and Prasa Rajbhandari, N. (2006) Applications of reduced tillage in hills of central Nepal. *Soil and Tillage Research* 88, 16–29.

Bajracharya, R. (2001) Land preparation: An integral part of farming systems in the mid-hills of Nepal. *Nepal Journal of Science and Technology* 3, 15–24.

Balesdent, J., Chenu, C. and Balabane, M. (2000) Relationship of soil organic matter dynamics to physical protection and tillage. *Soil and Tillage Research* 53, 215–230.

Balla, M., Awasthi, K., Shrestha, P., Sherchan, D. and Poudel, D. (1999) *Degraded Lands in Mid-hills of Central Nepal: A GIS Appraisal in Quantifying and Planning for Sustainable Rehabilitation.* LI-BIRD, Pokhara, Nepal.

Basnet, K. (1992) Conservation practices in Nepal: Past and present. *Ambio* 21(6), 390–393.

Bastakotia, N., Scholberga, J. and Groota, J.C.J. (2011) Farm management strategies to enhance the farm performance in changing climate: A case study in the mid-hills of Nepal. Tropentag 2011, Conference on International Research on Food Security, Natural Resource Management and Rural Development. 5–7 October 2011. University of Bonn.

Beare, M., Hendrix, P. and Coleman, D. (1994) Water-stable aggregates and organic matter fractions in conventional and no-tillage soils. *Soil Science Society of America Journal* 58, 777–786.

Begum, F., Man Bajracharya, R., Sharma, S. and Staula, B.K. (2010) Influence of slope aspect on soil physic-chemical and biological properties in the mid hills of central Nepal. *International Journal of Sustainable Development and World Ecology* 17(5), 438–443.

Blaikie, P.M., Cameron, J. and Seddon, D. (1980) *Nepal in Crisis: Growth and Stagnation at the Periphery*. Clarendon, Oxford, UK.

Brown, R. and Rosenberg, N. (1997) Sensitivity of crop yield and water use to change in a range of climatic factors and CO2 concentrations: a simulation study applying EPIC to the central USA. *Agricultural and Forest Meteorology* 83, 171–203.

Brown, S. (1997) Soil fertility, nutrient dynamics and socio-economic interactions in the Middle Mountains of Nepal. PhD thesis, Interdisciplinary Studies in Resource Management Science, University of British Columbia, Vancouver.

Brown, S., Schreier, P. and Shah, L. (1999) Lavkulich modeling of soil nutrient budgets: an assessment of agricultural sustainability in Nepal. *Soil Use and Management* 15, 101–108.

Carson, B. (1992) The land, the farmer and the future: A soil management strategy for Nepal. ICIMOD Occasional Paper No 21. ICIMOD, Kathmandu.

Chakrabarti, D. (1969) Investigation on erodibility and water stable aggregates of certain soils of eastern Nepal. *Journal of the Indian Society of Soil Science* 17, 391–514.

Chaney, K. and Swift, R. (1984) The influence of organic matter on aggregate stability in some British soils. *Journal of Soil Science* 35, 223–230.

Chaudhary, R.P. (2000) Forest conservation and environmental management in Nepal: A review. *Biodiversity and Conservation* 9, 1235–1260.

Chhetri, N. (2011) Climate sensitive measure of agricultural intensity: Case of Nepal. *Applied Geography* 31, 808–819.

CIA (Central Intelligence Agency) (2013) *The World Factbook 2013–14*. CIA, Washington, DC. Available at: https://www.cia.gov/library/publications/the-world-factbook/index.html (accessed 6 November 2014).

CIMMYT (2001) *Quality Protein Maize: Improved Nutrition and Livelihood for the Poor. Maize Research Highlights, 1999–2000*. CIMMYT, Mexico.

CSB (1999) *Statistical Year Book of Nepal 1997/1998*. HMG, National Planning Commission Secretariat, Thapathali, Kathmandu.

Dahal, R. (2006) *Geology for Technical Students*. Bhrikuti Academic Publications, Kathmandu.

Desbiez, A., Matthews, R., Tripathi, B. and Ellis-Jones, J. (2004) Perceptions and assessment of soil fertility by farmers in the mid-hills of Nepal. *Agriculture, Ecosystems and Environment* 103, 191–206.

Doolette, J. and Smyle, J. (1990) Soil and moisture conservation technologies: Review of literature. World Bank Technical Paper No 127 on Watershed Development in Asia. World Bank, Washington, DC, pp. 35–69.

FAO (Food and Agriculture Organization) (2000) *Manual on Integrated Soil Management and Conservation Practices*. FAO Land and Water Bulletin No 8. International Institute of Tropical Agriculture and FAO, Rome.

Franzluebbers, A. and Arshad, M. (1997) Soil microbial biomass and mineralizable carbon of water-stable aggregates. *Soil Science Society of America Journal* 61, 1090–1097.

Garcıa-Torres, L., Benites, J., Martınez-Vilela, A. and Holgado-Cabrera, A. (2003) *Conservation Agriculture: Environment, Farmers Experiences, Innovations, Socio-economy, Policy*. Kluwer Academic Publishers, Boston, Massachusetts.

Gartaula, H. and Niehof, A. (2013) Migration to and from the Nepal terai: Shifting movements and motives. *The South Asianist* 2, 29–51.

Gautam, R. (2006) *Research Biodiversity. Home Gardens in Nepal*. Proceedings of a National Workshop, 6–7 August 2004, Pokhara, Nepal. Local Initiatives for Biodiversity, Research, and Development, Pokhara, Nepal.

Government of Nepal Ministry of Agricultural Development (2012) Statistical information on Nepalese agriculture. Government of Nepal Ministry of Agricultural Development, Agribusiness Promotion and Statistics Division, Agri Statistics Section, Singha Durbar, Kathmandu.

Gurung, K., Tulachan, P. and Gauchan, D. (2005) *Gender and Social Dynamics in Livestock Management: A Case Study from Three Ecological Zones in Nepal.* Gender Component-LPP Project. Livestock Techonology Change, Livelihood Impacts and Policy Lessons, Nepal. Center for Mountain Research Development, Kathmandu.

Halbrendt, J., Gray, S.A., Crow, S., Radovich, T., Kimura, A.H., *et al.* (2014) Differences in farmer and expert beliefs and the perceived impacts of conservation agriculture. *Global Environmental Change* 28, 50–62.

Hill, I. (1999) Forest management in Nepal, economics and ecology. Technical Paper 445. World Bank, Washington, DC.

Hobbs, P. (2007) Conservation agriculture: What is it and why is it important for future sustainable food production? *Journal of Agricultural Science* 145, 127–137.

Horwith, B. (1985) A role for intercropping in modern agriculture. *BioScience* 35(5), 286–291.

Jirku, V., Kodesova, R., Nikodem, A., Muhlhanselova, M. and Zigova, A. (2013) Temporal variability of structure and hydraulic properties of topsoil of three soil types. *Geoderma* 2104-205, 43–58.

Joshi, K., Conroy, C. and Witcombe, J. (2012) Agriculture, seed, and innovation in Nepal: Industry and policy issues for the future. Project Paper, December 2012. International Food Policy Research Institute, Washington, DC.

Kafle, G., Limbu, P., Pradhan, B. and Fang, J. (2009) Piloting ecohealth approach for addressing land use transition, climate change and human health issues. NGO Group Bulletin on Climate Change. LI-BIRD, Pokhara, Nepal, p. 3.

Kalra, Y.P. (1995) Determination of pH of soils by different methods: Collaborative study. *Journal of AOAC International* 78, 310–321.

Kang, B., Reynolds, L. and Atta-Krah, A. (1990) Alley farming. *Advances in Agronomy* 43, 315–359.

Khatri-Chhetri, A. and Maharjan, K.L. (2006) Food insecurity and coping strategies in rural areas of Nepal. *Journal of International Development and Cooperation* 12(2), 25–45.

Knowler, D. and Bradshaw, B. (2007) Farmers' adoption of conservation agriculture: A review and synthesis of recent research. *Food Policy* 32, 25–48.

Koch, H.-J., Dieckmann, J., Buchse, A. and Marlander, B. (2009) Yield decrease in sugar beet caused by reduce tillage and direct drilling. *European Journal of Agronomy* 30(2), 101–109.

Laryea, K., Cogle, A. and Pathak, P. (1994) Tillage problems in rainfed soils: relevance to sustained productivity. In: Biswas, T.D., Narayanasamy, G., Yadav, J.S.P., Dev, G., Katyal, J.C. and Sidhu, P.S. (eds) *Soil Management for Sustainable Agriculture in Dryland Areas.* Bulletin No 16. Indian Society of Soil Science, New Delhi, pp. 119–134.

LRMP (Land Resource Mapping Project) (1985) Land System Map, 1:50,000 Scale Series. Topographical Survey Branch, Department of Survey, HMG/Nepal and Keinting Earth Science, Kathmandu.

Ludi, E. (2009) Climate change, water and food security. Technical report. Overseas Development Institute, London.

McCune, B. and Mefford, M. (2011) PC-ORD. Multivariate Analysis of Ecological Data. Version 6. MjM Software Design, Gleneden Beach, Oregon.

Maglinao, A. (2000) *Indicators of Sustainable Land Management for Slopeland Farms.* Extension Bulletin 484 of the Food and Fertilizer Technology Center, Tapei City.

Malla, G. (2008) Climate change and its impact on Nepalese agriculture. *The Journal of Agriculture and Environment* 9, 62–71.

Maskey, R., Sharma, B. and Joshi, M. (2003) Human dimensions in sustainable land use management in degraded land areas of Nepal. A paper prepared for presentation at the Open Meeting of the Global Environmental Change Research Community, Montreal, Canada, 16–18 October 2003.

Metz, J. (1995) Development in Nepal: Investment in the status quo. *GeoJournal* 35(2), 175–184.

MoAC (Ministry of Agriculture and Cooperatives) (2011) Statistical information on Nepalese agriculture. Ministry of Agriculture and Cooperatives, Government of Nepal, Kathmandu.

Nelson, G., Rosegrant, M., Koo, J., Robertson, R., Sulser, T., *et al.* (2009) Climate change: Impact on agriculture and costs of adaptation. Technical report, International Food Policy Research Institute, Washington, DC.

Neupane, R. and Thapa, G. (2001) Impact of agroforestry intervention on soil fertility and farm income under the subsistence farming system of the middle hills, Nepal. *Agriculture, Ecosystems and Environment* 84, 157–167.

NPHC (2012) National Population and Housing Census 2011. Government of Nepal National Planning Commission Secretariat Central Bureau of Statistics Kathmandu.

Ojeniyi, S. and Akanni, D. (2007) Effect of goat manure on some soil properties and growth, yield and nutrient status of tomato. *University of Khartoum Journal of Agricultural Sciences* 15(3), 396–406.

Pandey, P., Randey, H. and Nakagawa, M. (2009) Assessment of rice and maize based cropping systems for rural livelihood improvement in Nepal. *The Journal of Agriculture and Environment* 10, 57–64.

Paudel, B., Udawatta, R. and Anderson, S. (2011) Agroforestry and grass buffer effects on soil quality parameters for grazed pasture and row-crop systems. *Applied Soil Ecology* 48, 125–132.

Peacock, C. (1996) Manure. In: *Improving Goat Production in the Tropics: A Manual for Development Workers*. Oxfam in association with FARM-Africa, Oxford, UK, pp. 319.

Peck, J. (2010) Multivariate analysis for community ecologists: Step by step using PC-ORD. MjM Software Design, Gleneden Beach, Oregon.

Perfect, E., Kay, B., van Loon, W., Sheard, R. and Pojasok, T. (1990) Factors influencing soil structural stability within a growing season. *Soil Science Society of America Journal* 54, 173–179.

Pica-Ciamarra, U., Tasciotti, L., Otte, J. and Zezza, A. (2011) Livestock assets, livestock income and rural households: Cross-country evidence from household surveys. Joint paper of the World Bank, FAO, AU-IBAR, ILRI, with support from the Gates Foundation, Rome.

Pilbeam, C., Tripathi, B., Sherchan, D., Gregory, P. and Gaunt, J. (2000) Nitrogen balances for households in the mid-hills of Nepal. *Agriculture, Ecosystems and Environment* 79, 61–72.

Pilbeam, C., Mathema, S., Gregory, P. and Shakya, P. (2005) Soil fertility management in the mid-hills of Nepal: Practices and perceptions. *Agriculture and Human Values* 22, 243–258.

Poudel, S. and Kotani, K. (2013) Climatic impacts on crop yield and its variability in Nepal: Do they vary across seasons and altitudes? *Climatic Change* 116, 327–355.

Prasad, R. and Brook, R. (2005) Effect of varying maize densities on intercropped maize and soybean in Nepal. *Experimental Agriculture* 41, 365–382.

Puget, P., Chenu, C. and Balesdent, J. (2000) Dynamics of soil organic matter associated with particle-size fractions of water-stable aggregates. *European Journal of Soil Science* 21, 595–605.

Purvis, M. and Grainger, A. (2013) *Exploring Sustainable Development*. Earthscan, London.

Quinton, J. and Catt, J. (2004) The effects of minimal tillage and contour cultivation on surface run-off, soil loss and crop yield in the long-term Woburn Erosion Reference Experiment on sandy soil at Woburn, England. *Soil Use and Management* 20, 343–349.

Rasul, G. and Thapa, G. (2003) Shifting cultivation in the mountains of South and Southeast Asia: Regional patterns and factors influencing the change. *Land Degradation and Development* 14, 495–508.

Raut, N., Sitaula, K., Bajracharya, B. and Man, R. (2010) Agricultural intensification: Linking with livelihood improvement and environmental degradation in mid-hills of Nepal. *The Journal of Agriculture and Environment* 11, 83–94.

Regmi, B., Aryal, K., Shrestha, P., Subedi, A., Gautam, R., *et al.* (2004) Participatory identification of integrated agricultural technological packages suitable for shifting and sloping land cultivation areas of western hills of Nepal. Report. LI-BIRD, Pokhara, Nepal.

Schroeder, R. (1985) Himalayan subsistence systems: Indigenous agriculture in rural Nepal. *Mountain Research and Development* 5(1), 31–44.

Schwab, N. (2012) Agroforestry's influence on soil conditions and phytodiversity in a mid-hill village of Nepal. In: Stahr, K. *et al.* (eds) Sustainable Land Use and Rural Development in Mountain Areas: Program and Abstracts. University of Hohenheim, Stuttgart, Germany, pp. 139–140.

Sharma, B. (1996) *Sustainable Management of Soils in Rainfed Cropping Systems in the Mid-hills of Nepal: Role of Livestock, Fodder and Tree Crops*. APROSC, Kathmandu.

Sharma, D.P. (2011) Understanding the Chepangs and shifting cultivation: A case study from rural village of central Nepal. *Dhaulagiri Journal of Sociology and Anthropology* 5, 247–262.

Sharma, L.P. (1997) Eco-farming – An imperative to environmental sustainability and farming practices in Nepal. *Tribhuvan University Journal* 20(1), 49–55.

Shen, J., Yuan, L., Zhang, J., Li, H., Bai, Z., *et al.* (2011) Phosphorus dynamics: From soil to plant. *Plant Physiology* 156, 997–1005.

Shrestha, B., Maskey, S., Shrestha, R., Tripathi, B., Khadka, Y., *et al.* (2000) Soil fertility management: farmers' practices and perception in the hills of Nepal. Lumle Technical Paper No 2000/4. Lumle Agriculture Research Station, Pokhara, Nepal.

Shrestha, B., Williams, S., Easter, M., Paustian, K. and Singh, B. (2009) Modeling soil organic carbon stocks and changes in a Nepalese watershed. *Agriculture, Ecosystems and Environment* 132, 91–97.

Siddoway, F. and Barnett, A. (1976) Water and wind erosion control aspects of multiple cropping. *American Society of Agronomy Special Publication* 27, 317–335.

Sitaula, B., Sankhayan, P., Bajracharya, R. and Singh, B. (2005) A systems analysis of soil and forest degradation in a mid-hill watershed of Nepal using a bio-economic model. *Land Degradation and Development* 16, 435–446.

So, H., Kirchhof, G., Bakker, R. and Smith, G. (2001) Low input tillage/cropping systems for limited resource areas. *Soil and Tillage Research* 61, 109–123.

Swinton, S., Lupi, F., Robertson, G. and Hamilton, S. (2007) Ecosystem services and agriculture: Cultivating agricultural ecosystems for diverse benefits. *Ecological Economics* 64, 245–252.

Thapa, B. (1989) Farming systems in the middle hills of Nepal. PAC Technical Paper No 114. Pakhribas Agricultural Centre, Dhankuta, Nepal.

Thapa, B. (1994) Farmers' ecological knowledge about the management and use of farmland tree fodder resources in the mid-hills of eastern Nepal. PhD thesis, University of Wales, Bangor, UK.

Thapa, G. (1996) Land use, land management and environment in a subsistence mountain economy in Nepal. *Agriculture, Ecosystems and Environment* 57, 57–71.

Thierfelder, C., Cheesman, S. and Rusinamhodzi, L. (2012) Benefits and challenges of crop rotations in maize-based conservation agriculture (CA) cropping systems of southern Africa. *International Journal of Agricultural Sustainability* 11, 108–124.

Thoms, C.A. (2008) Community control of resources and the challenge of improving local livelihoods: A critical examination of community forestry in Nepal. *Geoforum* 39, 1452–1465.

Tiwari, K., Nyborg, I., Sitaula, B. and Paudel, G. (2008) Analysis of the sustainability of upland farming systems in the middle mountains region of Nepal. *International Journal of Agricultural Sustainability* 6, 289–306.

Tiwari, T., Brook, R. and Sinclair, F. (2004) Implications of hill farmers' agronomic practices in Nepal for crop improvement in maize. *Experimental Agriculture* 40, 397–417.

Troeh, F. and Thompson, L. (1993) *Soils and Soil Fertility*. McGraw-Hill, New York.

Tuladhar, J. (1994) Soil survey of Lumle Center. LARC Working Paper 94/15. Lumle Agricultural Research Center, Kaski, Nepal.

Upadhayay, M., Sharma, R., Yadav, S., Gupta, R. and Billore, S. (1990) Studies on population densities of component crops in sorghum/pigeonpea intercropping. *Indian Journal of Agronomy* 35, 60–66.

Wyatt-Smith, J. (1982) The agricultural systems in the hills of Nepal: The ratio of agricultural land to forest land and the problem of animal fodder. APROSC Occasional Paper No 1. APROSC, Kathmandu.

10 Preferences for Conservation Agriculture in Developing Countries: a Case Study on the Tribal Societies of India and Nepal

Cynthia Lai,* Catherine Chan, Aliza Pradhan, Bikash Paudel, Brinton Foy Reed and Jacqueline Halbrendt

University of Hawai'i at Mānoa, Honolulu, Hawaii, USA

10.1 Introduction

In many agricultural regions of the world, farmers are experiencing the effects of climate change and its subsequent effects on soil productivity, which lead to reduced agricultural productivity (FAO, 2012). For smallholder subsistence farmers who reside in developing countries, the effects of climate change coupled with population pressures are of even greater impact, due to existing marginalized land conditions (i.e. poor soil fertility, moisture retention, and erosion), as well as lack of capital, institutional support, and access to resources and information (Lai *et al.*, 2012a). With increasing population and decreasing land fertility, agricultural research in the 1960s and 1970s focused on agricultural intensification and increasing per capita food production (Conway and Barbier, 1990). The new technologies, innovations, and increased agricultural productivity that emerged from this period are recognized as the "Green Revolution". Although the resulting chemical fertilizers, pesticides, and breeding programs for high-yielding varieties provided increased yields, the successes were short-lived, as they failed to provide sustainable solutions to existing land degradation and soil fertility problems, particularly for the smallholder subsistence farmer (Conway and Barbier, 1990).

Given the low agricultural outputs and land degradation that smallholder farmers (i.e. those farming on less than 2 ha) face in conventional farming systems,

*Corresponding author; e-mail: laicynth@hawaii.edu

an opportunity exists for the introduction of improved farm management and cultivation practices. While former Green Revolution technologies have proven useful for increasing agricultural yields on larger, mechanized farms, this is not true for farmers with small landholdings, low income, poor access to capital, and limited technical capacity (Conway and Barbier, 1990; Lai *et al.*, 2012b). More relevant to the conditions of these smallholder farmers are conservation agriculture (CA) practices. CA practices aim to conserve natural resources through integrative management of existing soil, water, and biological resources with innovative technologies that promote soil health and enhance agricultural productivity. To be specific, CA practices refer to a general set of practices that include concepts of minimum tillage and intercropping in order to enhance soil health (FAO, 2000). Studies have verified that CA practices promote soil health by minimizing soil erosion and maintaining adequate surface cover throughout the year (ECAF, 2001). By combining the practices of reduced tillage (minimal soil disturbance) with permanent or semi-permanent organic soil cover to protect the soil from sun, rain, and wind, farmers are able to rebuild the soil as well as maintain and enhance its capacity for self-sustainability (FAO, 2012).

Multiple studies have investigated the costs and benefits of farmers adopting CA practices (Knowler and Bradshaw, 2007; FAO, 2012). Table 10.1 synthesizes the results of these studies. Market factors are those factors directly related to the farmer's sales, profits, and expenses, while non-market factors are those factors that are unrelated to the financial aspects of the on-farm operations (i.e. biodiversity). As one of the goals of smallholder farmers is to increase agricultural productivity for livelihood and income, it is important to understand not only the market and non-market costs and benefits but also their subsequent short- and long-term temporal impacts, and at what spatial or geographical scales those impacts are experienced (Table 10.1). These geographical impacts are important, as they identify the divergence between locally appropriable benefits and costs and benefits and costs accruing to a wider, nation region.

In theory, subsistence farmers are more likely to be concerned with the short-term and local impacts of CA adoption, as these will influence them more directly than will the less direct long-term impacts (FAO, 2001). For individual farmers, who are often risk-averse and focused on their own livelihood (Roumasset *et al.*, 1979), these short-term impacts (increased soil fertility, yield, and profits) may not outweigh the potential market and non-market costs of implementation (start-up costs, short-term problems, risk propensity, and time commitment) (FAO, 2001). On the other hand, given the numerous long-term impacts of CA (surface hydrology, reduced sediment loads, and increased carbon sequestration), the farmer seeking optimal outcomes may realize that through trade-offs between the various costs and benefits of CA adoption the net impacts of CA are more desirable than those of conventional practices (Roumasset *et al.*, 1979). As such, farmers are more willing to accept the initial short-term start-up costs for the potential of greater yields, profits, and reduced labor and the long-term and sustainable benefits of CA down the road. In regards to geographical impact, the distinction of local and national appropriable benefits and costs of CA may help to narrow down the scope of policy programs and objectives (Knowler and Bradshaw, 2007). For example, local appropriable costs and benefits may help focus incentive programs on mediating costs through training and subsidies

Table 10.1. Summary of potential market and non-market costs and benefits of conservation agriculture by temporal and geographical impact. (From UNFAO, 2001.)

		Temporal impact		Geographical impact	
		Short	Long	Local	National
Potential benefits					
Market	Increased soil fertility and moisture retention → yield increase	x	x	x	x
	Reduced labor for specific farming practices	x	x	x	x
	Increased profit due to yield increase	x	x	x	x
Non-market	Stabilized soil → reduced erosion downstream sedimentation		x		x
	Reduced toxic contamination of surface water and groundwater		x		x
	Reduced flooding, recharge of aquifers as a result of better infiltration		x		x
	Reduced air pollution resulting from mechanized tillage, reduction of CO_2 emissions to the atmosphere (carbon sequestration)		x		x
	Conservation of terrestrial and soil-based biodiversity		x		x
Potential costs					
Market	Purchase of specialized planting equipment	x		x	
	Short-term problems (i.e. pests) due to weeding	x		x	
	Application of additional herbicides	x		x	
Non-market	Perception of high risk to farmers because of technological unfamiliarity	x		x	
	Time requirement for farmers to acquire new skills and practices	x		x	

at the individual farmer scale. Alternatively, national-scale appropriable costs and benefits may guide national and even global policy and research and development towards supporting CA (Knowler and Bradshaw, 2007).

Due to its potential short- and long-term positive impacts, CA has been practiced on over 100 million ha (Mha) worldwide (Derpsch and Friedrich, 2009). However, despite the proven empirical evidence and financial analyses of the net benefits of CA adoption, the uptake and adoption of CA practices has been slow, particularly in developing countries (Anderson and D'Souza, 2014). CA adoption may be daunting to farmers who have sustained their families via traditional and conventional practices for generations. In India, only about 20% of CA technologies generated by agricultural research is even temporarily adopted by smallholder farmers, likely due to farmer uncertainty, risk aversion to new practices, implementation failures, and/or inapplicability of technologies (Feder *et al.*, 1985; Chambers, 1991). The direct benefits of CA to the farmer are not immediate, but rather require time and dedication in order to achieve improved yields, soil structure, and soil organic matter, as well as reduced soil erosion, chemical reliance, and labor (Hobbs, 2007; Hobbs *et al.*, 2008; Friedrich *et al.*, 2009; FAO, 2012). Based on existing environmental and land conditions and their dependence on agricultural productivity for their livelihood, when faced with new CA practices under uncertain environments, farmers face complex decisions in relation to adoption. Often, a farmer's decision to adopt a new practice cannot be explained by a single objective but by a consideration of multiple objectives. Moreover, agricultural decisions to adopt new farming practices never have sure outcomes, particularly for subsistence farmers who reside in rural regions of developing countries where poverty, lack of infrastructure and government support, coupled with the uncertainty of nature, further complicate their decisions (Roumasset *et al.*, 1979). For these reasons, and under this type of subsistence context (marginalized land conditions, poverty, lack of infrastructure and government support, and uncertainty of nature), it is important to go beyond biophysical evaluations of CA and include farmers' individual perspectives on, comprehension of, and preferences for these practices compared to their traditional ones. This type of socio-ecological research endeavor is a primary objective and stepping stone toward successful implementation and adoption.

10.1.1 Understanding the farmers' context: case studies from India and Nepal

As a response to increasing soil and water resource degradation in tribal and ethnic societies of developing countries, various programs and projects have offered strategies and solutions for addressing the management of sloping agricultural lands, with a particular focus on soil conservation and minimizing land degradation. The Sustainable Management of Agroecological Resources for Tribal Societies (SMARTS) project represents one of these types of initiatives. Based in the north central plateaus of India and the mid-hills of Nepal, the SMARTS project focuses research on the implementation of CA treatments in agriculture-based tribal villages, with a specific focus on minimum tillage and intercropping to improve household income. While some initiatives may be more focused on the biophysical research outcomes (i.e. soil and water resource health) of CA, the SMARTS project further incorporates the traditional and ecological knowledge of subsistence farmers. In addition, it collaborates farmers with agricultural research and extension professionals in order to develop the most optimal and applicable conservation practices to increase adoption rates and agricultural sustainability.

To understand the context in which the farmers of these rural and developing regions make decisions, it is important to understand the critical factors influencing the adoption of new CA practices. In order to do so, we focused on two tribal populations of subsistence farmers in India and Nepal. These tribal farmer societies were selected as they represented vulnerable populations that could directly benefit from CA adoption. The two case study regions share similar household, economic, environmental, and social conditions, as well as farm and farmer characteristics

(Table 10.2). In both locations, subsistence farming on sloping land is practiced, although Nepal's (25% slope) land area is steeper than India's (2–5% slope). Villages are small, with around 30–40 households per village. In both countries, the average household income level is below the poverty line. Most farmers are illiterate and practice monocropping systems on an average 0.6 ha. The main cropping systems are maize followed by either mustard (in India) and millet or legumes (in Nepal). Off-farm agricultural input costs are minimal to none for these farmers, as they typically use saved seeds and on-farm manure for fertilization. The majority of household income in these tribal regions comes from the sale of crops, and the remaining income comes from livestock and/or off-farm employment (i.e. mining).

India

Smallholder farmers in the tribal areas of Odisha state, India, have struggled to produce adequate crop yields under the constraints of marginal land, low inputs, and

Table 10.2 Socio-economic profiles of the two tribal areas for the study of farmer preferences for conservation agriculture.

	India	Nepal
Context	Tribal Subsistence Rainfed	Tribal Subsistence Rainfed
Name of village	Tentuli	Thumka Gorkha Hyakrang
Physical geography	Upland 2–5% slope	Upland 25% slope
Economic conditions	Lack of access to capital, credit, finance	Lack of access to capital, credit, finance
Illiteracy	75%	80%
Environmental conditions	Marginal, rainfed land Poor soil fertility, moisture retention, degradation and erosion Seasonal intense rainfall → soil runoff	Marginal, rainfed land Poor soil fertility, moisture retention, degradation and erosion Seasonal intense rainfall → soil runoff
Average level of education	Primary	High school
Average households per village	40	30
Average annual household income	US$420	US$601
Current farming practices	Monocrop systems Conventional (plow) tillage	Monocrop systems Conventional (plow) tillage
Average farm size (ha)	0.6	0.6
Main crops by importance	Maize, mustard, rice	Maize, millet, rice
Cropping season	June–September October–January Maize/mustard/fallow	March–July July–November Maize/legume/fallow

continuous monocrop farming systems (Lai *et al.*, 2012b). The district of Kendujhar consists of multiple tribal villages (30–100 households; on average, five to seven members per household) and represents one of the poorest districts in the state of Odisha, in terms of both economic and environmental resources (Lai *et al.*, 2012a). It has very low education levels, minimal resources, and is being influenced by developmental and population pressures. In terms of agriculture, farming is conducted on marginal, rainfed land. It suffers from poor soil fertility, moisture retention, limited irrigation, susceptibility to erosion and other effects induced by climate change, poverty, and traditional farming practices. The village of Tentuli represents one of the tribal societies in the district of Kendujhar and is the focus of this case study. According to baseline surveys conducted by the SMARTS project in 2011, farmers in Tentuli engage in subsistence farming and practice conventional plow tillage, grow mainly maize and mustard, and have farm sizes of approximately 0.6 ha. The cropping season is approximately June–September for maize and October–January for mustard. Off-farm employment includes mining for male householders, while women focus on household responsibilities. There are opportunities to labor on other farms as well. With on- and off-farm employment, farming households make an annual income of about US$420 (Lai *et al.*, 2012a).

Nepal

Nepal's hill agricultural system is characterized by upland, rainfed, and maize-based farming, where the degradation of soil fertility is apparent in the region's comparatively low and decreasing crop yields, which have caused severe food deficiencies. Soil degradation, primarily due to erosion, affects 14.7 Mha of Nepal's land area (FAO, 1994). Intense rainfall on sloping lands accelerates soil runoff, decreasing soil fertility and resulting in a long-term decline in soil productivity and environmental moderating capacity. Moreover, this area is continuously degraded by farmers' unsustainable agricultural practices, due to their lack of access to new technology and information (i.e. CA), which would enable them to improve on existing traditional practices. Subsistence farmers in this particular study reside in villages on highly sloping land. These farmers practice conventional tillage of mainly maize and millet on an average farm size of 0.6 ha and have an annual income of US$601 (Reed *et al.*, 2014).

Influences of farmer CA preferences on decision making

Although farmers from both regions have many similarities, it was expected that there would be context-specific differences in their CA preferences due to potential cultural and site-specific characteristics. General differences in on-farm practices between the two societies under similar environmental and production contexts may imply conceptual and perhaps traditional differences in on-farm preferences of agricultural practices. For example, both countries follow different agricultural farming practices within their shared maize-based systems. In India, farmers plow, followed by broadcasting of maize seed; while in Nepal, farmers also plow, but then place maize seeds into the furrows left by the plow. In addition, on-farm differences such as cropping season, main farm crops, and land slope differ between the regions (Table 10.2); these may also influence farmer preferences for new agricultural practices and technologies.

Specific farmer goals, criteria, and preferences ultimately influence farmers' decisions to adopt CA (Alphonce, 1997). In order to understand farmers' perceptions and preferences of novel agricultural technology and CA treatments, we must understand their initial preferences of CA treatments prior to on-farm trials and, after working directly with them on-farm, identify any potential changes in preferences and what factors or reasons may have influenced those changes. Furthermore, it is important to compare research and extension professionals' preferences of the selected criteria and treatments to those of the farmers with whom they work, as the professionals are the ones to develop and disseminate the CA treatments and information. The decision-making process of research and extension professionals may be different from farmers, and as such may highlight significant perception gaps that might explain why farmers are struggling to adopt. Professionals are trained to be analytical and apply a systematic scientific method, whereas farmers are looking for what works within their dynamic environment (Carr and Wilkinson, 2005). As the farmer is the key locus of decision making when it comes to the adoption of novel CA practices, it is important that his of her preferences and indigenous knowledge is incorporated into those of the research and extension professional.

10.2 Methods

In order to explore farmer preferences for CA, we followed a three-step process. Part one included the introduction of CA and CA treatments to tribal societies through workshops. Part two consisted of the identification of farmer goals and criteria as it related to the adoption of new practices via focus group discussions and subsequent evaluation of conventional and CA treatments via exposure to on-station and on-farm trials and socio-economic analyses. Finally, part three included determining farmers' and research and extension professionals' preferences for CA criteria and treatments utilizing the analytical hierarchy process (AHP). Before and after on-farm trials, 40 Indian and 47 Nepalese farmers were surveyed to assess whether or not their initial preferences changed over time after working directly with the treatments.

10.2.1 Introduction of conservation agriculture

As with any new agricultural practices, adoption is unlikely to be immediate, but rather incremental. The transition from conventional to conservation agriculture is a complex undertaking, particularly for subsistence farmers. There are various avenues of adoption and varying constraints affecting what farmers in this context can achieve realistically through technology transfer, given their lack of experience with CA, the marginalized conditions, and the time frame of a single agricultural season or of the extension program. As such, CA production system treatments that are simple and are adapted from existing practices are more likely to be preferred by farmers, and subsequently adopted, than more complex practices (Guerin, 2001). In order to gain a better understanding of existing farmer practices and traditions, data from in-field agronomic and socio-economic survey results, multiple village visits, and initial ground-truthing data were compiled. These data were used to develop three maize-based CA production system treatments after multiple consultations with local experts and agricultural technicians (Table 10.3). In these cases, legumes were selected for intercropping due to their soil health benefits and higher value in the market. The three CA treatments were introduced to farmers via an interactive workshop hosted by the SMARTS research and extension professionals prior to on-farm trials in 2011.

India

The most common upland crop in the tribal regions of India is maize (*Zea mays*), which is grown between the months of June–September during the rainy season, followed by mustard (*Brassica juncea* L.), an oilseed grown in the post-rainy season between the months of October–January. The alternative CA production systems (treatments) introduced consisted of different combinations of minimum tillage and intercropping of maize and a legume (Table 10.3). To achieve minimum tillage, the land was tilled once prior to sowing, rather than twice, as with conventional tillage. For intercropping, cowpea (*Vigna unguiculata*, a legume) was planted between rows of maize. The inter-row spacing for maize was standard for a maize monocrop treatment, with no reduction of maize plants in the intercropped treatment. Treatment one (T1) represented the existing practice of conventional tillage (control treatment). Treatment two (T2) incorporated conventional tillage with intercropping, but kept the existing practice of plowing and incorporated intercropping of a legume (Herrera and Harwood, 1973). Treatment three (T3) introduced the practice of minimum tillage in order to minimize and reduce soil disturbance. Treatment four (T4) represented the most innovative practice, as it provided a solely CA approach of minimum tillage and legume intercropping to increase soil nutrients and resilience by moving away from a monocrop system.

Nepal

The treatments in Nepal and India consisted of different combinations of minimum tillage and intercropping. In Nepal, strip tillage was used and a legume (cowpea (*V. unguiculata*) in 2011 and black gram (*Vigna mungo*) for 2012–2014) was intercropped with millet (*Eleusine coracana*) rather than maize (*Z. mays*) (Table 10.3).

Table 10.3. Description of the conservation agriculture (CA) production systems (treatments) and conventional technologies in India and Nepal.

Country	CA treatment (T)
India	T1. Control: conventional tillage/maize
	T2. Conventional tillage/maize–cowpea intercrop
	T3. Minimum tillage/maize
	T4. Minimum tillage/maize–cowpea intercrop
Nepal	T1. Control: conventional tillage/maize followed by millet
	T2. Conventional tillage, maize followed by legume
	T3. Conventional tillage/maize followed by millet–legume intercrop
	T4. Strip tillage/maize followed by millet–legume intercrop

In contrast to India, where intercropping was completed in the first season, in Nepal, millet and legume intercropping was practiced in the second season after the maize crop. T1 represented the traditional Nepalese mid-hill farmers' practice of conventional plow tillage and maize followed by millet. In T2, conventional tillage was practiced, with maize followed by legume as the sole crop. T3 and T4 represented the introduction of two completely new agricultural practices for these Nepal farming communities. For T3, farmers practiced their traditional conventional tillage but with maize followed by an intercropping of millet and a legume. T4 included minimum (strip) tillage with maize followed by millet and legume intercropping.

10.2.2 Farmer preferences for CA practices

In India and Nepal, farmer focus group discussions followed the introduction of CA and the selected CA treatments in order to investigate what major factors influenced farmers' decisions to adopt. It was determined via focus group discussions that the farmers' main goal when selecting new on-farm practices was to improve income. Based on previously published information (Lai *et al.*, 2012b; Reed *et al.*, 2014), review of the literature, and focus group discussions, farmers from both countries identified profit, yield, labor, and soil environmental benefit as important CA outcomes that directly related to maximizing their income.

Using data and information from experimental on-station trials from each country, an economic and comparative analysis was conducted of the eight treatments across the two countries, as they related to the selected criteria (Table 10.4). Production system yield data were derived from the on-station plot outputs over the entire production year and converted to maize yield equivalent (MYE). To illustrate, for Nepal, yields of millet and legumes were converted to the MYE using the market price at the time of the study. The same was done for

Table 10.4. Conservation agriculture (CA) treatments and their outcomes from farmer trials in India and Nepal under the SMARTS project.

Country	CA treatment	Profit (US$/ha/ year)	Labor required (person days/ha/ year)	Yield (MYE[a] t/ha/year)	Soil environmental benefit (rank order)[b]
India	T1	518 (0)	257 (0)	7.21 (0)	1
	T2	516 (–0.34)	432 (+68)	8.56 (+18.7)	2
	T3	455 (–12)	279 (+8.5)	6.48 (–10.1)	3
	T4	660 (+27)	440 (+71)	9.11 (+26.4)	4
Nepal	T1	129 (0)	373 (0)	2.76 (0)	1
	T2	1142 (+892)	330 (–11.5)	3.60 (+30.4)	2
	T3	542 (+427)	448 (20.1)	3.47 (+25.7)	3
	T4	617 (+478)	400 (7.2)	3.13 (+13.4)	4

Parentheses indicating percent increase or decrease from Treatment 1 (Lai *et al.*, 2012b; Reed *et al.*, 2014).
[a]MYE, maize yield equivalent.
[b]Rank of 1 through 4, 4 indicating the most optimal soil health-enhancing outcome.

India for maize and mustard, converting the outputs into a single maize equivalent unit. Profit on 1 ha was calculated for a whole year. Profit was derived from the difference between the market price and cost of production per unit multiplied by the total yield of each treatment. The labor required was derived by determining the person days (8 h/day) required for each treatment for maize, legume, millet (Nepal), or maize, cowpea, mustard (India) on a per hectare basis for a year. In order to facilitate comparisons between treatments, results are reported as the percent increase or decrease from the conventional treatment (T1). Although no long-term analysis has been completed to determine the soil benefits, an extensive literature review identified the potential soil benefits for each treatment. Based on the literature review, the soil benefit for each treatment was ranked 1–4, with 4 indicating the optimal soil health-enhancing outcome.

India

In terms of the Indian farmers' preferred outcome, as well as local field station crop production data, T4 provided the best outcome of the CA treatments for farmers, as it provided optimal values for three (yield, profit, and soil environmental benefit) of the four outcome criteria. In terms of profit, T4 provided the highest yield due to the intercrop of cowpea with maize, providing a 27% increase in maize equivalent yields compared to conventional practices (T1). The T4 practices of minimum tillage and intercropping additionally provided the best outcome for soil environmental benefit utilizing practices of minimum tillage to reduce soil disturbance and intercropping with a legume to increase soil nutrients (Herrera and Harwood, 1973; Jaganathan *et al.*, 1974; Kalra and Gangwar, 1980; Anderson and D'Souza, 2004; Hobbs, 2007; Hobbs *et al.*, 2008). In terms of labor required, conventional practices required less labor than the CA treatments. Treatments that introduced minimum tillage increased labor hours invested in plowing and harvesting. In intercropping treatments, labor due to sowing and harvesting increased. T2 resulted in less labor compared to the other two CA treatments, which might have been a result of cowpea taking up more space between the maize rows and outcompeting the weeds. T2 was also the second best in terms of yield. Profit for T2 was nearly equivalent compared to conventional practices (–0.34% from T1). T3 required the least labor of all the CA treatments (though still more than conventional), but also produced the lowest yield and profit of the four treatments.

Nepal

Based on Nepalese farmers' preferred outcome as well as local field station crop production data, T2 provided the best overall outcome in terms of profit, yield, and labor. T4 provided the best option for soil quality improvement, the second most profit of all the treatments and third best yield performance. The highest maize yield was found under T2, most likely due to the introduction and subsequent positive effects of growing legume crops that fix nitrogen. T2 also had the lowest labor requirement due to the lack of nursery management and transplanting labor, as was required by millet. Because of lower labor requirements and higher yields of legume crops (which had higher market value), T2 was most profitable.

10.2.3 Farmers' preferences: analytic hierarchy process approach

In order to investigate farmers' preferences of these treatment options and the criteria/objectives the farmers used to make adoption decisions, the analytic hierarchy process (AHP) approach was used. AHP is a multicriteria decision-making tool developed by Thomas Saaty in the 1970s (Saaty, 2008). It is preferred over alternative methodologies because of its applicability to group settings, ability to organize and analyze complex decisions into simple pairwise comparisons for evaluation, and ability to reflect the most optimal "choice" based on an identified goal and understanding of the problem (Saaty, 2008). This is ideal in a case such as this, in which farmers' preferences for farming practices and technologies cannot be explained by a single objective, but rather by a compromise between multiple objectives (criteria) to achieve their goal. To gather the data and conduct the AHP, one must designate a hierarchy of decision making (Fig 10.1) that includes three major steps (Braunschweig, 2000):

1. Determine the ultimate goal of the preferred choice decision (Level 1: improved income).
2. Select the objectives/criteria to be used to decide among the choice options (Level 2: profit, labor saving, yield, and soil environmental benefit).
3. Lay out the given choice options (treatments) (Level 3).

Once the hierarchy has been developed and organized (goal, objectives/criteria, and alternatives), systematic pairwise comparisons are conducted, such that each element in the tier below is compared to each other element in the same tier with respect to one element in the tier above. These pairwise comparisons are made on a 5-, 7- or 9-point scale to indicate strength of preference. In order to obtain a ranking of the overall priorities, an eigenvector value must be calculated. The eigenvector is calculated by: (i) squaring the pairwise matrices; (ii) adding all the rows of the matrices and normalizing them; and (iii) multiplying the weight of each objective with the rankings of the alternatives with respect to the same objective to obtain the overall preferences. Besides determining

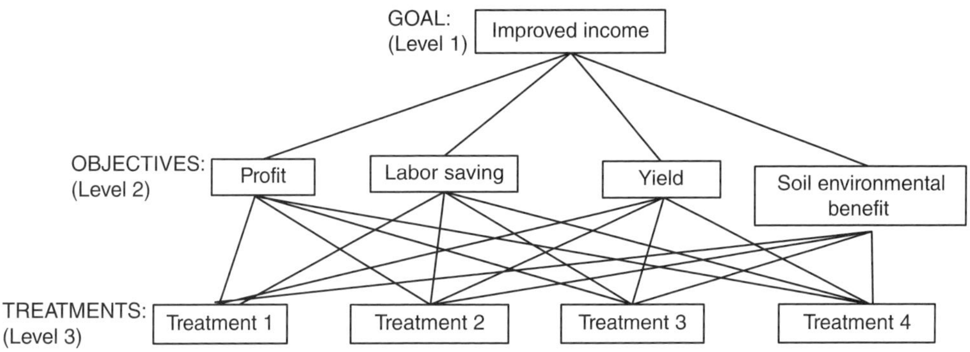

Fig. 10.1. An AHP diagram for India showing the three levels and the relevant pairwise comparisons.

eigenvalues for priorities, the priorities may also be computed via the following equation (Braunschweig, 2000):

$$P_l = \sum_{M=1}^{M} P_{lm} V_m \qquad \text{with} \qquad \sum_{l=1}^{L} P_{lm} = 1 \qquad \text{and} \qquad \sum_{m=1}^{M} V_m = 1$$

where P_l is the final priority of treatment l; P_{lm} is the priority of treatment l with respect to criterion m; V_m is the weight of objective m; l is treatment (1… L), (Treatments 1–4); m is objective (1 … M), (criteria: profit, labor saving, yield, and soil environmental benefit).

Similar to determining the eigenvector value of a pairwise comparison matrices, the equation illustrates that for each alternative/treatment, the local (by level) priorities are multiplied by the corresponding objective weight, and the results are summed up to obtain the global (overall) priority of the project with respect to the ultimate goal. For this study, Expert Choice Software was utilized to compute the local and global priorities of 40 Indian and 47 Nepalese farmers. The farmers were surveyed both before and after on-farm trials to assess whether or not their initial preferences changed over time after working directly with the treatments.

10.3 Results

10.3.1 Farmers' preferences: CA criteria and treatments

Farmers in both India's and Nepal's tribal societies had the opportunity to implement on-farm trials of the selected CA treatments for approximately 2 years. While participants in both case studies experienced similar access to resources, information, and support of CA treatments, the two countries responded differently to CA treatment preferences before and after the on-farm trials.

India

Farmers prioritized the following criteria for selecting treatment options in descending order of importance prior to on-farm trials as follows: yield (0.329), profit (0.274), soil (0.213), and labor (0.184) (Table 10.5). The most pressing concern

Table 10.5. Farmer preferences of criteria for selecting CA treatments before 2011 and after 2013 implementing on-farm trials. (From Lai *et al.*, 2012b; Reed *et al.*, 2014.)

	India				Nepal			
	Before		After		Before		After	
Criterion	Weight	Rank	Weight	Rank	Weight	Rank	Weight	Rank
Profit	0.274	2	0.190	3	0.097	3	0.232	3
Soil	0.213	3	0.426	1	0.594	1	0.303	2
Yield	0.329	1	0.311	2	0.213	2	0.347	1
Labor	0.184	4	0.073	4	0.096	4	0.117	4

for farmers of this case study in India was productivity, followed by profit, then soil. Farmers did not deem labor availability of high importance, as there was generally a sufficient labor supply. After 2 years of farm trials with CA treatments, farmers' preferences changed, indicating a deeper understanding of CA, particularly as it related to soil environmental benefit, to crop productivity and income. After practicing CA, yield was ranked second in importance over profit and labor, respectively (Table 10.5). Thus, farmers in India shifted their preferences following on-farm trials, placing more importance on the criteria of soil environmental benefit (0.426) and yield (0.311) over profit (0.190) and labor (0.073).

In terms of CA treatments before on-farm trials, Indian farmers strongly preferred T4 (minimum tillage intercropped with legume; 0.347) and T2 (conventional tillage intercropped with legume; 0.366), compared to T1 (conventional tillage; 0.141) (Table 10.6). Farmers selected treatments that best supported increased yield and profit. T4 and T2, specifically, had greater yields and profit compared to conventional practice (T1), because these conservation treatments included intercropping of maize with cowpea. Since cowpea had a high market value, and maize yield was not decreased in these treatments, profit was maximized. In these tribal villages, farmers had never intercropped before. The addition of another crop resulted in additional overall incomes (output of two crops instead of one, higher market value of cowpea) and enhanced soil fertility, due the nitrogen-fixing components of cowpea (Herrera and Harwood, 1973). Before CA treatment interventions, farmers preferred T2 (conventional tillage, maize intercropped with legume). After practicing CA for 2 years, farmers preferred T4, a practice that embraced CA practices of minimum tillage and continuous cover crop to maximize soil moisture. However, both before and after CA trials, their treatment preferences clearly highlighted their preferences for greater yields and profits (i.e. T2 and T4). This consistent preference of greater yields and profits both before and after trials, coupled with the observed shift in treatment preferences (before: T2; after: T4) indicates that the outcome of greater yields and profits is an underlying factor for farmer adoption compared to the actual agricultural technology applied. Nonetheless, T4 represented the most optimal of the treatments based on the presented on-station objectives of yield, profit, and soil environmental benefit (Table 10.4), and based on farmer preferences after 2 years of on-farm trials (Table 10.6).

Table 10.6. Farmer preferences in India and Nepal for CA treatments prior to on-farm trials and after on-farm trials.

	India				Nepal			
	Before		After		Before		After	
CA treatment (T)	Weight	Rank	Weight	Rank	Weight	Rank	Weight	Rank
T1	0.141	4	0.065	4	0.09	4	0.157	4
T2	0.366	1	0.182	2	0.439	1	0.371	1
T3	0.147	3	0.160	3	0.23	3	0.248	2
T4	0.347	2	0.592	1	0.241	2	0.224	3

Nepal

Prior to on-farm trials, Nepalese farmers prioritized soil environmental benefit (0.594), followed by yield (0.213), profit (0.097), and labor (0.096) (Table 10.5). The almost twofold greater priority placed on soil health over yield, the next preferred criteria, reflects the farmers' wish to improve soil health. After 2 years of farm trials with CA treatments, farmers' preferences changed and shifted from soil environmental benefits to yield as the most preferred criteria. This change in farmer preference since the on-farm trials may have been due to the low cowpea yield following on-farm trials of T2 (conventional tillage; maize followed by legume), indicating that, in terms of short-term needs, farmers did not see reduction of crop yield as worth the trade-off of soil environmental benefits. This implies that even though farmers may understand the significance of soil health as it relates to sustainable agriculture, when it comes to implementing new practices to accommodate their subsistence livelihood, sufficient yield represents a critical factor.

Farmer preferences of CA treatments before and after on-farm trials further support their perceived need for increased yields. Prior to on-farm trials, Nepalese farmers indicated a strong preference for T2 (conventional tillage; maize followed by a legume) followed by T4 (strip tillage; maize followed by millet–legume intercrop), T3 (conventional tillage; maize followed by millet–legume intercrop), and T1 (conventional tillage) (Table 10.6). Similar to Indian farmers, Nepalese farmers selected treatments that best reflected outcomes that supported increased yield and profit. Therefore, farmers preferred treatments that included intercropping of an additional crop, particularly a legume, thereby enhancing soil nutrients, achieving higher yields. Following on-farm trials, preference for T2 remained a priority, but T4 was now outweighed by T3, due to greater yields than found in T4, indicating farmer treatment preferences shifted to yield as a critical CA criterion. Based on the shifts in farmer preferences from before to after on-farm trials for both criteria and treatments, it seems that despite their intentions towards improved soil health, farmers' immediate need for profit and yield outweighs all other criteria.

10.3.2 Interpreting and understanding farmer versus professional preferences

Prior to introducing CA to these tribal regions, farmers had minimal exposure to environmental education and had minimal to no knowledge of CA. Research and extension professionals, on the other hand, should have background knowledge of CA and its specific short- and long-term benefits. Therefore, this project and select case studies represented a unique opportunity for research and extension professionals to utilize their science-based knowledge and disseminate the innovative agricultural practices of CA to farmers in marginalized land areas.

As research and extension professionals develop and introduce CA practices to farmers, it is important that farmers' and professionals' preferences for treatments align for the most effective implementation and outcomes of CA treatments. Therefore, following the CA introduction workshops and focus group discussions, farmers conducted on-farm trials that were monitored closely by research and extension professionals so that the professionals could also work directly with the treatments, witness their implementation and impacts, and subsequently develop their own personal CA treatment preferences. Given the differences in perspective and background knowledge between tribal farmers and research and extension professionals, it important to investigate possible perception gaps between farmer and professional preferences for CA criteria and treatments before and after on-farm trials.

In India, tribal farmers and research and extension professionals indicated different preferences for CA criteria and treatments prior to on-farm trials, but similar preferences following on-farm trials (Table 10.7). Before on-farm trials, farmers prioritized yield (0.329), while research and extension professionals prioritized profit (0.356). This was further supported by each group's CA treatment preferences prior to on-farm trials, in that research and extension professionals selected T4 as their preferred CA treatment, as it offered the optimal outcome in terms of greatest soil environmental benefits and highest profit compared to conventional practices (T1). On the other hand, farmers selected T3 as their preferred CA treatment, due to its optimal outcome in yield and profit compared to conventional practices.

In order for farmers to make the decision to adopt a new practice, the practice must achieve improved incomes compared to current practices. According to farmers, yield, profit, labor required, and soil environmental benefit are the objectives that must be considered when adopting a new practice given a set of options (Lai *et al.*, 2012b). Generally speaking, both stakeholder groups preferred criteria representing immediate gains from CA treatments. In fact, both farmers and professionals have overlooked the main objectives and significance of CA, to enhance soil health and structure for sustained agricultural productivity. As such, farmers and professionals are prioritizing the wrong objectives in order to achieve the outcomes they desire, improved sustainable income through CA.

Initially, farmers and professionals selected yield and profits, respectively, as their most weighted of the four criteria when selecting for CA treatments. During on-farm conservation agriculture production systems (CAPS) trials, farmers were able to gain first-hand experience and understanding of the specific costs and benefits of CA treatments, given their observations. At the same time, professionals were

Table 10.7. Farmer and research/extension preferences for CA criteria and treatments, with respect to the goal of maximizing farmer income, before and after on-farm trials in India. (From Lai *et al.*, 2012b.)

	Before				After			
	Farmers		Research and extension		Farmers		Research and extension	
	Weight	Rank	Weight	Rank	Weight	Rank	Weight	Rank
Soil benefit	0.213	3	0.271	2	0.426	1	0.370	1
Yield	0.329	1	0.245	3	0.311	2	0.322	2
Labor	0.184	4	0.128	4	0.073	4	0.062	4
Profit	0.274	2	0.356	1	0.190	3	0.246	3
T1	0.141	4	0.103	4	0.065	4	0.067	4
T2	0.147	3	0.175	3	0.182	2	0.211	2
T3	0.366	1	0.25	2	0.16	3	0.156	3
T4	0.347	2	0.472	1	0.592	1	0.566	1

able to observe farmer behavior and application of the preferred treatments and disseminate and interpret CA information on a more engaging level. After approximately 2 years of CA practice, and despite initial differences, farmers and research and extension professionals were able to align their preferences regarding treatment criteria as it related to the goal of maximizing farmers' income (Table 10.7). Farmers shifted their preferences to soil environmental benefits as their preferred criteria (0.426), followed by yield (0.311), profit (0.190), and labor (0.073). Given the tangible evidence of the impacts of CA practices (increased agricultural productivity) following on-farm trials, farmers selected T4 as their preferred CA treatment. Professionals' preferences indicated a shift similar to that of the farmers after 2 years of on-farm trials. While professionals initially indicated profit as the preferred criteria for evaluating CA treatments, they too later preferred soil benefit (0.370) followed closely by yield (0.322), profit (0.246), and labor (0.062). This shift in preferences towards an emphasis on soil health implies that after 2 years of on-farm trials, farmers and professionals are able to witness the benefits of CA practices (minimum tillage and intercropping) and understand that soil health is the initial catalyst to achieving the short-term gains of yield and profit that come with CA adoption.

On-farm trials in India truly represent a success story in terms of CA awareness and dissemination. According to unpublished results under the SMARTS project by Pradhan *et al.*, "... requests from additional village farmers to implement CA treatments during the third year of the study suggest that they recognize these benefits and are willing to modify conventional practices to achieve them" (Pradhan *et al.*, 2013, unpublished results). This particular tribal case study yielded not only tangible benefits in terms of improved crop yields but also in increasing farmer awareness of CA at the local as well as at the policy level. It also represents a successful exchange of information and technologies between research and extension professionals as it relates to professionals introducing a new practice and tribal farmers accepting and applying it.

10.4 Conclusion

Based on the results from India and Nepal of farmer preferences of CA treatments compared to conventional practices, there is high potential for CA in smallholder and subsistence farming regions. Farmers are interested in trying new practices, particularly ones that increase immediate gains compared to their existing practices. On the other hand, farmers are hesitant to abandon their conventional tillage practices for minimum tillage, despite the multiple expected benefits of increased agricultural productivity and soil health. Farmers are worried about the perceived risk involved in trying new techniques, given their limited experience and knowledge of CA. Therefore, if research and extension professionals are interested in implementing CA treatments that include minimum tillage, they must improve their training and workshop delivery, enhancing their ability to disseminate CA information by showing tangible benefits to farmers.

Furthermore, lack of adoption is more often than not due to a mismatch of CA technologies to the farmer context (Giller *et al.*, 2009). According to results comparing farmer and professional preferences of CA treatments, farmers and professionals may have different preferences for CA objectives and treatments. By increasing understanding of farmer needs and preferences for CA practices, professionals can better ensure that they are introducing the most applicable and relevant technologies within the subsistence farmer context. While both professionals and farmers are often focused on short-term gains when applying CA technologies, improved soil health represents the major precursor to increased agricultural productivity. As such, professionals should ensure that workshops are focusing on disseminating the soil benefits of CA treatments and that farmers are understanding the positive relationship between agricultural productivity (yield) and soil health.

Understanding farmer preferences at the individual level is important not only for the purposes of increasing farmer adoption rates but also for increasing policy and government support. If researchers and government can understand farmers' preferences and their criteria for technology better, this would help to ensure efficient planning, successful implementation, effective intervention, and ultimately increase the potential for adoption by participating farmers. This movement has already been catalyzed in India, where local district administration and District Departments of Agriculture have been approved to implement maize–cowpea intercropping with reduced tillage across 2,000 ha of potential maize area in the district of Kendujhar, Odisha (SANREM CRSP, 2012). Additionally, a project proposal has recently been approved to replicate minimum tillage with intercropping across 1,000 ha of farmlands (SANREM CRSP, 2012).

Conservation agriculture represents a relatively new phenomenon for these tribal subsistence farmers, and abandoning conventional and traditional practices for a new and unfamiliar practice is a complex decision. In order to develop policies that support all those involved (farmers, researchers, and extension agents), all stakeholders require a thorough understanding of farm-level conditions and practices in the short and long term: what criteria farmers use to determine agricultural practices and which of these criteria matter the most to those farmers. The AHP offers an approach that attempts to formalize and make transparent the decision-making process of these farmers and establishes criteria and assigns weights to understand better their preferences behind selecting new on-farm practices, particularly CA practices.

References

Alphonce, C. (1997) Application of the analytic hierarchy process in agriculture in developing countries. *Agricultural Systems* 53, 97–112.

Anderson, J. and D'Souza, S. (2014) From adoption claims to understanding farmers and contexts: a literature review of conservation agriculture (CA) adoption among smallholder farmers in southern Africa. *Agriculture, Ecosystems and Environment* 187, 116–132.

Braunschweig, T. (2000) Priority setting in agricultural biotechnology research: Supporting public decisions in developing countries with the analytic hierarchy process. *ISNAR Research Report* 16, 1–110.

Carr, A. and Wilkinson, R. (2005) Beyond participation: Boundary organizations as a new space for farmers and scientists to interact. *Society and Natural Resources* 18, 255–265.

Chambers, R. (1991) Scientist or resource-poor farmer – whose knowledge counts? Keynote address. Seminar on Crop Protection for Resource-Poor Farmers. Institute of Development Studies, University of Sussex, UK.

Conway, G. and Barbier, E. (1990) *After the Green Revolution: Sustainable Agriculture for Development*. Earthscan, Sterling, Virginia.

Derpsch, R. and Friedrich, T. (2009) Global overview of conservation agriculture adoption. Paper presented at the 4th World Congress on Conservation Agriculture, New Delhi, India, 4–7 February 2009.

ECAF (European Conservation Agriculture Federation) (2001) Conservation agriculture in Europe. Available at: http://www.ecaf.org/index.php?option=com_content&task=view&id=53&Itemid=52 (accessed 30 July 2013).

FAO (Food and Agriculture Organization) (1994) Land Degradation in South Asia: 1st Severity, Causes and Effects upon the People. World Soils Report 78. FAO, Rome.

FAO (2000) Soil management and conservation for small farms. FAO Soils Bulletin 77. FAO, Rome.

FAO (2001) *The Economics of Conservation Agriculture*. Food and Agriculture Organization Land and Water Development Division, Rome, Italy.

FAO (2012) What is conservation agriculture? Food and Agriculture Organization of the United Nations. Last updated January 2012. Available at: http://www.fao.org/ag/ca/1a.html (accessed 10 November 2014).

Feder, G., Just, R. and Zilberman, D. (1985) Adoption of agricultural innovations in developing countries: A survey. *Economic Development and Cultural Change* 33, 255–299.

Friedrich, T., Kassam, A. and Taher, F. (2009) *Adoption of Conservation Agriculture and the Role of Policy and Institutional Support*. International Consultation on No-Till with Soil Cover and Crop Rotation, Shortandy, Kazakhstan, 8–10 July, 2009.

Giller, K., Witter, E., Corbeels, M. and Tittonell, P. (2009) Conservation agriculture and smallholder farming in Africa: The heretics' view. *Field Crops Research* 114, 23–34.

Guerin, T. (2001) Why sustainable innovations are not always adopted. *Resources, Conservation and Recycling* 34(1), 1–18.

Herrera, W.T. and Harwood, R.R. (1973) Crop inter-relationship in intensive cropping system (mimeo.) IRRI, Los Banos, Philippines.

Hobbs, P.R. (2007) Conservation agriculture: What is it and why is it important for future sustainable food production. *The Journal of Agricultural Science* 145, 127–137.

Hobbs, P.R., Sayre K. and Gupta, R. (2008) The role of conservation agriculture in sustainable agriculture. *Philosophical Transactions of the Royal Society* B 363, 543–555.

Jaganathan, N.T., Morachan, Y.B. and Ramiah, S. (1974) Studies on the effect of maize and soybean association in different proportions and spacing on yield. *Journal of Potassium Research* 10(2), 386–391.

Kalra, G.S. and Gangwar, B. (1980) Economics of intercropping of different legumes with maize at different levels of nitrogen under rainfed conditions. *Indian Journal of Agronomy* 25, 181–185.

Knowler, D. and Bradshaw, K. (2007) Farmers' adoption of conservation agriculture: a review and synthesis of recent research. *Food Policy* 32(1), 25–48.

Lai, C., Chan-Halbrendt, C., Halbrendt, J., Shariq, L., Roul, P., *et al.* (2012a) Comparative economic and gender, labor analysis of conservation agricultural practices in tribal villages in India. *International Food and Agribusiness Management Review* 15(1), 73–86.

Lai, C., Chan-Halbrendt, C., Halbrendt, J., Naik, D. and Ray, C. (2012b) Farmers' preference of conservation agricultural practices in Kendujhar, Odisha using analytic hierarchy process. In: Mulvaney, M.J., Reyes, M.R., Chan-Halbrendt, C., Boulakia, S., Jumpa, K., Sukvibool, C. and Sombatpanit, S. (eds) *Conservation Agriculture in Southeast Asia and Beyond*, Special Publication No 7. World Association of Soil and Water Conservation, Beijing, pp. 85–100.

Reed, B., Chan-Halbrendt, C., Tamang, B.B. and Chaudhary, N. (2014) Analysis of conservation agriculture preferences for researchers, extension agents, and tribal farmers in Nepal using analytic hierarchy process. *Agricultural Systems* 127, 90–96.

Roumasset, J., Boussard, J. and Singh, I. (1979) *Risk, Uncertainty and Agricultural Development.* Southeast Asian Regional Center for Graduate Study and Research in Agriculture (SEARCA), Philippines and the Agricultural Development Council, New York.

Saaty, T.L. (2008) Decision making with the analytic hierarchy process. *International Journal of Services Sciences* 1(1), 83–98.

SANREM (Sustainable Agriculture and Natural Resource Management) CRSP (Collaborative Research Support Program) (2012) SANREM CRSP: 2012 Annual Report. Available at: http://www.oired.vt.edu/sanremcrsp/wp-content/uploads/2013/12/SANREM_AnnualReportFY12.pdf (accessed 4 August 2014).

UNFAO (2001) The Economics of Conservation Agriculture. Rome, Italy. Available at: http://www.fao.org/docrep/ 004/y2781e/y2781e00.htm (accessed 2 November 2013).

11 Empowering Women through Conservation Agriculture: Rhetoric or Reality? Evidence from Malawi

JANE MAHER,[1*] PAUL WAGSTAFF[2] AND JOHN O'BRIEN[2]

[1]*Department of Geography, School of Natural Sciences, Trinity College Dublin, Dublin, Ireland;* [2]*Concern Worldwide, Dublin, Ireland*

11.1 Introduction

Malawi is a landlocked country in southern Africa, 1,500 km from a seaport. It is one of the poorest countries in sub-Saharan Africa (SSA), with a GDP per capita of US$805 in 2011, and ranks 170th out of 185 on the Human Development Index, an indicator that combines life expectancy, education, and income as a measure of development (UNDP, 2013). In Malawi, agriculture is the primary economic sector, representing approximately 37% of the country's GDP and employing about 80% of the labor force in 2010 (African Development Bank, 2012). Approximately 80% of Malawi's population lives in rural areas (World Bank, 2012); 90% of these people are smallholder farmers that rely on rainfed subsistence farming techniques (IFAD, n.d.). Systematic plowing of agricultural land has intensified in recent years due to land scarcity, which has resulted in significant soil degradation and declining yields (Scherr and Yadav, 1996). As in much of SSA, erratic rainfalls heightened by climate change and a growing population have led to food insecurity in Malawi. Food insecurity threatens much of rural Malawi, with more than 40% of the rural population living in poverty and vulnerable to seasonal food crises. The occurrence of poverty and hunger is exacerbated by a reliance on *ganyu*, an informal labor market system, through which the poor provide daily agricultural labor on larger farms and agricultural estates in exchange for nominal wages.

*Corresponding author; e-mail: maherj8@tcd.ie

11.1.1 Women and agriculture in Malawi

Women in Malawi generally fare worse than their male counterparts on most social and economic indicators. Their unequal status is shaped by the interlocking factors of poverty and discriminatory treatment in the family and public life. Gender inequality in Malawi reduces women's access to productive resources, development opportunities, and decision-making processes, which creates a cycle of vulnerability and poverty.

As in Malawi, agriculture is underperforming in much of Africa. One reason for this underperformance is the serious gender gap women face in access to productive resources. They often lack access to land tenure, extension services, credit, improved crop variations, and access to markets, and experience lower levels of human capital (Quisumbing and Pandolfelli, 2009). Globally, women are a vital part of agriculture and food production, making up two-thirds of the agricultural workforce (Ayoade, 2011) and producing 50% of the world's food (OECD, 2011). Often, men have access to improved technologies for farming, while women still use traditional techniques that are labor-intensive, as well as time- and energy-consuming (Carr and Hartl, 2010). Providing women with the same access and control over resources can result in better agriculture and human outcomes (Farnsworth, 2010). In fact, providing women with the same quantity and quality of inputs that men typically receive, while also improving agricultural education, could increase national agriculture outputs in SSA by up to 20% (World Bank, 2005).

The persistent gender gap in agriculture hinders women's productivity. Closing the gender gap would produce significant gains for society by increasing agricultural productivity, reducing poverty and hunger, and promoting economic growth (FAO, 2011). Evidence suggests that women are more likely than men to spend their income on the well-being of their families, including more nutritious foods, school fees, and health care (Bunch and Mehra, 2008). Hence, many see the economic empowerment of rural women as a central strategy in the attainment of the Millennium Development Goals, a set of goals agreed by the world's countries and development institutions to meet the needs of the world's poorest, by 2015. This includes eradicating extreme hunger, achieving universal primary education, improving child and maternal health, and gender equality (Carr and Hartl, 2010).

11.1.2 Conservation agriculture in Malawi

Conservation agriculture (CA) is a farming system that has been promoted by numerous international organizations as a panacea to agricultural problems in smallholder farming in the tropics (Giller *et al.*, 2009). The overall goal of CA is to make better use of agricultural resources, compared with conventional agricultural techniques, through the integrated management of soil, water, and biological resources (Knowler and Bradshaw, 2006). The FAO (2012) defines CA as "an approach to managing agro-ecosystems for improved and sustained productivity, increased profits and food security while preserving and enhancing the resource base and the environment". There are three core principles to CA: no tillage, covering soil, and crop rotation.

Proponents of CA suggest it is an option for adaptation and mitigation of the threats of food insecurity and climate change, because it reduces vulnerability to both excessive rainfall and drought, while increasing food production (FAO, 2008). As rainfall becomes more unpredictable, CA can give farmers more security. Field preparation and sowing can occur before the start of the wet season, and so crops can take advantage of maximum growing time once rains fall. Under conventional agriculture, tilling land begins once the rains start to fall, potentially sacrificing valuable time. This can be a significant factor, as it is estimated that for every day planting is delayed after the first possible planting date, maize yields can decrease by up to 2% (Maal, 2011).

However, CA is not without its criticisms, especially in SSA. The benefits of CA accumulate over time, and yields increase over a longer period. Giller *et al.* (2009) state that CA is an inappropriate farming system for the vast majority of resource-constrained smallholder farms, due to the immediate need for food for survival. As a result, smallholder farmers often attribute more value to the immediate benefits of agriculture than those realized in the future. Giller *et al.* (2009) also argue that CA often goes against traditional and cultural activities such as burning crop residues for land clearing and controlling rodent and pest populations. Where crop residues are not burnt, they are often sold or used as livestock feed, particularly among agropastoralist communities.

In SSA, CA is being promoted by many international organizations as an answer to agricultural problems such as soil degradation and water availability, and has been incorporated into regional agricultural policies by the New Partnership for African Development (NEPAD) and the Alliance for a Green Revolution in Africa (AGRA) (Kassam *et al.*, 2009). In addition, many non-governmental organizations (NGOs) have introduced CA projects. In 2004, Concern Worldwide (CW) began promoting CA as a viable alternative to traditional farming practices. By encouraging CA usage, CW hopes to assist smallholder farmers in addressing their poor production outcomes and prevent environmental degradation, while recognizing their resource constraints. CW's CA program focuses on reducing soil disturbance through the use of permanent planting basins (zaï holes), technology from the Sahel that has proven successful in semi-arid areas. Soil is covered through the use of crop residues, with additional grass mulch as required. The main carbohydrate crop (maize) is rotated with a legume crop (groundnuts, cowpeas, and soybean); a third crop is chosen by the farmer, usually a cash crop (soybean, sunflower) or a food reserve (sorghum).

11.1.3 Women and CA

There have been a limited number of studies on the impact of CA on women, and there is a need to examine the relationship between women and CA, particularly since gender issues in agriculture have not been treated as an integral part of policy and programming (FAO, 2011). Those few studies that do exist show that CA can reduce the time women spend on agricultural activities (Milder *et al.*, 2011). In Zambia, women both reduced their workload and spread their work out over a longer period by using CA. In fact, the farming calendar shifted earlier into the dry season and women expended 60% less energy on land preparation (Maal, 2011). Furthermore, research has shown CA can have positive impacts for women through the improved nutritional status of household members, the ability to sell surplus produce, and the resulting increase in income, which can be invested in school fees, clothes for children, and agricultural inputs (Maal, 2011). However, as weed density is often said to increase under CA, Giller *et al.* (2009) expresses a concern that CA may shift the burden of work from men, who traditionally carry out tilling, on to women, who are typically responsible for weeding.

In order to address the lack of existing research on CA impacts on women in SSA, we examined four areas of women's lives: time and labor spent on agricultural endeavors; agricultural production and its effect on household food security; decision making in the home; and social capital. We sought to determine whether CA was a beneficial method of agriculture for women to practice, in light of these factors, as well as to investigate the role CA could play in empowering women through agriculture. Concern Worldwide (CW) and Concern Universal (CU), two NGOs that implement CA projects in Malawi, assisted with the project. Both organizations' CA projects are gender neutral in implementation; that is, gender is not taken into consideration when selecting those to participate in the project. However, since the CA project started in 2004, CW has noted that women adopt CA at a faster rate than men, and also have a greater interest in learning the techniques. CA is implemented at the community level with the assistance of lead farmers. Lead farmers are selected by CW extension workers, trained in CA and facilitation skills, and then given responsibility for training other CA beneficiaries.

11.2 Methodology

To measure if CA is a beneficial farming system for women, we employed an analytical framework of assessing the impact CA has on women's time and labor, agricultural production and household food security, decision making in the home, and social capital. Semi-structured, one-on-one interviews with CA practicing farmers, focus groups, and the disaggregation of harvest data were used to address the research questions. The study population was defined as women in Malawi who have been practicing CA for at least 1 year and have harvested crops under a CA farming system. A control population of men practicing CA for at least 1 year and women practicing only conventional agriculture was also asked to participate. The control populations were chosen to enable gender-sensitive comparisons to be drawn between the experiences of women practicing CA and men practicing CA, as well as between women practicing and not practicing CA.

During project planning, a random sampling method was chosen to select the study participants. However, on arrival at the study site, it was determined that random sampling could not be carried out. Households practicing CA were often spread out across a wide area, and thus random sampling would have proven very time-consuming and reduced the sample size significantly. Therefore, participants were selected non-randomly, based on availability and willingness to take part. Initially, this was of concern, because non-random sampling has often provided a weak basis for generalizing about a population (Walliman, 2005). However, to be eligible for CA projects promoted by CW and CU, households had to be in the poor or very poor quartile. Therefore, it was felt that this method of sampling would still result in a representative population of the poor farming community.

Due to availability, 80 female CA farmers and 33 control (male CA and female conventional only) farmers were interviewed. The study subjects were interviewed to gain information on land use, decision making in the home, social status, labor, and difference in yields. During the project planning stage, the use of a structured household survey was intended to be the main method of primary data collection. However, once in Malawi, this proved to be an inflexible method of data collection that could result in losing valuable information. As Fife (2005) states, the complexity of lives cannot be understood through a multiple-choice questionnaire. Rather, in order to measure the complexity of the impact CA has on women's lives, we decided to use semi-structured, one-on-one interviews with open-ended questions. This allowed the interviewer to judge the quality of the questions answered, probe further into areas that arose, and allowed the interviewee to shape his or her response (Walliman, 2005). Key informants such as extension staff at partner organizations and lead farmers were also interviewed to shed light on CA adoption rates, people's understanding of CA, and their views on women practicing CA.

Eighteen focus groups were carried out, consisting of female CA farmers, male CA farmers and female conventional farmers. Focus group size was limited to between five and eight participants, as larger groups could limit the detail obtained. Focus groups were carried out using participatory research appraisal (PRA) tools. For example, the Daily Activity Clock (DAC), a tool that illustrated the different kinds of daily tasks carried out in a day, was particularly useful to compare workloads between different groups in communities. Other PRA tools used were the seasonal calendar, used to explore seasonal changes in workload and food availability, and the labor requirement chart, used to determine the time and effort spent on different labor activities. The above PRA tools were selected to explore the potential difference practicing CA makes on daily life and the activities carried out.

Data were collected over 6 weeks in June and July 2012 in four districts in central Malawi, Dowa, Lilongwe, Nkhotakota, and Ntcheu. CA has been practiced in Dowa, Lilongwe, and Nkhotakota since 2010, with the support of CW. In Ntcheu, where it is promoted and implemented by CU, it has been used since 2008. This district was therefore chosen as a site to gain an insight into the impact of women practicing CA over a longer period. Data were collected with the assistance of a field facilitator provided by the partner organization. Daily data collection started with meeting a lead farmer, who assisted in ensuring beneficiaries were available for interviews, with minimal disturbance to daily activities.

The resulting data were analysed in Microsoft Excel. T-tests were used to determine whether a variation between two groups was significant. ANOVA tests were used when there were multiple variables, in order to limit errors that might occur by carrying out multiple T-tests (Carlberg, 2011). Though for most agronomy research a significance level (P value) of 0.05 or less was considered significant, it was felt that for social research with a small sample size, anything ≤ 0.1 could be considered significant.

11.3 Results and Discussion

11.3.1 Time and labor

To discover the impact CA has on female farmers in Malawi, both women who were practicing CA and women who were not were asked to carry out a labor requirement chart indicating how long it took to carry out each activity in conventional agriculture and CA. Analysis of the information provided by the respondents highlighted that throughout the year the labor requirements for conventional agriculture were consistently higher. On average, labor demands for CA were reduced by 34 days compared to labor demands for conventional agriculture (Fig. 11.1). In addition to this, interviews and focus group discussions confirmed further that CA postively impacted on time and labor requirements. Women stated that CA was less intensive and saved time, because work was spread throughout the year.

This reduction in labor demands and intensity aligns with the findings presented by Maal (2011) and seems to discount the concerns presented by Giller *et al.* (2009, 2011) that CA increases the burden of work further on to women, as in Malawi, tilling is frequently done with a hand-held hoe, an acceptable method for women to carry out. Furthermore, it is often stated that not tilling the soil increases weeds (Giller *et al.*, 2009). No increase in weed density when using CA was shown. This could be attributed to sufficient soil cover through thick layers of mulch, not only crop residues but also grass cut from outside the farm, and the use of planting basins, which reduces weed density when practicing CA (Maal, 2011). Moreover, this research shows that women feel the weeding requirements for CA are less when compared to conventional agriculture. Women have stated that weeding is less intensive, as it can be done by hand, whereas in conventional agriculture, weeding requires a hoe.

To understand the above reduction in labor demands, the DAC was used to show a female CA famer's average day during both the wet and dry seasons compared to a female conventional farmer. The results depict how the average day has changed for

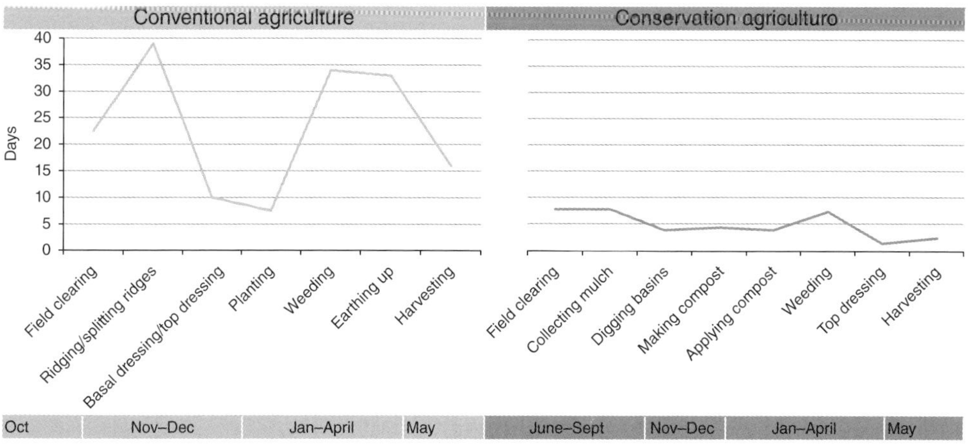

Fig. 11.1. Labor requirement chart for women (conventional agriculture and CA).

women who are practicing CA, largely as a result of pushing the agricultural calendar into the dry season. Female CA farmers state that July–October is the busiest time in the CA calender, when land preparation activities such as collecting mulch, making planting basins, and composting take place. During the wet season, with the majority of work on CA plots completed, women report not having to tend their CA plots. This creates a 3–5 h window of free time, compared to the daily activities of those practicing conventional agriculture only. The majority of women (71%) reported spending this time working on conventional agricultural land. Others reported spending free time carrying out household chores or working in business (29%).

Participants were asked to discuss the type of, if any, secondary source of income they earned. This allowed the research to investigate whether potential free time created by practicing CA allowed women to become more involved in income-generating activities. In all four districts studied, the main source of secondary income for female CA farmers and non-CA farmers was *ganyu* (casual labor) and selling products such as vegetables, firewood, fish, and livestock (Table 11.1). An interesting and positive finding showed that female CA farmers were more involved in selling products than relying on *ganyu*, compared to female non-CA farmers. This suggests that CA is beneficial in more ways than a simple shift of time and energy to other forms of agriculture.

Ganyu is used as a coping mechanism during times of food shortages. In Malawi, it is the most important source of livelihood after own-farm production. Conflict often arises between time dedicated to own-farm production and *ganyu*, perpetuating a vicious cycle of food insecurity (Whiteside, 2000). *Ganyu* is an activity that helps meet a household's short-term needs, but can have negative long-term implications. For example, the reduction in dependency on *ganyu* can result in more time for own-farm production and, hopefully, long-term sustainable development of these households (Whiteside, 2000). Discussions in focus groups revealed that women practicing CA carried out less *ganyu*, because there was more food available within their homes.

Other sources of income included owning businesses, metalwork, carpentry, tailoring, and employment in the poultry industry. Men had the most diverse range of income-generating activities (Table 11.1). When participating in income-generating activities, women's access is often limited to low-income activities (African Development Fund, 2005). It can be presumed that men remain the breadwinners in most households, whether CA practicing or not.

Table 11.1. Statistical differences in secondary source of income (SSI).

	Female CA farmers (n = 80) (%)	Female non-CA farmers (n = 18) (%)	Male CA farmers (n = 15) (%)	F	*P* value
Earning SSI	75	61	90	3.86	0.06
Ganyu	41	63	11	2.36	0.1
Selling products	55	36	44	7.02	0.01
Other	4	–	55	3	0.1

11.3.2 Returns on agricultural production

Key informant interviews with CW extension workers revealed that when both practice CA, men harvest more than women farmers. The total harvest in kilogram per hectare of land cultivated (Fig. 11.2) confirmed this for both maize and soybean. However, women produced more groundnuts. It is probable that men spend more time looking after maize, the staple crop, and soybean, a cash crop that can generate significant revenue. Groundnuts have always been a woman's crop, and women have traditionally always had control over the use and sale of groundnuts, while the men focus on maize and tobacco. Female CA farmers harvest more on all counts than female non-CA farmers, indicating that CA has a positive effect on women's agricultural production (Fig. 11.2). In addition to this, the harvest data show that female lead farmers are harvesting more than male farmers, with the exception of soybean. The results show a trend of significance for maize and soybean (P value = 0.1); however, the difference in yields for groundnuts is not significant.

Key informant interviews revealed that extension workers believed low education and high rates of illiteracy were responsible for the differences between male and female yields. However, results showed that there were no clear causal linkages between higher harvest and increased education. Male CA farmers had the highest level of education on average, with 8 years of education. Female non-CA farmer attended, on average, 6 years of schooling and female CA farmers had, on average, 5 years of education. Furthermore, there have been conflicting studies regarding the role of education in the uptake of CA and harvest rates (Knowler and Bradshaw, 2006). This research indicates that education may not be a determining factor in agricultural output.

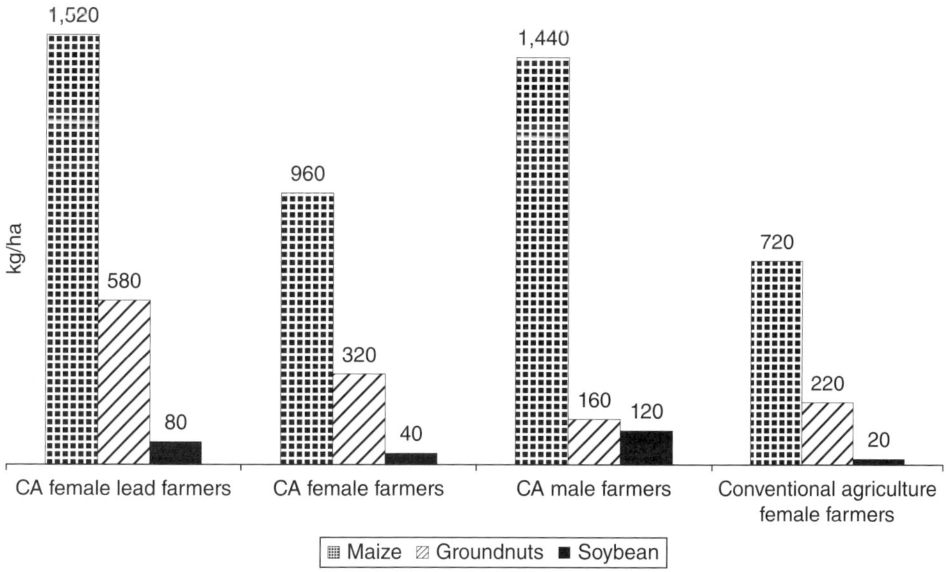

Fig. 11.2. Harvest data (kg/ha).

Harvest data were disaggregated by the sex of the lead farmer. This disaggregation aimed to show whether the sex of the person providing CA extension services affected the harvest produced. Women who receive extension services from a female lead farmer harvest more maize and soybean compared with those who have a male lead farmer. Milder *et al.* (2011) state women often have less access to extension services. Where extension services are provided by a man, it is sometimes culturally unacceptable that a woman attends such training. Although this was not observed during this study, an interesting finding did emerge: female lead farmers often trained their husbands, and together they trained beneficiaries. Female lead farmers then trained females, while their husbands trained men in CA techniques. Given the above findings, gender-specific extension services could lead to higher yields. Therefore, CA projects with gender-specific extension services could potentially allow for significant effects on future production and yields.

11.3.3 Household food security

In order to assess whether CA enhanced household food security, interview respondents were asked if they saw an improvement in the amount of food available in their home. The majority (66%) of female CA farmers stated that they saw an improvement. Over a quarter of CA farmers (28%) felt that there was only a small improvement in food availability. The two reasons cited for this were poor seed varieties (17%) and poor rainfall (11%). A small percentage (6%) saw no change or improvement in the amount of food available in his or her home. The average food availability (Fig. 11.3) showed the periods when food was available in the home and the times when there was no food available in the home of female CA farmers and female non-CA farmers. There was an increase of 1 month's food availability in the homes of those practicing CA compared to those practicing conventional agriculture. Despite this, both CA and non-CA households struggle to attain food security throughout the year.

As highlighted earlier, female CA farmers report a reduced dependency on *ganyu*, and this may have contributed positively to household food security. The majority of participants report having more food available in their homes. However, female CA farmers in Dowa reported the use of poor maize varieties, which experienced rot, leading to no observable improvement in household food

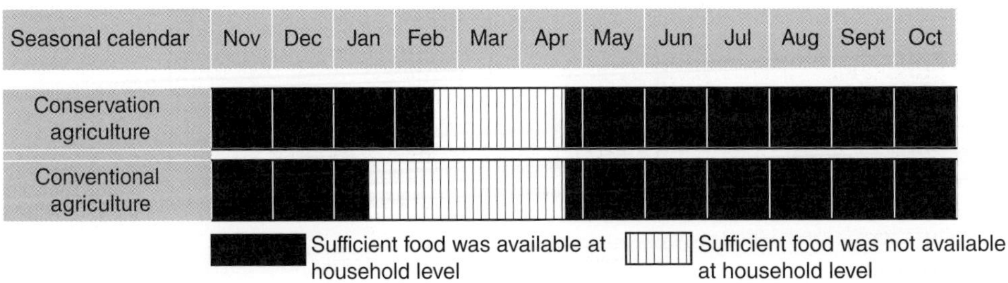

Fig. 11.3. Food security calendar.

security. Similarly, female CA farmers in Nkhotakota reported that soybean was planted in waterlogged soil, resulting in poor yields. In Ntcheu, where women had been practicing CA for longer, CA farmers stated that poor rainfall that year had affected yields and that *ganyu* might be required, while in other years none was carried out.

11.3.4 Decision making within the household

In Malawi, there are two types of land ownership: patrimonial and matrimonial. Patrimonial land ownership refers to land that is owned and inherited through the paternal line, and matrimonial land ownership refers to land owned and inherited through the maternal line. However, even in matrimonial land ownership, the husband assumes responsibility for household decisions (Maal, 2011). Due to strong cultural, social, and political norms that favor men in Malawi, we did not expect CA to impact women's roles in making household decisions. However, our results indicate that there is a statistically significant difference in decision making at the household level between female CA farmers and female non-CA farmers (Table 11.2). Interviews with female non-CA farmers revealed that the only significant contribution they made to decision making related to determining the use of crops, if they were not sold, and food expenditure (Table 11.2). Female CA farmers appeared to have a greater overall involvement in making decisions at the household level.

Another interesting finding was that women had a greater say over the use of crops produced through CA; 45% stated that they made the decision on whether a CA crop was sold or kept. This was much higher than the 14% of women practicing conventional agriculture who reported involvement in such a decision. This result suggests that women may gain greater involvement in and control of household decisions as a result of CA. Female CA farmers noted a change since attending agricultural and CA training, indicating the impact that CA could have on decision making in households. Through acquiring new agricultural techniques, their input in decision making in the household increased.

It is widely acknowledged that women are more likely to reinvest income into their family, and empowering women is a well-proven strategy for improving children's well-being (FAO, 2011). The significant difference seen in female decision making for household expenditure indicates that participation in CA is having a positive effect on women's roles within the household (Table 11.2).

Table 11.2. Statistical significance of female decision making in CA households and non-CA households.

	Female CA farmers (%)	Female non-CA farmers (%)	F	*P* value
Agricultural decision making	13	0	8	2.19×10^{-5}
Crop-use decision making	29	17	5.33	0.06
Expenditure decision making	26	8	2.89	0.09

Women in Ntcheu attributed their children's school attendance and completion, as well as improved household food security, to the success of CA practices.

11.3.5 Social capital

The International Food Policy Research Institute (IFPRI, 2011) states that one of the challenges in the adoption of CA is the attitude that many people have towards those practicing CA. It is often stated that CA is the "lazy man's" method of farming, and people often mock those who practice it (Carr and Hartl, 2010, p. 30). Giller *et al.* (2009) also state that cultural, economic, and social aspects prevent CA from being a success in SSA. Its adoption can prove particularly challenging in agropastoralist communities where the mulch intended for soil cover is often needed for livestock feed. In such communities, livestock are extremely valuable and a sign of wealth. As such, their feed often takes preference over environmental concerns such as soil cover. However, in the study sites in Malawi, there was a very low level of livestock ownership; hence, this was a lesser problem. Another challenge faced is the burning of mulch. This is often done to clear fields, control pests, and also to hunt for mice, which are a vital source of protein in many parts of Malawi. It is these cultural, economic, and social realities that are cited as undermining the potential expansion in social capital among women practicing CA.

Information was obtained during interviews about social status and involvement in groups within the community. The results showed that CA has had a positive effect on the social capital of women practicing CA. Women report that CA has helped them to form close bonds with other women practicing CA. However, 55% of female CA farmers reported experiencing jealousy due to practicing CA. The main reasons cited for any jealousy experienced were: inputs received (35%), because of the greater yields produced under CA (26%), and others wanting to join (9%). Nevertheless, many women (45%) did not experience any jealousy, and often stated that they felt admiration from other members of the community because of their involvement in CA. Even with over half of the women experiencing jealousy because of practicing CA, most (95%) still felt that CA has had a positive effect on their social status. Some stated that jealousy turned to admiration after greater yields were produced. Women practicing CA often felt that they were perceived as better farmers and female non-CA farmers often came to seek advice on how to start practicing CA. The 5% of women who felt that CA had affected their social status negatively was due to a poor harvest resulting from rotten maize.

The attendance of women at CA training has also helped to change attitudes within the home. In response to the increased agricultural production attributable to CA, men often appreciate and practice the new techniques that women are practicing. Furthermore, 40% of women practicing CA felt that in their homes, their opinion was valued more since starting CA practices.

The participation of female CA farmers and female non-CA farmers in groups and committees at the community level was examined and showed that female CA farmers (73%) were more involved in groups and committees, and a greater proportion also held leadership positions (65%) within organizations compared to female non-CA farmers (58% participation and 20% representation, respectively).

These positions include secretary, treasurer, and chairperson. The differences seen in membership and leadership positions are statistically significant. Men have the lowest participation in groups in the community (60%). However, they retain the majority of leadership positions, with nearly three-quarters of men sampled holding leadership positions. Overall, the impact of CA has been shown to be largely positive, with women experiencing increased social status, growing influence in the home, and confidence and admiration within the community.

11.4 Conclusion

This research sought to examine whether CA could play a role in the empowerment of women through agriculture. Our results indicate that there are indeed clear positive impacts for women who practice CA, in areas of time, labor, agricultural production, food security, decision making, social status, and confidence. It has helped to create a sense of time and control for women. Furthermore, the study results indicate that women have become more involved in decision-making processes at the household level and within the community after adopting CA practices. Our findings indicate that CA, as a farming system, can certainly contribute to diminishing the gender gap in agriculture.

However, there is a still a lot to be done to help empower women. Culturally, there are still challenges and issues that women have to overcome. The persistent gender gap seen in agriculture is due largely to women's restricted access to resources. To close the gender gap and really empower women, women need to be incorporated further into agricultural development programs through greater inclusion at all levels, including project planning, implementation, and as project beneficiaries. Currently, in spite of the improvements this research shows CA can bring to women's lives and livelihoods, there is a lot of targeted work required to close the gender gap successfully, especially in terms of changing the social and cultural norms that place women in a disadvantaged position.

References

African Development Bank Group (2012) *Malawi Country Strategy Paper 2013–2017*. ORSB Department, AfDB, Lilongwe.

African Development Fund (2005) Republic of Malawi, Multi-Sector Gender Profile. Available at: http://www.afdb.org/fileadmin/uploads/afdb/Documents/Project-and-Operations/malawi.pdf (accessed 8 August 2012).

Ayoade, A.R. (2011) Relevance of women in agriculture programme in Oyo State, Nigeria. *International Journal of Agriculture, Environment and Biotechnology* 4(3), 249–254.

Bunch, S. and Mehra, R. (2008) *Women Help Solve Hunger. Why is the World Still Waiting?* International Center of Research on Women, Washington, DC.

Carlberg, C., (2011) *Statistical Analysis: Microsoft Excel 2010*. Pearson Education, Indianapolis, Indiana.

Carr, M. and Hartl, M. (2010) *Lightening the Load Labour-saving Technologies and Practices for Rural Women*. International Fund for Agriculture Development (IFAD) and Practical Action Publishing Ltd, Warwickshire, UK.

FAO (Food and Agriculture Organization) (2008) Climate Change and Food Security: A Framework Document. Available at: http://www.fao.org/forestry/15538-079b31d45081fe9c3dbc6ff34de4807e4.pdf (accessed 18 March 2012).

FAO (2011) The State of Food and Agriculture 2010–2011. Available at: http://www.fao.org/docrep/013/i2050e/i2050e.pdf (accessed 21 March 2012).

FAO (2012) What is Conservation Agriculture? Available at: http://www.fao.org/ag/ca/1a.html (accessed 10 May 2012).

Farnsworth, C.R. (2010) Gender Aware Approaches in Agricultural Programmes: A study of Sida-supported Agricultural Programmes. *Sida Evaluation* 2010:3. Available at: http://www.oecd.org/countries/burkinafaso/46145893.pdf (accessed 10 May 2012).

Fife, W. (2005) *Doing Fieldwork, Ethnographic Methods for Research in Developing Countries and Beyond.* Palgrave Macmillan, New York.

Giller, K.E., Witter, E., Corbeels, M. and Tittonell, P. (2009) Conservation agriculture and smallholder farming in Africa: the heretics' view. *Field Crop Research* 114 (1), 23–34.

Giller, K.E., Corbeels, M., Nyamangara, J., Triomphe, B., Affholder, F., *et al.* (2011) A research agenda to explore the role of conservation agriculture in African smallholder farming systems. *Field Crop Research* 124(3), 468–472.

IFAD (International Fund for Agricultural Development) (n.d.) Rural Poverty in Malawi. Available at: http://www.ruralpovertyportal.org/web/guest/country/home/tags/malawi (accessed 9 April 2012).

IFPRI (International Food Policy Research Institute) (2011) *Approaches to Implementation of Conservation Agriculture Among Promoters in Malawi.* International Food Policy Research Institute and Ministry of Agriculture and Water Development, Malawi.

Kassam, A., Friedrich, T., Shaxson, F. and Pretty, J. (2009) The spread of conservation agriculture: Justification, sustainability and uptake. *International Journal of Agricultural Sustainability* 7(4), 292–320.

Knowler, D. and Bradshaw, B. (2006) Farmers' adoption of conservation agriculture: A review and synthesis of recent research. *Food Policy* 32(1), 25–48.

Maal, B. (2011) *Women, Gender and Conservation Agriculture in Malawi.* Norwegian Agency for Development Cooperation (NORAD), Oslo.

Milder, J.C., Majanen, T. and Scherr, S.J. (2011) Performance and Potential of Conservation Agriculture for Climate Change Adaptation and Mitigation in Sub-Saharan Africa. Eco Agriculture Discussion Paper No 6. Eco Agriculture Partners, Washington, DC.

OECD (Organisation for Economic Co-operation and Development) (2011) Women's Economic Empowerment Issues Paper April 2011. DAC Network on Gender Equality. Available at: http://www.oecd.org/dataoecd/50/60/47561694.pdf (accessed 15 March 2012).

Quisumbing, A.R. and Pandolfelli, L. (2009) Promising Approaches to Address the Needs of Poor Female Farmers: Resources, Constraints, and Interventions. IFPRI Discussion Paper 00882. Available at: http://www.ifpri.org/sites/default/files/publications/ifpridp00882.pdf (accessed 15 April 2012).

Scherr, S.J. and Yadav, S. (1996) *Land Degradation in the Developing World: Implications for Food, Agriculture and the Environment to 2020.* International Food Policy Research Institute, Washington, DC.

UNDP (United Nations Development Programme) (2013) *Human Development Report 2013, The Rise of the South: Human Progress in a Diverse World.* UNDP, New York.

Walliman, N. (2005) *Your Research Project.* 2nd edn. SAGE Publications, London.

Whiteside, M. (2000) Ganyu Labour in Malawi and its Implications for Livelihood Security Interventions: An Analysis of Recent Literature and Implications of Poverty Alleviation. Agriculture Research and Extension Network Paper, No 99. Overseas Development Institute (ODI), London.

World Bank (2005) Gender and "Shared Growth" in Sub-Saharan Africa. *Briefing Notes on Critical Gender Issues in Sub-Saharan Africa.* Available at: http://siteresources.worldbank.org/EXTABOUTUS/Resources/GenderGrowth.pdf (accessed 7 May 2012).

World Bank (2012) Rural population data. Available at: http://data.worldbank.org/indicator/SP.RUR.TOTL.ZS (accessed 9 April 2012).

12 Gendered Implications of Introducing Conservation Agriculture (CA): A Case Study in the Hill Region of Nepal

Jacqueline Halbrendt,* Bikash Paudel and Catherine Chan

University of Hawai'i at Mānoa, Honolulu, Hawaii, USA

12.1 Introduction

As one of the poorest countries in the world, and experiencing rising populations, Nepal is at a high risk of food crisis. The majority of Nepal's population lives on marginal land in rural areas, where food security is low and continuing to decrease (FAO, 2012; World Bank, 2012). Much of Nepal's poverty is concentrated in the hill region, where farming communities depend on sloping, degraded fields for sustenance and face seasonal food scarcity (FAO, 2007; Tiwari *et al.*, 2008; Shively *et al.*, 2011). Conservation agriculture (CA) practices have long been proposed as a potential remedy for such issues; nevertheless, these practices have been introduced on a limited basis only and have seldom met with success. A combination of social, economic, cultural, and environmental factors may have contributed to difficulties in promoting the adoption of long-term sustainable agricultural practices such as CA (Paudel and Thapa, 2004). Research has shown that traditional practices often persist, despite development efforts by government extension or non-governmental organizations (NGOs) to introduce new practices (Yadav, 1987; Bunch, 1999; Cochran, 2003). Factors such as gender, education level, and economic status have each been identified as important indicators of a willingness to learn new farming practices (Kessler, 2006; Knowler and Bradshaw, 2007).

CA practices are especially beneficial on sloping agricultural lands prone to degradation and erosion, particularly for smallholder farming systems with low inputs available, such as in Nepal's hill region. This region is particularly vulnerable to degradation, as research has shown the potential for increased soil loss on sloping agricultural lands (Shrestha *et al.*, 2004). More than one-third of

*Corresponding author; e-mail: jhalbren@hawaii.edu

Nepal's total agricultural land is located in the hill region, which must support 44% of the country's population of 29.8 million people (Thapa and Paudel, 2002). Thus, with declining productivity per capita, the hill region has become a major focal area for reducing food security vulnerability. Current farming systems consist of practicing full tillage and sole-crop farming year round. Practicing minimum tillage can reduce the risk of degradation by minimizing soil disturbance and retaining soil structure. Field experiments in Nepal and elsewhere have shown significant evidence that minimum tillage can reduce both runoff and soil loss (Tiwari *et al.*, 2009). Adopting an intercropping system with a leguminous crop benefits soil fertility through the fixation of nitrogen, provision of soil organic matter, and creation of additional soil cover (Thapa, 1996).

Though the adoption of CA has proven effective at increasing crop productivity on sloping lands and improving the livelihoods of subsistence farmers in Africa and Asia (Pimentel *et al.*, 1995; Lal, 2001; Hobbs *et al.*, 2008; Derpsch and Friedrich, 2009; Kassam *et al.*, 2009; Jun *et al.*, 2010), adoption is low in Nepal. Reasons for this include a lack of understanding among researchers of the preferences and incentives of farmers and the benefits of CA systems (Carr and Wilkinson, 2005; Probst *et al.*, 2007). Hence, an increased understanding of farmer preferences regarding the implementation of CA systems can help researchers and extension agents understand better which CA practices may appeal to farmers and have the greatest potential for long-term adoption. Within this, it is important to understand how conservation agriculture production system (CAPS) adoption affects all groups and subsets of people. However, research on the gender-based perceptions and implications of CA practices and how such practices may be beneficial or detrimental is lacking. Interventions that create undue burdens of labor on the local community or a particular gender may impede efforts to increase agricultural production. Moreover, while women consist of approximately half of the population in rural subsistence farming communities and bear much of the burden of household and agricultural labor, there is a lack of research measuring gender preferences for and the labor effects of agricultural interventions. Given that women share the workload in many subsistence farming communities, it is important to recognize the impact that various interventions will have on women's overall burden of labor and livelihood. Household labor capacity can be a limiting factor for making changes to agriculture in subsistence communities, and the inability for individuals to increase their agricultural labor hours can contribute to failure to adopt practices despite their potential to increase food security, improve soil conditions, and maintain livelihoods. This is particularly relevant for rural women, who already take on a disproportionate workload for both agricultural and household duties (Gurung *et al.*, 2005; Blackden and Wodon, 2006). Identifying and mitigating such impacts can be a determining factor in sustaining long-term agricultural development and improving food and nutritional security among poor, marginalized communities in the hill region of Nepal.

It is also important to assess gender-based access to information regarding CA practices when evaluating the feasibility of long-term adoption. Evidence shows that the degree of farmers' access to information influences conservation decisions positively (Bekele and Drake, 2003; Bandiera and Rasul, 2006). In fact, there are several barriers women face in receiving information about new technologies. For example, social taboos, such as women being discouraged from being outside their village alone, still exist in rural areas and limit the level of outside information that women can access. Moreover, most marketing agents, credit agents, and extension workers in Nepal are men, further limiting women's access to information due to social taboos (Suvedi and McNamara, 2012). Additionally, lower levels of education (57% literacy for women compared with 75% for men in Nepal; CBS, 2012) also act as a barrier to information access. Oftentimes, it is the women themselves that do not think that acquiring new information is their role in the household and they defer to the men to access new information. Therefore, it is clear that despite their invaluable role in household agriculture, women are often left out of discussions regarding new technologies. Hence, it is important to identify the information networks through which women farmers get information about CA practices and identify the gaps and areas for strengthening adoption of CA.

The Chepang ethnic group, living in Nepal's central hill region (Fig. 12.1), are a perfect example of both the socio-economic and gender issues often encountered when promoting CA. As one of the most marginalized tribal communities in Nepal in terms of geographic location and socio-economic status (Piya *et al.*, 2013), they face the challenges of cultivating marginal lands, limited access to resources, and a high burden of labor for women, making them a good candidate for the implementation of CA. Villages are isolated on mountainsides without direct access to road networks and markets, limiting opportunities for income generation, and few, if any, receive agricultural extension services. While outmigration is prevalent in much of Nepal (CBS, 2012), it is not as common among the Chepang people, due to its prohibitive cost. However, some Chepang men do work as seasonal agriculture labor in nearby villages or migrate to cities as construction workers. As a result, women's roles in agriculture are increasing in Chepang societies. Women provide more than half of the workforce and increasingly must take on larger roles in household decision making. It has been estimated that few

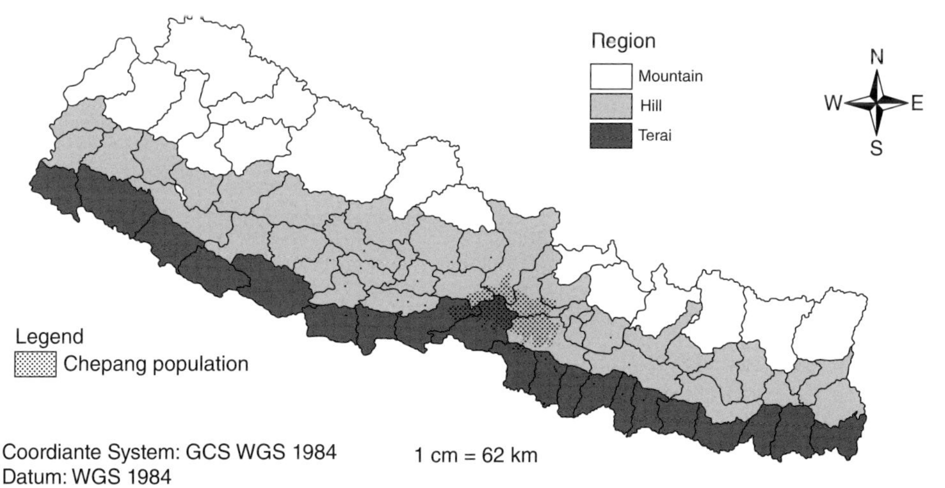

Fig. 12.1. Native areas of Chepang tribal communities in Nepal. (From Gurung, 1998.)

Chepang households are self-sufficient in agricultural production; hence, food scarcity exists for approximately 6 months of the year (Piya *et al.*, 2011). In general, women are the most vulnerable to food insecurity in the family. Traditionally, women cook food and serve all other members of the family before serving themselves. Therefore, if there is less food than required, women are more often affected than men. It is expected that preferences for new agricultural interventions will be different by gender due to differing roles in household and agricultural activities, access to information, and labor requirements.

12.2 Methodology

12.2.1 Profile of study sites

Three Chepang villages, located near the Trisuli River, were selected to study CA. The selected villages were Thumka, in the Gorkha district, Hyakrang, in the Dhading district, and Kholagaun, in the Tanahun district. Village size ranged from 26 households in Kholagaun to 30 households in Hyakrang, and 42 households in Thumka. These communities were selected due to their high risk of food insecurity resulting from marginal agricultural lands, small landholdings, and high malnutrition. Agriculture in this area is characterized by subsistence, rainfed farming systems, with typically less than 0.7 ha of arable land per household and limited opportunities for income generation. Farmers use traditional continuous cultivation methods of terracing, plowing with draft power, and monocropping in a maize-based agricultural system along the hillsides of Nepal. Since there were two growing seasons in the study villages, there were different crop rotations for different production systems. All production systems grew maize in the first season, then millet or cowpea in the second.

12.2.2 Research methods

Baseline socio-economic study

A socio-economic study was conducted in 2011 to develop a baseline understanding of local conditions. The survey assessed factors such as income, household size, education, farm size, farming practices, and off-farm wage earning. A total of 37 households from the villages were surveyed to develop a socio-economic profile of the villages.

Gender-based labor analysis

In order to conduct an activities analysis by gender, individual households in the three Chepang communities were surveyed in June 2012. Male and female heads of each household were surveyed separately in face-to-face interviews to assess gender participation in various agricultural activities and how much time was dedicated to these activities. Time-use surveys can be employed to estimate agricultural activities by gender and serve to measure the distribution of labor in the household and community (Beteta, 2006). Three major upland crops were the focus of the survey: maize, millet, and legumes, with the typical conventional cropping pattern of maize followed by a relay crop of either millet or a legume such as cowpea. Survey analysis measured the labor hours required for both conventional cultivation approaches for a complete cropping season: (T1) maize followed by millet, and (T2) maize followed by legumes. The survey specifically assessed the labor hours required at five stages of cultivation: plowing, fertilizer application, sowing/transplanting, weeding, and harvesting. A total of 77 surveys were collected in June 2012, with an even distribution of male and female heads of household, and comprising 79% of local households.

Field experiments were also conducted to measure the labor shifts from the introduction of two CA practices, intercropping (IC) and strip tillage (ST). In addition to T1 and T2, the experimental plots also had two CA systems: (T3) full tillage, maize followed by a millet–cowpea intercrop; and (T4) strip tillage, maize followed by millet–cowpea intercrop (Table 12.1). Experimental plots were established on eight representative farmer fields in each village, and the labor required for each treatment was recorded by gender. Changes in labor distribution were measured by finding the difference in labor hours by shifting from farmers' conventional practices (T1 and T2) to conservation practices (T3 and T4). Data for the CA experiments were converted to percent change in labor hours per treatment and applied to the total farm data for maize, millet, and legumes to account for differences between the field size on the total farm and the smaller CA experimental plots.

Determining farmer preferences of CA

To explore farmers' preferences for CA systems, an analytic hierarchy process (AHP) survey was conducted in the three villages in 2013. By this time, farmers had either practiced some form of CA or observed their neighbors doing so. AHP was selected due to its ability to take contradictory viewpoints and strength of preference into account, its broad applicability, and its widespread use as a multi-objective decision-making tool in a group setting (Alphonce, 1996; Vaidya and Kumar, 2006; Valniunas *et al.*, 2009). The first step in the AHP analysis is to determine the hierarchical structure of the goal, followed by objectives to meet the goal and alternative options to achieve the goal (in this case, the on-farm treatments) (Fig. 12.2). The overall goal of adopting CA was defined as improving farmer income. This was determined using farmer focus groups and a literature review. Similarly, the objectives deemed most important to meeting the goal were determined by farmer focus groups. These were: increasing profit, labor savings, yield, and soil quality. All treatments included in the on-farm trials were included

Table 12.1. List of conventional and CA practices selected for on-farm trial.

Treatment	Cropping system and type of tillage
Treatment 1 (T1)	Maize followed by millet, full tillage
Treatment 2 (T2)	Maize followed by legume, full tillage
Treatment 3 (T3)	Maize followed by millet–legume intercrop, full tillage
Treatment 4 (T4)	Maize followed by millet–legume intercrop, strip tillage

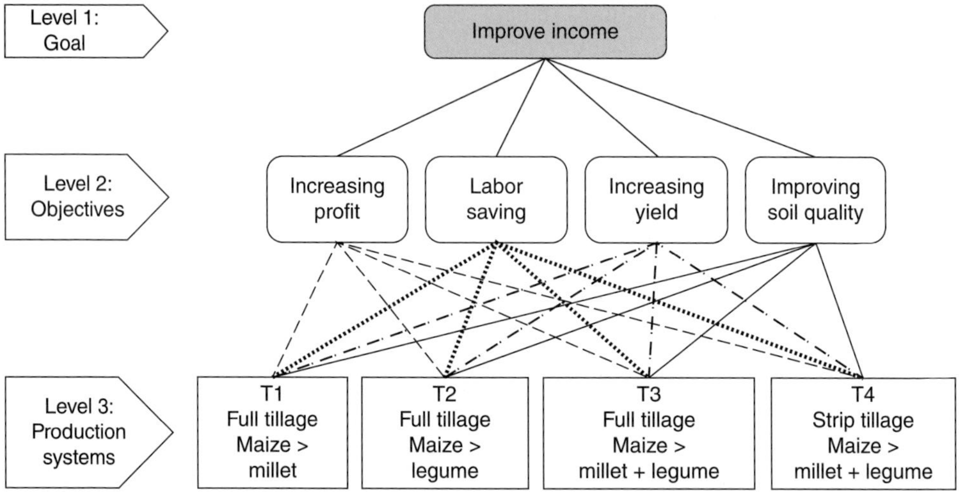

Fig. 12.2. Diagram of goal, objectives, and production systems at different levels of the analytic hierarchy process.

as options for CA production systems (Table 12.1). After the AHP hierarchy structure was developed, the next step was to sample male and female household decision makers randomly in each village. About 25% of total households ($n = 47$) were surveyed. After the AHP data collection, results were analyzed using Expert Choice software (Version 11.5.829) and results were discussed in groups with respondents.

Information network analysis

To assess the information networks through which farmers obtained information about CA practices, we conducted a survey in the three study sites in 2012. Altogether, 148 farmers (80 male and 68 female) from 87 households (89% of total households) in the villages took part in the survey. The study assessed the status of information networks through which the farmers received their knowledge of CA practices. The study used the social network analysis (SNA) method in which information was collected through gender-disaggregated personal interviews. A maximum of one male and one female member from each household was surveyed.

The SNA questionnaire included two key questions:

Question 1. Can you name 5–10 people (including other farmers) from whom you receive information about new agriculture technologies?
Question 2. Did you also seek advice for any of the CA practices (i.e. minimum tillage, crop rotation, and cover crop/mulching) from the person/s mentioned in Question 1?

Information sources were first identified from Question 1. Next, sources that did not provide information about CA practices (identified from Question 2) were deleted. Thus, the information networks of CA practices were determined. For simplicity, the network was termed an “information network of CA”. The visual representation of the network was created by NetDraw software and data were analyzed using UCINET (Borgatti *et al.*, 2013).

In addition to visual observation, statistical tools were applied to compare networks based on attributes of the "nodes" (i.e. farmers). Comparison of network density is the most popular method to compare networks. Network density is a unitless number ranging between 0 and 1, which shows the ratio of existing ties in the network to the total number of possible ties (Borgatti *et al.*, 2013). For example, if there are three people in a group, and only two know each other, then the density of the network would be 1/3, where 1 is the number of existing ties and 3 is the number of all possible ties among the three people.

$$\text{Network density} = \frac{\#\text{ existing ties}}{\#\text{ all possible ties}}$$

An analysis of variance (ANOVA) of a structural block model was used to test the difference among the density of networks characterized by advice seeking from "male-to-male", "male-to-female", "female-to-male", and "female-to-female" connections. An ANOVA density model for a structural block model was used to compare the difference in networks by gender. The ANOVA of a structural block model tests whether the patterns within and between group ties differ across groups (Hanneman and Riddle, 2005); hence, it was applicable for comparing the differences between genders.

12.3 Results and Discussion

12.3.1 Socio-economic profile of project sites

Average household size ranged from nine per household in Thumka to six people per household in Kholagaun (Table 12.2). Average landholdings in Thumka and Hyakrang were similar (0.68 ha/household), but lower in Kholagaun (0.43 ha/household). This difference in farm size may be the result of greater land divisions among extended family homes in Kholagaun as compared to the other two villages. Most heads of household were illiterate, though primary education was more prevalent among young people. Rice and maize were the staple crops consumed, though these households often purchase rice due to the low availability of suitable land for rice production. Thus, maize was the primary staple grown by land-poor households. The average annual household income ranged from US$401 in

Table 12.2. Selected socio-economic factors of the three study villages.

Village (number of households)	Average annual income (US$)	Average household size (% of women)	Average level of education	Average farm size (ha)	Major staple crops grown
Thumka (44)	500	9 (55.2%)	Primary	0.68	Rice, maize
Hyakrang (32)	401	7 (47.6%)	Primary	0.68	Rice, maize
Kholagaun (27)	584	6 (56.5%)	Primary	0.43	Rice, maize

Hyakrang to US$584 in Kholagaun. This variation in annual income may be due in part to the remoteness of the villages, crop yields, distance to markets, distance from off-farm employment opportunities, access to external resources, and availability of exchange labor. Approximately 32% of total average income in Thumka and Kholagaun came from crop sales. Off-farm employment opportunities, such as construction and mining, were occasionally available in nearby towns. However, most farmers only sought such employment in the dry season, when their labor was not required on their farm. A few villagers were skilled laborers (such as carpenters) in addition to farmers, with compensation sometimes taking the form of exchanged agricultural labor.

12.3.2 Distribution of agricultural labor by gender and changes required by the adoption of conservation agriculture interventions

Results indicated that for both farmers' conventional practices (T1 and T2), males provided 47% of the total labor, while females provided 53%. With the introduction of CA systems, T3 and T4, a similar trend was maintained, with males providing 45% and 46% of the total labor and females 55% and 54%, respectively. Overall trends showed that while men managed most of the plowing, women did the majority of fertilizer application, weeding, and harvesting activities. While sowing and transplanting were predominantly female tasks in both conventional farmer practices, the work was distributed more evenly in the CA systems, where time needed for sowing increased due to intercropping (Fig. 12.3). Changes in labor requirements by gender were measured by finding the difference in percent labor for each activity as cultivation practices shifted from conventional farmer practices (T1, T2) to CA practices (T3, T4). For the percent difference in labor by gender as farmer practices switch from conventional fields to CA, all changes showed a shift in labor that resulted in a greater proportion of total labor for women. Shifting from millet to full tillage (FT) millet–legume (T1–T3) resulted in 1.96% more labor for women, while shifting from millet to strip tillage (ST) millet–legume (T1–T4) had the least increase of labor for women among the studied treatments, with only 0.70% more labor for women (Fig. 12.3). Changing fields from legume cultivation to FT millet–legume (T2–T3) resulted in 2.17% more labor for women and shifting from legume to ST millet–legume (T2–T4) showed a 0.91% increase in labor (Fig. 12.4).

While both CA options showed a total increase in percent labor for women, it is more beneficial for women to shift from millet to a MT millet–legume practice, since there would be less of an increase in labor; 0.7% as compared with 1.96% increased labor (Fig. 12.3). Physically tasking activities such as weeding were also reduced with this practice. However, in terms of men's labor, both practices reduced the total percent labor, and shifting to intercropping would have the greatest benefit in terms of labor and opportunities for off-farm wage earning. In shifting from the conventional farmer practice of FT maize–legume to CA systems, there was a greater overall shift in labor. Nevertheless, shifting to an ST millet–legume (T2–T4) required less change in terms of total labor, as the change in labor was 0.91% compared with the full tillage intercrop system (Fig. 12.4). For all of the CA options, it was also important to note that there were

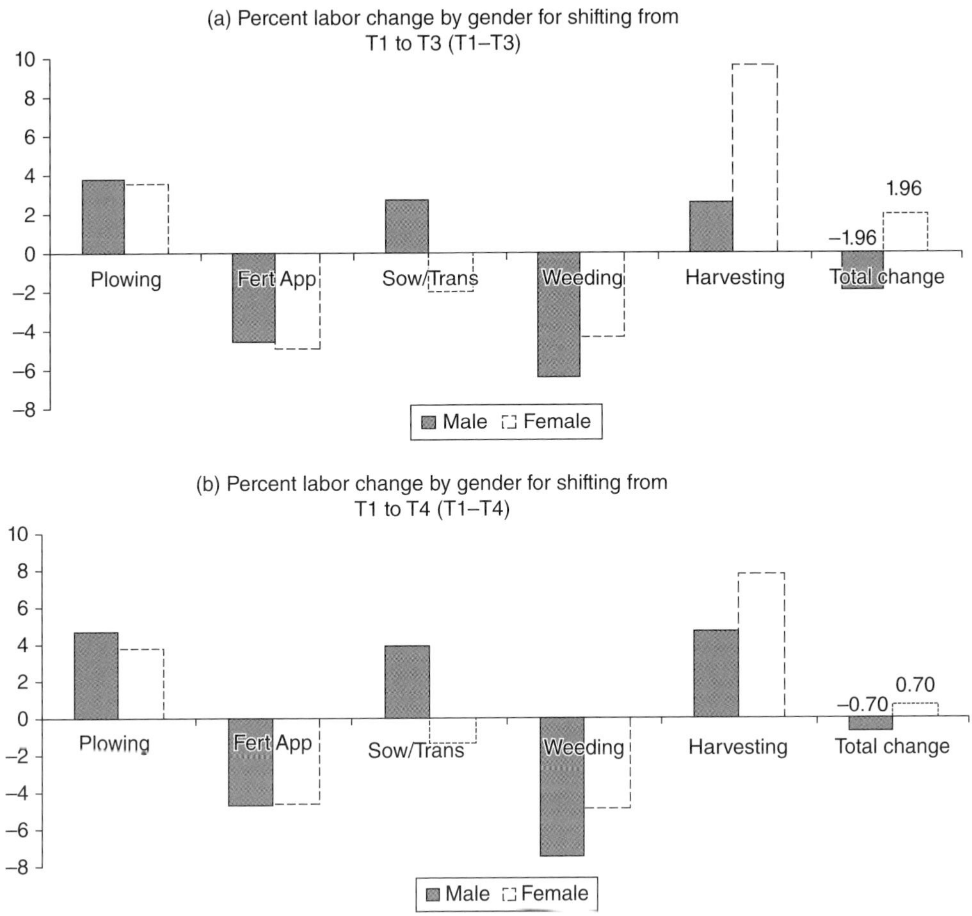

Fig. 12.3. Labor changes by gender for shifting from millet cultivation to: (a) CA practices of FT millet–legume intercrop (T3); and (b) ST with millet–legume intercrop (T4) practices. Values above zero indicate an increase in percent total labor, while values below zero show a decrease in labor requirements.

wide shifts in labor required for each activity. All options showed decreases in weeding and fertilizer application, while plowing and sowing/transplanting were increased. However, increased demands for harvesting were predominantly taken care of by women.

12.3.3 Male versus female preference for conservation agriculture interventions

AHP data analysis showed that T2 was the system most preferred by both males and females (providing 39% and 33% of weights, respectively) in terms of their perception of its potential to increase household income (Fig. 12.5). T2 was followed by T3, with a 24% weight by males and 30% by females. Both males and

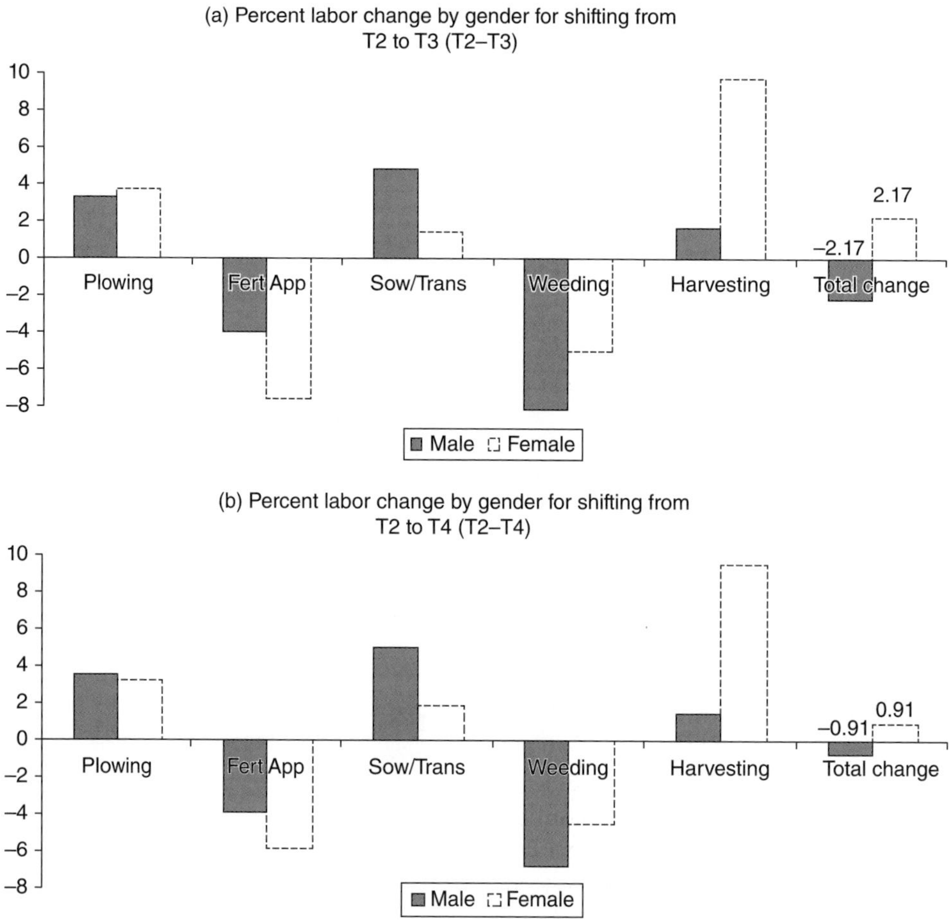

Fig. 12.4. Labor changes by gender for shifting from legume cultivation to: (a) CA practices of FT millet–legume (T3); (b) ST millet–legume (T4).

females selected T4 as the third preference (21% of males, 19% of females). Both males and females also consistently rated the traditional system (T1) as the least preferred system (18% and 16%, respectively). Thus, there was a nearly unanimous preference for CA systems over the traditional production system between the genders.

These preferences were based on the importance given to different objectives for contributing to enhancing the overall goal of household income. In general, there was an agreement between males and females regarding the most and least important objectives. Both males and females gave highest importance to crop yields (33% and 40%, respectively) (Fig. 12.6). Thus, results suggest a shared understanding among farmers that yield is the most critical objective for farming. Since farmers have short-term needs for food production, prioritizing yield is perfectly reasonable. Farmers need to increase yield and production to feed their families. This need should promote the adoption of technologies, such as intercropping, which can increase total crop production. Similarly, there was

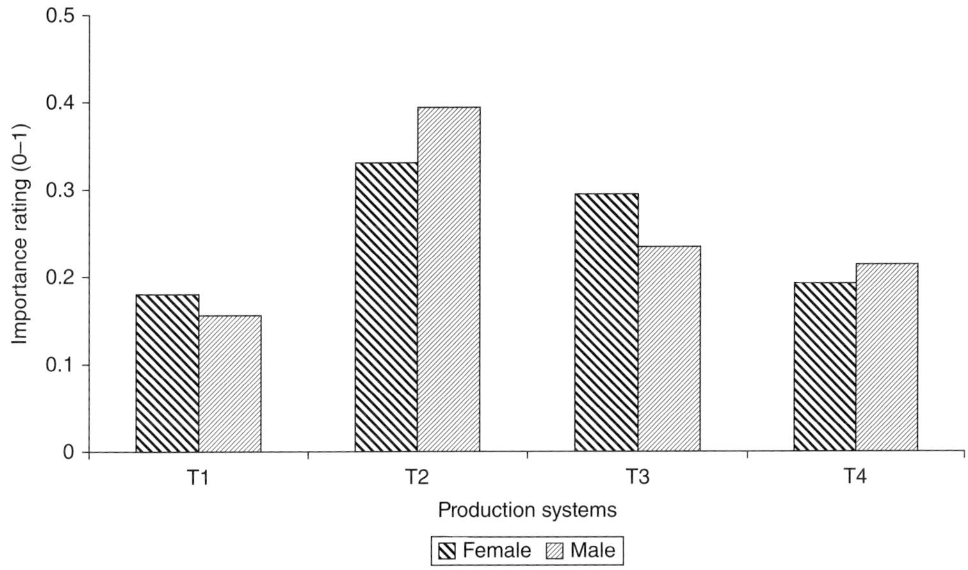

Fig. 12.5. Preferences for different production systems, by gender.

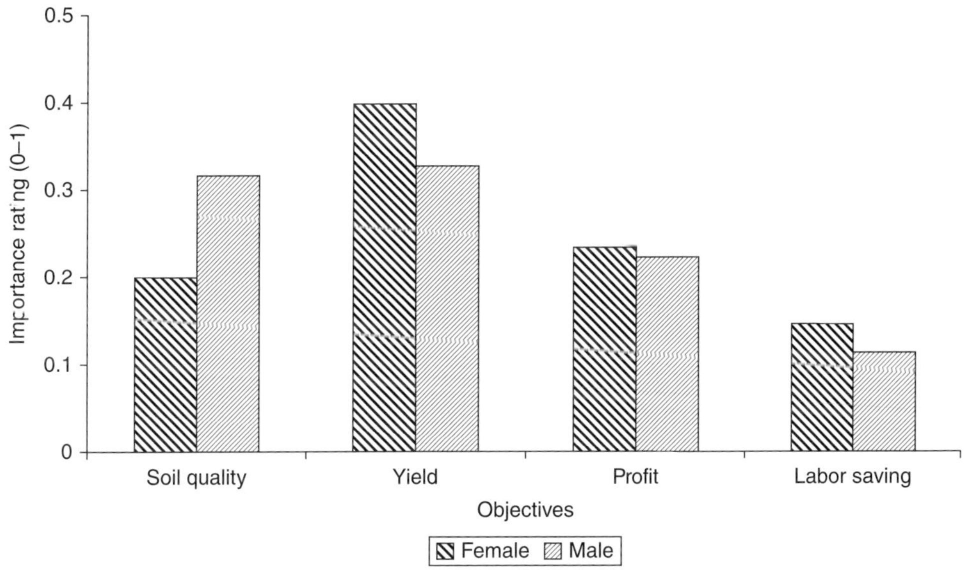

Fig. 12.6. Preferences among farmers for different objectives to enhance household income, by gender.

a shared belief between males and females that labor saving was the least important objective (12% and 15%, respectively) (Fig. 12.6). This indicates that farmers do not consider labor savings a critical motivation for the adoption of CA systems. Though one of the most frequently stated benefits of minimum tillage system is labor saving, this benefit does not seem to appeal to farmers and is unlikely to enhance CA adoption.

Despite a shared understanding for the most and least important objectives, males and females differed in their perception regarding the second most important objective. Soil quality was the second most important objective (32%) followed by profit (23%) for males, while profit was the second most important objective (24%) followed by soil quality (20%) for females. This may be attributed to agriculture being the main source of income for females, while males have more opportunities for off-farm wage earning. For this reason, females may have given higher importance to profit than males. On the other hand, generally, the role of buying fertilizers and other external inputs from market is a man's job. When soil quality is degraded, it requires a higher amount of fertilizer. This may be a reason that males give higher importance to soil quality. Since there is not an agreement about the importance of soil quality between males and females, this may present a challenge for the adoption of technologies such as strip tillage that improve the soil quality but do not always increase the short-term profit.

As we have shown, T2, the full tillage of maize in the first season followed by legumes in the second, was the most preferred system by both males and females. In terms of soil quality, this treatment is not ideal as the land is plowed twice for land preparation and increased risk of soil degradation may occur. However, since T2 produced nearly twice as much legume yield as other treatments and the market price of cowpea and black gram were much higher than millet (US$0.88 and US$1.22/kg, respectively, for legumes, US$0.38/kg for millet), T2 produced the highest profit. Similarly, the crop yields were also highest for this system. Nevertheless, with a combined higher yield, profit, and leguminous crop soil benefit, this treatment option was most preferred by both genders. It should be noted that T4 is the only CA system that incorporates two CA principles, legume intercropping and minimum tillage, both of which have strong support in the scientific literature. Since T4 is the third priority after T2 and T3, it seems that farmers do not prefer strip to full tillage. Additionally, there are no significant gender differences in terms of perceptions of tillage. This lower preference for strip tillage may be due in part to lack of knowledge and first-hand experience with the long-term advantages of minimum tillage for soil quality. Even male farmers, who preferred soil quality over short-term profit, selected T2 and T3 over T4. This disconnect in preference for treatments which do not promote soil quality, while claiming soil quality as a priority, indicates a clear lack of information on the benefits of CA practices. Although farmers who took part in the study have either implemented CA systems or observed other farmers doing so over the last few years, they have yet to witness any longer-term soil benefits. Therefore, fully disseminating CA's long-term advantages might influence farmers' willingness to adopt strip tillage-based CA systems.

12.3.4 Gender-based CA information networks

Farmers' adoption of new agriculture technology can be encouraged by effectively disseminating information on the technology's benefits. However, Nepal's agriculture extension system has not placed a focus on CA technologies and the Chepang study group has been historically underserved in terms of access to extension services. As a result, informal networks and connections are the major sources of farmers' information. This was supported by the results of the CA information network analysis. Although combined CA systems that integrate the three practices of minimum tillage, continuous soil cover, and crop rotation are relatively new for farmers, many Chepang farmers are aware of CA practices individually. The CA information network was mapped by determining the route of information flow among farmers within villages and from outside sources. The analysis indicated that direct farmer-to-farmer communication was the main source of information for both male and female farmers. Extension personnel from different NGOs, agriculture input suppliers, and crop buyers were the primary outside sources of information. The results confirmed that the government extension system was weak in the study communities and neither male nor female farmers mentioned them as a frequent information source.

Despite these similarities, there was a clear gender difference between CA information networks for males and females. Males were more frequently represented in CA information networks than females. Fewer female farmers were connected to the network than male farmers. In addition, female farmers were connected by more indirect and long-tailed connections, while males' networks were more direct than females (Fig. 12.7). While females had more frequent connections with other farmers from the same villages (indicated by circles in the figure), male farmers had additional frequent connections with information sources from outside the village (indicated by inverted triangles in the figure). Furthermore, in most cases, males were the end node of the information (information sources), indicating that most advisors of new agriculture technologies were male. The few female sources of outside information (primarily affiliated with NGOs) were more often approached by female farmers for advice.

There are various reasons for women's weaker information network. Outside the village, the main sources for information were extension agents from GOs and NGOs, agricultural input distributors, crop buyers, teachers, local leaders, and farmers from other villages. However, it is usually the male household members who interact with these sources and travel to markets, since there is a social taboo on women traveling alone outside the village. Moreover, because of the male-dominant nature of their society, women feel uncomfortable interacting with unknown males. Additionally, females are not encouraged to seek out information sources, as it is not seen as a woman's role; procuring new information and technologies is generally regarded as a man's role. The sum of these social structures was that women had a weaker information network that was confined primarily to their village. However, in spite of women's weaker CA information network, in general, women's social networks for cultural, social, and religious matters are believed to be stronger than men's. Therefore, once women feel enabled to seek out information regarding agriculture technology, their traditional networks may prove to be an effective means for technology transfer.

In addition to strength of network, there was also a difference in the composition of the male and female networks. There were many male farmers in the women's CA technology network, indicating that women were open to approaching male farmers within their villages for information. However, there were

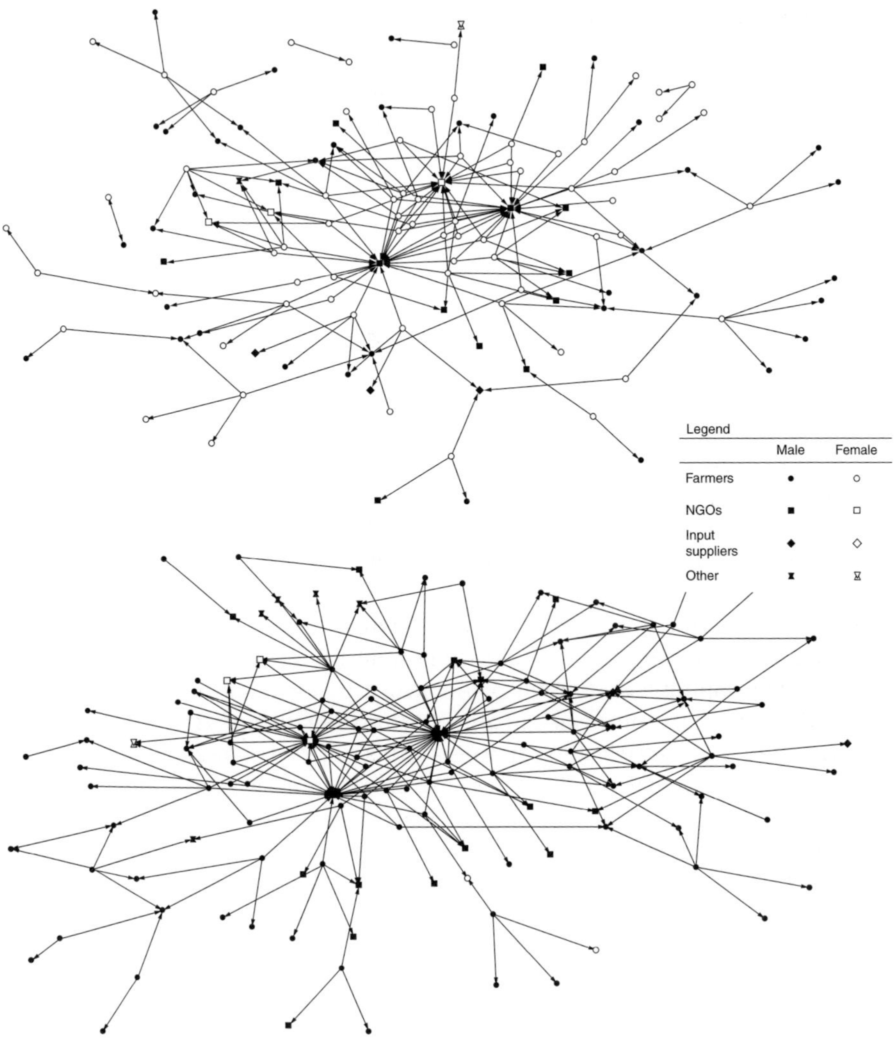

Fig. 12.7. CA information network of females (top) and males (bottom) for information regarding CA practices. The arrow shows the direction of information dispersal. (Data source – primary.)

almost no female farmers as information sources for male farmers, indicating that male farmers were reluctant to ask female farmers for advice, or saw them as poor sources of information.

The results of network density comparison with the ANOVA density model showed that there was no significant difference between the density of male-to-male networks (1% of all possible ties) and female-to-female networks (0.6% of all possible ties) (Table 12.3). Nor was the density of female-to-male ties (1.2% of all possible ties) different from the homophilic (within same gender) networks. However, the density of male-to-female networks was significantly lower than all other types of ties (0.3% of all possible ties, $P = 0.04$), signifying that male

Table 12.3. Density of CA technology network by gender.

Network of ties (by gender)		
Asked by	Asked to	Density (%)
Male	Male	1.0[NS]
(n = 148)	Female	0.3*
Female	Male	1.2[NS]
(n = 83)	Female	0.6[Base]
	R^2	0.1*

*, significance at the probability of 0.05; NS, non-significant according to the ANOVA test of structural block model.

farmers rarely asked female farmers for advice on new agriculture technologies. This behavior is part of the traditional cultural belief that men are primarily responsible for finding new information. Therefore, the female information network is closely related to the overall position of females in society; hence, a multidisciplinary approach to the empowerment of women is required to increase their sphere of information.

Considering the increasing role of women in decision making regarding the adoption of new technologies, women's weak information network is a challenge for encouraging the adoption of CA practices. Therefore, it is important that female farmers are informed and educated about CA systems to increase the rate of adoption and transfer of information. This can be achieved by tapping into females' traditional networks (e.g. social, religious, and cultural networks) within and outside the villages, which are believed to be as strong as male's networks (Subramaniam, 2004). Female farmers can adapt and customize CA practices according to their needs by discussing these technologies within their network. Additionally, it is important to engage women in the process of testing CA systems and disseminating the results, which can help foster a learning environment within the traditional networks. At the macro level, it is evident that there is a lack of female agricultural experts who can help rural, uneducated female farmers to adopt sustainable agriculture technologies. Therefore, females should be encouraged to join agriculture education programs and government extension services.

While it is clear that female Chepang farmers are challenged by a weaker CA information network than their male counterparts, the limited information in their networks presents an additional problem. Although women have knowledge about individual CA practices such as crop rotation, land cover, and minimum tillage, many may not be aware of the CA systems in which all three practices are implemented together to generate synergistic effects. Hence, information about CA systems can be delivered more effectively by ensuring that extension workers, NGOs and GOs, and market players such as input sellers and crop buyers have an integrated understanding of overall CA practices and that they actively work to engage both men and women in information dissemination.

It is important to remember, however, that a woman's capacity for adopting a CA system does not operate in isolation. Aside from the lack of knowledge of and access to information about CA systems, there are cultural norms that may hinder women's power to alter decisions regarding adopting new technologies. Therefore, in addition to providing technical information, it is important to empower women socially and economically. This can, in part, be achieved through the active engagement of women in agriculture projects, savings groups, and marketing activities.

12.4 Conclusions and Implications

The results of this research have shown that there are a number of important considerations in terms of gender that may affect the feasibility of CA adoption. The implications of integrating sustainable CA practices, such as intercropping and minimum tillage, can have different effects on labor demands. Depending on access to wage-earning opportunities and the existing distribution of both household and agricultural labor, these effects may impact men and women differently. Such changes to agricultural practices can result in an inequitable or impractical redistribution of labor among the genders, and may create a barrier to their adoption. Therefore, the gendered implications of introduced practices are vital to consider in planning agricultural interventions. Additionally, one should consider the results of this research within the larger scope of the household and community context. Other agricultural activities and household obligations may or may not restrict the flexibility of absorbing shifts in labor due to adopting conservation practices. The shifts in individual activities from more or less physically demanding tasks can further affect the acceptability of a new practice by a community. Finally, opportunity costs in terms of the potential for wage earning are important to consider in the overall benefits of agricultural labor saving for subsistence farming communities.

Regarding farmers' preferences for CA interventions, both men and women seem to be in consensus on the relative importance of enhancing crop yield. The genders differed, however, on whether profit or soil quality was seen as the next most important objective. The study showed that CA systems were preferred to traditional production systems, regardless of a farmer's gender. This implies that farmers are looking for alternative technologies, providing an opportunity for CA implementation. Both male and female farmers show a clear preference for integrating legumes into the farming system, either through sole cropping after maize or through intercropping with millet. At the same time, farmers showed no preference for strip tillage-based CA systems over conventional tillage, which implied a knowledge gap among farmers concerning the full benefits of strip tillage over the long term.

While information on CA practices seemed to be lacking for all Chepang farmers, female farmers were particularly challenged when it came to getting information on CA practices from outside sources. To resolve such issues and facilitate successful technology transfer, our first recommendation is that researchers, extension agents, and farmers conduct regular meetings engaging women to discuss CA systems, their potential benefits and practical feasibility, and to identify farmer concerns and barriers to adoption. Ongoing trial plots in farmer's fields can further be used to facilitate discussion of CA systems' benefits with women farmers. Second, there should be more emphasis on education and training for women farmers regarding the practices and benefits of CA. Third, emphasis should be placed on utilizing females' traditional networks for effective dissemination of CA information among women farmers. Finally, as there is a lack of female extension workers, the government should focus on the long-term development of female extension personnel, thereby resulting in a more gender-neutral and -inclusive agriculture extension system.

It is important to note that these results are locally specific, depending on culture and environmental conditions, and are expected to differ by location. Nevertheless, the framework provided here could be used to measure potential labor shifts elsewhere. On the whole, assessing gendered impacts and integrating gender sensitivity is a vital aspect of the planning and implementation process of agriculture development projects. As shown here, gender inclusion in the introduction of agriculture technologies can aid in developing CA projects, promoting community equity, and can increase the likelihood of the adoption of yield-improving practices.

Acknowledgements

The authors would like to give appreciation and thanks to the participating survey respondents and enumerators, and our partners in Nepal, LI-BIRD and IAAS. This project was made possible through the funding of USAID/Feed the Future Food Security Innovation Lab: Collaborative Research on Sustainable Agriculture and Natural Resource Management.

References

Alphonce, C. (1996) Application of the analytic hierarchy process in agriculture in developing countries. *Agricultural Systems* 53, 97–112.

Bandiera, O. and Rasul, I. (2006) Social networks and technology adoption in northern Mozambique. *The Economic Journal* 116(514), 869–902.

Bekele, W. and Drake, L. (2003) Soil and water conservation decision behavior of subsistence farmers in the Eastern Highlands of Ethiopia: A case study of the Hunde-Lafto area. *Ecological Economics* 46(3), 437–451.

Beteta, C.H. (2006) What is missing in measures of women's empowerment? *Journal of Human Development* 7(2), 221–241.

Blackden, C.M. and Wodon, Q. (eds) (2006) *Gender, Time Use, and Poverty in Sub-Saharan Africa* (Vol 73). World Bank Publications, Washington, DC.

Borgatti, S.P., Everett, M.G. and Freeman, L.C. (2013) *Ucinet for Windows: Software for Social Network Analysis*. Analytic Technologies, Harvard, Massachusetts.

Bunch, R. (1999) Reasons for non-adoption of soil conservation technologies and how to overcome them. *Mountain Research and Development* 19(3), 213–220.

Carr, A. and Wilkinson, R. (2005) Beyond participation: Boundary organizations as a new space for farmers and scientists to interact. *Society and Natural Resources* 18(3), 255–265.

CBS (Centre Bureau of Statistics) (2012) Nepal Population Census, 2011. Centre Bureau of Statistics, Nepal Planning Commission, Nepal.

Cochran, J. (2003) Patterns of sustainable agriculture adoption/non-adoption in Panamá. Thesis, McGill University, Montreal, Canada.

Derpsch, R. and Friedrich, T. (2009) Global overview of conservation agriculture adoption. Paper presented at the 4th World Congress on Conservation Agriculture, 4–7 February 2009, New Delhi.

FAO (Food and Agriculture Organization) (2007) FAO/WFP Food Security Assessment Mission to Nepal. An FAO global information and early warning system on food and agriculture work food programme special report: 25 July 2007. FAO, Rome.

FAO (2012) Asia's women in agriculture, environment and rural production: Nepal. FAO Sustainable Development Department Factsheet. Available at: ftp://ftp.fao.org/sd/sdw/sdww/Nep.pdf (accessed 17 November 2014).

Gurung, H.B. (1998) *Nepal, Social Demography and Expressions*. New Era Publication, Kathmandu.

Gurung, K., Tulachan, P.M. and Gauchan, D. (2005) Gender and social dynamics in livestock management: A case study from three ecological zones in Nepal. LPP Project, Center for Mountain Research Development Report, Kathmandu.

Hanneman, R. and Riddle, M. (2005) Introduction to social network analysis. Available at: http://www.faculty.ucr.edu/~hanneman/nettext/ (accessed 12 November 2014).

Hobbs, P.R., Sayre, K. and Gupta, R. (2008) The role of conservation agriculture in sustainable agriculture. *Philosophical Transactions of the Royal Society* 363, 543–555.

Jun, F., Mingan, S., Quan Jiu, W., Jones, S.B., Reichardt, K., *et al.* (2010) Toward sustainable soil and water resources use in China's highly erodible semi-arid loess plateau. *Geoderma* 155(1), 93–100.

Kassam, A., Friedrich, T., Shaxson, F. and Pretty, J. (2009) The spread of conservation agriculture: Justification, sustainability and uptake. *International Journal of Agricultural Sustainability* 7(4), 292–320.

Kessler, C.A. (2006) Decisive key-factors influencing farm households' soil and water conservation investments. *Applied Geography* 26, 40–60.

Knowler, D. and Bradshaw, B. (2007) Farmers' adoption of conservation agriculture: A review and synthesis of recent research. *Food Policy* 32, 25–48.

Lal, R. (2001) Managing world soils for food security and environmental quality. *Advances in Agronomy* 74, 155–192.

Paudel, G.S. and Thapa, G.B. (2004) Impact of social, institutional and ecological factors on land management practices in mountain watersheds of Nepal. *Applied Geography* 24, 35–55.

Pimentel, D., Harvey, C., Resosudarmo, P., Sinclair, K., Kurz, D., *et al.* (1995) Environmental and economic costs of soil erosion and conservation benefits. *Science* 267(5201), 1117–1123.

Piya, L., Maharjan, K.L., Joshi, N.P. and Dangol, D.R. (2011) Collection and marketing of non-timber forest products by the Chepang community in Chitwan district of Nepal. MPRA Paper No 35389. Available at: http://mpra.ub.uni-muenchen.de/35389/ (accessed 12 November 2014).

Piya, L., Maharjan, K.L. and Joshi, N.P. (2013) Determinants of adaptation practices to climate change by Chepang households in the rural mid-hills of Nepal. *Regional Environmental Change* 13(2), 437–447.

Probst, K., Hoffman, V. and Christinck, A. (2007) *Farmers and Researchers: How Can Collaborative Advantages be Created in Participatory Research and Technology Development? Module: Knowledge and Innovation Management*. University of Hohenheim, Stuttgart, Germany, pp.120–137.

Shively, G., Gars, J. and Sununtnasuk, C. (2011) A Review of Food Security and Human Nutrition Issues in Nepal. Purdue University Department of Agricultural Economics Staff Paper 11-05.

Shrestha, D.P., Zinck, J.A. and Van Ranst, E. (2004) Modelling land degradation in the Nepalese Himalaya. *Catena* 57(2), 135–156.

Subramaniam, M. (2004) (eds) *The Power of Women's Informal Networks: Lessons in Social Change from South Asia and West Africa*. Lexington Books, Lanham, Maryland.

Suvedi, M. and McNamara, P. (2012) Strengthening the Pluralistic Agricultural Extension System in Nepal. MEAS Rapid Scoping Mission, USAID. Available at: https://dl.dropboxusercontent.com/u/15810717/Country%20Reports/MEAS%20Country%20Report%20NEPAL%20-%20Dec%202011.pdf (accessed 12 November 2014).

Thapa, G.B. (1996) Land use, land management and environment in a subsistence mountain economy in Nepal. *Agriculture, Ecosystems and Environment* 57, 57–71.

Thapa, G.B. and Paudel, G.S. (2002) Farmland degradation in the mountains of Nepal: A study of watersheds 'with' and 'without' external intervention. *Land Degradation and Development* 13, 479–493.

Tiwari, K.R., Sitaula, B.K., Nyborg, I.L.P. and Paudel, G.S. (2008) Determinants of farmers' adoption of improved soil conservation technology in a middle mountain watershed of central Nepal. *Environmental Management* 42, 210–222.

Tiwari, K.R., Sitaula, B.K. and Bajracharya, R.M. (2009) Runoff and soil loss responses to rainfall, land use, terracing and management practices in the middle mountains of Nepal. *Acta Agriculturae Scandinavica, Section B* 59(3), 197–207.

Vaidya, O.S. and Kumar, S. (2006) Analytic hierarchy process: An overview of applications. *European Journal of Operational Research* 169, 1–29.

Vainiunas, P., Zavadskas, E.K., Peldschus, F., Turskis, Z. and Tamosaitiene, J. (2009) Model of construction design projects' managers qualifying by applying analytic hierarchy process and Bayes rule. In: Grasserbauer, M., Sakalauskas, L. and Zavadskas, E.K. (eds) *Knowledge-Based Technologies and OR Methodologies for Strategic Decisions of Sustainable Development*, 5th International Vilnius Conference, Lithuania, 30 September 2009. The Association of European Operational Research Societies, Vilnius, Lithuania, pp. 148–153.

World Bank (2012) *Nepal Country Overview*. Available at: http://www.worldbank.org/en/country/nepal (accessed 12 November 2014).

Yadav, R.P. (1987) Agricultural research in Nepal: Resource allocation, structure, and incentives. International Food Policy Research Institute, Research Report 62, September 1987.

Index